CW00569121

Environmental Electrochemistry

ACS SYMPOSIUM SERIES **811**

Environmental Electrochemistry

Analyses of Trace Element Biogeochemistry

Martial Taillefert, EDITOR
Georgia Institute of Technology

Tim F. Rozan, EDITOR
University of Delaware

American Chemical Society, Washington, DC

Library of Congress Cataloging-in-Publication Data

Environmental electrochemisty : analyses of trace element biogeochemistry / Martial Taillefert, editor, Tim F. Rozan, editor.

p. cm.—(ACS symposium series ; 811)

Includes bibliographical references and index.

ISBN 0–8412–3774–3 (alk. paper)

1. Environmental chemistry—Congresses. 2. Electrochemistry—Congresses. 3. Trace elements—Environmental aspects—Congresses. I. Taillefert, Martial, 1967 – II. Rozan, Tim F., 1964 – III. American Chemical Society. Division of Environmental Chemistry. IV. American Chemical Society. Meeting (220th: 2001 : Washington, D.C.). V. Series.

TD193.E553   2002
628.01'54137—dc21   2001053392

The paper used in this publication meets the minimum requirements of American National Standard for Information Sciences—Permanence of Paper for Printed Library Materials, ANSI Z39.48–1984.

PRINTED IN THE UNITED STATES OF AMERICA

Foreword

The ACS Symposium Series was first published in 1974 to provide a mechanism for publishing symposia quickly in book form. The purpose of the series is to publish timely, comprehensive books developed from ACS sponsored symposia based on current scientific research. Occasionally, books are developed from symposia sponsored by other organizations when the topic is of keen interest to the chemistry audience.

Before agreeing to publish a book, the proposed table of contents is reviewed for appropriate and comprehensive coverage and for interest to the audience. Some papers may be excluded to better focus the book; others may be added to provide comprehensiveness. When appropriate, overview or introductory chapters are added. Drafts of chapters are peer-reviewed prior to final acceptance or rejection, and manuscripts are prepared in camera-ready format.

As a rule, only original research papers and original review papers are included in the volumes. Verbatim reproductions of previously published papers are not accepted.

ACS Books Department

Contents

In-Situ Measurements
at the Sediment–Water Interface

Sediment Porewaters and Microbial Mats

New Technologies in Electrochemistry

Trace Metal Complexation

Indexes

Preface

Electrochemistry has been used extensively in the past three decades to determine the chemical composition of environmental samples from the water column, sediments, soils, biofilms, and microbial mats. These electrochemical methods have many advantages over other analytical techniques for environmental research: (1) the techniques are non-destructive, which minimize sample perturbation, (2) the data can be collected rapidly and reproducibly, (3) the detection limits have appropriate sensitivity for most environmental applications, (4) direct information on the chemical speciation can be obtained, (5) the instrumentation can be very compact, which is attractive for field deployment, and (6) the electrochemical sensors can be miniaturized, allowing for non-invasive in-situ sampling. For these reasons, electrochemical methods are extremely useful tools for understanding complex natural biogeochemical processes, however, the present literature lacks a comprehensive publication, which details the current multidisciplinary uses of electrochemistry in natural environments.

The goal of this American Chemical Society (ACS) Symposium Series publication is to demonstrate the usefulness of electrochemical methods in environmental research. The chapters are based on a Symposium on Electrochemical Methods for the Environmental Analysis of Trace Element Biogeochemistry that was organized at the 200th ACS National Meeting in Washington, D.C. in August 2000. The chapters present instrumental designs and techniques currently being employed by researchers and cover a wide range of environmental applications including trace metal measurements in the water column of freshwater and marine environments, redox chemical species at hydrothermal vents, in anoxic water bodies, sediments and microbial mats, major cations and anions in extraterrestrial systems, metal complexing properties of natural waters, and mineral–water interface processes.

The book is targeted to all scientist whose goal is to better understand trace element cycling in aquatic systems, including environmental chemists and engineers, geochemists, soil scientists, marine chemists, and environmental microbiologists who require tools to characterize the chemistry of microbial environments.

To cover this wide array of interests we have divided the book's 19 chapters into six sections: (1) development and application of electrochemical techniques for in-situ measurements in the water column of lakes, rivers, and oceans; (2) development and application of electrochemical techniques for on-line measurements in the water column of lakes, rivers, and oceans; (3) development and application of electrochemical techniques for in-situ measurements at the sediment–water interface in lakes and marine systems; (4) applications of electrochemical techniques for sediment and microbial mat measurements; (5) novel electrochemical technologies in development or destined to be utilized in extraterrestrial environments; and (6) recent applications of electrochemical techniques for metal complexation studies. Each section comprises a number of different research examples where a variety of analytical techniques (voltammetric, potentiometric, amperometric) have successfully been used to help understand biogeochemical processes in diverse environments. Chapters were categorized according to their main focus, and some may apply to more than one section. Therefore, we encourage the reader to consult each individual chapter.

Acknowledgments

The editors are grateful to the authors for their outstanding contributions, to the referees who contributed significantly to the quality of this book by reviewing each chapter, and finally to the acquisitions staff of the ACS Books Department. Special thanks to Anne Wilson and Kelly Dennis for their patience and support during the acquisition of this book and to Margaret Brown for editing and production.

Martial Taillefert
School of Earth and Atmospheric Sciences
Georgia Institute of Technology
Atlanta, GA 30332–0340

Tim Rozan
College of Marine Studies
University of Delaware
Lewes, DE 19958–1298

Introduction

Chapter 1

Electrochemical Methods for the Environmental Analysis of Trace Elements Biogeochemistry

Martial Taillefert[1] and Tim F. Rozan[2]

[1]School of Earth and Atmospheric Sciences, Georgia Institute of Technology, 221 Bobby Dodd Way, Atlanta, GA 30332
[2]College of Marine Sciences, University of Delaware, 700 Pilottown Road, Lewes, DE 19958

Since the 1970's, when the first *in situ* measurements of oxygen in the oceans were reported, the development of electrochemical methods for the analysis of trace element biogeochemistry in the environment has significantly improved. From conductimetry to measure salinity, to amperometric and potentiometric sensors that can measure a single analyte, to voltammetric sensors that can measure several species during the same scan, a variety of electrochemical techniques have been utilized to better understand biogeochemical processes in the environment. These techniques have been integrated into a variety of devices for laboratory experimentation and *in situ* deployment (or online measurements). The development of microsensors has significantly contributed to the application of these techniques in sediments, biofilms, and microbial mats, where data can now be collected at the micrometer scale. Electrochemical techniques have also been adapted to measure the chemical speciation of trace elements in natural environments following physical and chemical separations. Finally, the complexation properties of most naturally occurring ligands have been determined by electrochemical measurements performed on synthetic and natural samples.

This first chapter is intended to familiarize the reader with the electrochemical terminology, techniques, current applications, and future directions in environmental chemistry and biogeochemistry research.

Electrochemical Techniques and Sensors

Potentiometry

In potentiometry, the difference of potential between two electrochemical cells is measured with a high impedence voltmeter in the absence of appreciable currents. The difference in potential is measured between a reference electrode, usually a Ag/AgCl or a $Hg/HgCl_2$ electrode, and an indicator electrode. The boundary potential, or the difference in potential between the external and the internal solution at the indicator electrode, is sensitive to the activity of ions according to the Nernst equation.

Potentiometric sensors commonly used in the environment include membrane electrodes to determine the pH ($1-3$), or S^{2-} ($4-6$), molecular electrodes to determine pCO_2 ($3, 7$), and ion selective electrodes (ISEs) to determine NO_3^- ($8-10$), NO_2^- ($10, 11$), NH_4^+ ($10, 12$), Ca^{2+} (10), CO_3^{2-} (10), or free metal activities ($13, 14$). Unfortunately, interferences prevent the use of NO_3^- and NH_4^+ electrodes in marine environments. Finally, the detection limits of ISEs are usually not suitable for the determination of free metals in natural environments, but the recent development of polymer membrane ISEs ($15, 16$) is promising to increase their sensitivity.

Coulometry

Coulometry is based on the oxidation and reduction of an analyte for a sufficient period of time to assure its quantitative conversion to a new oxidation state. Two coulometric methods are used to analyze environmental samples: chronopotentiometry and chronoamperometry. In chronopotentiometry, an element is electrolytically preconcentrated at an electrode and the potential is recorded at a reference electrode as a function of time upon addition of a chemical reactant ($17, 18$) or application of a constant current (e.g., $19, 20$). The first technique, called potentiometric stripping analysis (PSA), is usually used to analyze trace metals in freshwater (21) and seawater (22), as well as for metal complexation studies (23). The second technique, called constant current stripping analysis (CCSA), has been used to analyze trace metal concentrations in aquatic systems ($19, 20$).

In chronoamperometry, an element is electrolytically preconcentrated at an electrode by applying a constant potential and the current of the reaction is simultaneously recorded at an auxiliary electrode as a function of time (24). To our knowledge, the application of chronoamperometry for the analysis of environmental samples is limited. A chronoamperometric method has been used to measure organic compounds in the environment (*25*), and recently, a biosensor has been developed to determine NH_4^+ by the intermediate of a chronoamperometric measurement of NADH involved with NH_4^+ in the reduction of an enzyme (*26*).

Amperometry

In amperometry, the current is measured at an auxiliary electrode when a constant potential is applied between a reference electrode, usually silver/silver chloride, and the indicative electrode.

The most common amperometric sensor is the oxygen electrode, or Clark electrode (*27, 28*), which has been routinely deployed to measure concentrations of oxygen in aquatic systems (e.g., *4, 28-34*). Another amperometric sensor has been built to detect N_2O in biofilms (*35*). Recently, bioelectrochemical sensors have been developed to measure NO_3^- in marine sediments (*36*) and CH_4 in sediments and biofilms (*37*). These biosensors use a N_2O microelectrode (*35*) and an oxygen microelectrode (*4*), respectively, as indirect electrochemical detector. The NO_3^- biosensor contains a microbial community between a membrane and the N_2O electrode which reduces NO_3^- and NO_2^- to N_2O. This biosensor does not suffer from chloride interference and can be used in marine environments. The CH_4 biosensor contains aerobic methane oxidizers which consume O_2. The decrease in O_2 is inversely proportional to the CH_4 concentration.

An amperometric microsensor has been developed for on-line measurements of H_2S (*38, 39*). Dissolved sulfide is detected indirectly by measuring the reoxidation current of ferrocyanide produced by sulfide reduction of ferricyanide. In contrast to potentiometric sensors, this electrode does not suffer from oxygen interferences because oxygen does not react with ferricyanide.

Voltammetry

In voltammetry, the potential is ramped between a working electrode and a reference electrode. At a particular potential, the analyte is oxidized or reduced at the working electrode and the current resulting from this reaction is measured at an auxiliary electrode. Voltammetric techniques are attractive to measure chemical species in the environment because they can detect several analytes in the same potential scan, they have low detection limits, and generally do not suffer from matrix problems (e.g., high salinity) (*40*). Voltammetry in the environment is mostly used with mercury electrodes, because their analytical window is ideal for the direct measurements of $O_{2(aq)}$, ΣH_2S (i.e., H_2S, HS^-, S^{2-}, and $S(0)$ in S_8 and S_x^{2-}), thiols, Fe^{2+}, Mn^{2+}, $S_2O_3^{2-}$, $S_4O_6^{2-}$, I^-, and other trace metals. The latter are either measured directly (i.e., Cu, Zn, Cd, and Pb) (e.g., *41-45*) or indirectly (e.g., Cr, Co, Ni, Cu, Pb, Cd, U) by adding an organic ligand that forms an electrochemically reactive complex (e.g., *42, 46-48*). However, other substrates have been developed to measure organic compounds and metals (e.g., *24, 49-51*) in the environment.

Different voltammetric methods may be used with mercury electrodes for aqueous sample analysis. In linear sweep voltammetry (LSV), a direct current potential is ramped, and the current is sampled as a function of time. This technique is not very sensitive and is generally used to measure dissolved oxygen only. Cyclic voltammetry (CV) is similar to LSV, except that the potential is scanned back to its initial value once it reaches its final value. This technique is generally used to study the electrochemical properties of chemical species. In differential pulse voltammetry (DP), the potential is ramped by using pulses and the current measured at the end of each pulse. The sensitivity and selectivity of this method make it suitable for trace metal analysis. Finally, in square wave voltammetry (SWV) the potential is scanned using square waves (a potential pulse of the same intensity is applied in the positive direction, then in the negative direction), and the current is sampled at the end of each pulse. Because of the polarity inversion, the currents are of opposite signs and their subtraction enhances the current intensity. SWV is the most popular technique because of its great sensitivity and speed. It is used to measure ΣH_2S, Fe^{2+}, Mn^{2+}, $S_2O_3^{2-}$, $S_4O_6^{2-}$, and I^- by scanning the potential in the positive direction (cathodic square wave voltammetry: CSWV). Alternatively, DP and SWV are both used to measure trace metals Cu, Cd, Pb, and Zn by scanning the potential in the negative direction after preconcentration by electrodeposition at the surface of the electrode (anodic stripping voltammetry: ASV). Finally, CSV and SWV are both used to measure Cr, Co, Ni, Cu, Cd, Pb, U and other elements by scanning the potential in the positive direction after preconcentration of the electrochemically reactive metal-organic complex at the electrode (adsorptive cathodic stripping voltammetry: AdCSV).

Environmental Applications

In situ measurements in water bodies

In situ measurements are needed to determine the chemical composition of water bodies because the chemical speciation can be altered when a sample is collected by exposure to the atmosphere or by contamination (*42, 52*). In addition, samples cannot always be collected and brought back to the laboratory for analysis. In these instances, e.g. extraterrestrial environments, *in situ* measurements, or remote sensing, is also required (see Chapter 16).

Instrumentation has been developed to deploy many different sensors in water bodies. The most commonly used sensors are amperometric oxygen electrodes and potentiometric pH electrodes deployed on CTD instruments (*28*), however, voltammetric systems have also been deployed and are presented in Chapters 2, 3, and 4.

On-line measurements in water bodies

As an alternative to *in situ* measurements, electrochemical techniques have been used over the last twenty years to analyze natural waters using on-line systems (e.g., *38, 44-46, 53, 54*). In the procedure for on-line measurements, a water column sample is pumped onboard ship into a flow cell, where the analysis is performed immediately. As the sample is not handled or exposed to the atmosphere, these analyses can be considered very close to *in situ* measurements.

On-line total measurements of trace metal concentrations in marine waters have been mainly performed by voltammetry with hanging mercury drop electrodes (HMDE) or mercury thin-film electrodes (MTFE) using ASV (*44, 45, 53*) and AdCSV (*46* and Chapter 5). When used, the mercury thin film was plated once before analysis (*44*) or *in situ* before each analysis (*45*). In addition, Mn^{2+}, Fe^{2+}, ΣH_2S, and $FeS_{(aq)}$ were measured on-line with a HMDE in the water column of a lake (*54*). Recently, a flow cell that accommodates a voltammetric microelectrode has been used in deep-sea environments (see Chapter 4). In addition, the only amperometric microsensor used on-line (see above) has been developed for water column analysis of H_2S (*38, 39*).

Finally, a new technology has emerged to quantify the "free" metal fraction in solution. This method uses supported liquid membranes (SLM) or membranes coated with a liquid hydrophobic chelator (*55* and see Chapter 6). The SLM is placed between the solution to analyze and an internal solution, and the chelator transports the "free" metal across the membrane to the internal solution, where it is analyzed.

In situ measurements at the sediment-water interface

Most biogeochemical processes in sediments occur very close to the sediment-water interface (sometimes within millimeters). This narrow spatial scale is extremely susceptible to sample perturbation during collection of a sediment core. *In situ* measurements minimize sediment-water interface distortions (mixing) and allow for submillimeter scale resolution. Such measurements have been routinely conducted to determine microprofiles of oxygen with Clark microelectrodes (*3, 29-32*), and pH (*2, 3, 56*) and pS^{2-} (*57*) with potentiometric microelectrodes. Recently, such measurements have been performed for the first time with voltammetric microelectrodes (*56*). A voltammetric microelectrode was mounted on a remotely operated vehicle (ROV) to measure O_2, ΣH_2S, Mn^{2+}, and Fe^{2+} on a (sub)millimeter scale down to 5 cm in the sediment.

Alternatively, benthic chambers have been deployed at the sediment-water interface to measure fluxes of O_2 with Clark electrodes (*32*), pH with potentiometric electrodes (*32, 58*), and H_2S with amperometric sensors (*58*). New studies with these devices are presented in Chapters 7, 8, and 9.

Porewaters in sediments and microbial mats

The spatial resolution of standard procedures to collect porewaters is not sufficiently high to study biogeochemical processes that occur at redox gradients in sediments and microbial mats. In addition, artifacts can occur during sampling and handling of porewaters. Electrochemical measurements in intact sediments or microbial mats represent a useful alternative to the conventional porewater extraction procedures. They have been performed since the 80s with amperometric and potentiometric microelectrodes after collection of a sediment core to measure O_2, pH, NO_3^-, and H_2S (e.g., *1, 4, 8*). In the 90s, these microelectrodes were improved or new ones designed to analyze O_2 and H_2S (*33*), H_2S (*39*), NO_2^- (*11*), NO_3^-, NO_2^-, NH_4^+, Ca^{2+}, and CO_3^{2-} (*59*), and pH and pCO_2 (*3, 7, 59*), sometimes *in situ* in microbial mats located in shallow systems (*33*). Voltammetric microelectrodes have been used for measurements of O_2, ΣH_2S, Fe^{2+}, Mn^{2+}, and $FeS_{(aq)}$ in marine sediments (*60-63*), microbial mats (*63*), and in biofilms (*64*). In addition, soluble organic-Fe(III) complexes have been found ubiquitously in marine sediments (*61, 62*) and in biofilms (*64*) where sulfides are not present.

Investigations with voltammetric microelectrodes in lake sediments are reported for the first time in Chapter 10 and 11. New findings on the nitrogen cycle in lake sediments are presented in Chapter 12. New measurements have also been conducted with voltammetric microelectrodes in marine sediments: the cycling of the main redox species is presented in Chapter 10, while the seasonal cycling of soluble organic-Fe(III) complexes is presented in Chapter 13. Finally, studies conducted in marine stromatolites are documented in Chapter 14 and in microbial mats in Chapter 12 and 15.

New technologies in electrochemistry

New technologies in electrochemistry include the recent development and application of bioelectrochemical sensors to monitor chemical compounds in sediments, microbial mats, or biofilms which cannot be directly monitored by electrochemical sensors. These sensors use microbial or enzymatic reactions in the presence of the analyte and a reactant in front of an electrochemical sensor which indirectly detects the reactant or the product activity change (e.g. *26, 36, 37*). The microbial growth parameters (i.e., temperature, ionic strength) have to be optimized and interference problems (i.e., presence of inhibitors) solved to ensure optimal response of these bioelectrochemical sensors.

A new array of microelectrodes has been built to measure microprofiles near the sediment-water interface (*65*, Chapter 2). This system is promising because it can simultaneously resolve spatial and temporal measurements at the sediment-water interface. However, the electrode array is big enough to disturb the sediment during its insertion and requires a long period of equilibration before the measurements. It has yet to be tested in real sediments.

Supported liquid membrane technology (Chapter 6) should be soon adapted to *in situ* measurements in natural waters and possibly sediments to detect free hydrated trace metal concentrations.

Novel technologies to measure remotely the chemical composition of extraterrestrial environments is waiting to be deployed (Chapter 16). This system is compact and self-operated, it will be able to determine the chemical composition of geological material on other planets.

Finally, nanoelectrodes have been integrated in an atomic force microscope to map the chemical composition at solid-water interfaces by considering both the solid topography and the solution near the surface (Chapter 17). This technique will provide information on the mechanisms of mineral dissolution and formation, catalyzed by chemical or biological reactions.

Trace Metal Complexation

Voltammetry is not only used to detect chemical species in the environment, but also to determine metal-ligand complexation parameters. Measurements are usually performed by titrating a natural solution with a metal and measuring the electrochemically labile metal in solution. Linearization of these metal titration curves can provide a conditional stability constant and a concentration of active complexation sites (*66, 67*) assuming a given stoichiometry. Such measurements have been performed by ASV with Cu (e.g., *68-72*), Cd (*68-71*), Pb (e.g., *69-71, 73*), and Zn (*69, 70*) or by competitive ligand-exchange ASV (CLE-ASV) with

Cu (*69, 74, 75*), Pb (*75*), Cd (*76-77*), and Zn (*78*) or competitive ligand-exchange CSV (CLE-CSV) with Cu (*72, 78, 79*) and Pb (*80*).

Alternatively, stability constants of metal complexes can be determined by the De Ford and Hume technique (*81*) if the redox reaction at the electrode is reversible (*82*). This technique measures the shift in potential at which a metal is reduced when it is complexed by a ligand. The potential shift is recorded during the ligand titration, and the stability constant computed from these titration curves (*81*). This method is usually used to measure stability constants of Cu, Pb, Cd, and Zn with simple ligands such as OH^-, Cl^-, CO_3^{2-} (*18*) or well defined organic ligands (*83*). Recently, it has been used to determine stability constants of Mn, Fe, Co, Ni, Cu, and Zn with HS^-, $S_2O_3^{2-}$, CN^-, and mercaptoethanol (*84, 85*).

Finally, pseudopolarograms have also been used to determine metal complexation in natural waters (*86-88*). This method is based on the observation that the reduction of a metal complexed by a ligand shifts towards more negative potentials than that of a free metal ion. To increase the detection limit of the technique to trace metal levels, pseudopolarograms can be generated by using ASV, where a half-wave is generated through a series of independent ASV experiments at increasingly negative potentials. From this half-wave, conditional stability constants can be calculated (*87, 88*).

Problems associated with these complexing properties measurements include: the deposition potential by ASV may discriminate between strong and weak electrochemically-labile complexes (*89*); adsorption of non-electroactive ligands at the electrode surface may prevent diffusion of the metal to the electrode or may complex the metal (*18, 90, 91*); the analytical window influences the class of ligands titrated (*18, 92, 93*); and the stoichiometry of complexes is often assumed rather than determined. These problems all affect the accuracy of the determined complexation parameters.

This book would not be complete without considering the use of voltammetric methods in the determination of metal complexation in natural waters. Chapter 18 reviews competitive ligand-exchange methods using voltammetry to determine complexing parameters in freshwater. Finally, recent evidence for the role of trace concentrations of dissolved sulfides in complexing trace metals in river waters is presented in Chapter 19.

Future Directions

Electrochemistry is a continuously growing field and new techniques and methods keep emerging from the literature. As a result, environmental chemists, chemical oceanographers, and microbiologists are constantly improving their techniques to study biogeochemical and environmental processes. With improvements in nanotechnology, electrodes will probably become smaller and more sensitive. In addition, commercially available *in situ* instrumentation has flourished recently, allowing us to adapt electrochemical sensors on moorings, remotely operated vehicles (ROV), rosettes or cables to monitor biogeochemical processes in the environment over a spatial and/or temporal scale. Electrochemical sensors have already been adapted on free landers deployed at the sediment-water interface and to be deployed on Mars (see Chapter 16). The next obvious step is to adapt these sensors on autonomous underwater vehicles (AUV) which could freely map the water column of lakes and oceans.

In addition, investigations in sediment porewaters should be performed in three dimensions and with time as sediments are highly heterogeneous (even on small spatial scales) and most biogeochemical processes are not at equilibrium.

Finally, progress in the theory of voltammetry and improvements of analytical methods to determine metal complexation by natural organic is essential to assess the toxicity of metal contamination. With the recent progress in separation technology, future studies will probably aim toward separating natural organic matter in smaller units that can each be characterized for their metal complexation properties.

References

1. De Jong, S. A.; Hofman, P. A. G.; Sandee, A. J. J. *Mar. Ecol. Prog. Ser.* **1988**, *45*, 187-192.
2. Wilson, T. R. S.; McPhail, S. D.; Braithwaite, A. C.; Koch, B.; Dogan, A.; Disteche, A. *Deep-Sea Res.* **1988**, *36*, 315-321
3. Cai, W-J.; Reimers, C. E. *Limnol. Oceanogr.* **1993**, *38*, 1762-1787
4. Revsbech, N. P.; Jørgensen, B. B.; Blackburn, T. H.; Cohen, Y. *Limnol. Oceanogr.* **1983**, *28*, 1062-1074.
5. Eckert, W.; Frevert, T.; Trüper, H. G. *Wat. Res.* **1990**, *24*, 1341-1346.
6. Peiffer, S.; Frevert, T. *Analyst* **1987**, *112*, 951-954.
7. De Beer, D.; Glud, A.; Epping, E.; Kühl, M. *Limnol. Oceanogr.* **1997**, *42*, 1590-1600.
8. De Beer, D.; Sweerts, J.-P. *Anal. Chim. Acta* **1989**, *219*, 351-356.
9. Jensen, K.; Revsbech, N. P.; Nielsen, L. P. *App. Environ. Microb.* **1993**, *59*, 3287-3296.
10. Müller, B.; Buis, K.; Stierli, R.; Wehrli, B. *Limnol. Oceanogr.* **1998**, *43*, 1728-1733.
11. De Beer, D.; Schramm, A.; Santegoeds, C. M.; Kühl, M. *Appl. Environ. Microbiol.* **1997**, *63*, 973-977.
12. De Beer, D.; Van den Heuvel, J. C. *Talanta* **1988**, *35*, 728-730.
13. Belli, S. L.; Zirino, A. *Anal. Chem.* **1993**, *65*, 2583-2589.

14. Mousavi, M. F.; Sahari, S.; Alizadeh, N.; Shamsipur, M. *Anal. Chim. Acta* **2000**, *414*, 189-194.

15. Sokalski, T.; Ceresa, A.; Zwickl, T.; Prestch, E. *J. Am. Chem. Soc.* **1997**, *119*, 11347-11348.

16. Bakker, E.; Pretsch, E. *Trends Anal. Chem.* **2001**, *20*, 11-19.

17. Jagner, D. In *Marine electrochemistry*; Whitfield, M.; Jagner, D., Eds.; John Wiley & Sons Ltd: Chichester, NY, 1981; Chap. 4.

18. Buffle, J. *Complexation Reactions in Aquatic Systems an Analytical Approach*; Halsted Press: a division of Wiley: New York, NY, 1988; p 692.

19. Wang, J.; Larson, D.; Foster, N.; Armalis, S.; Lu, J.; Rongrong, X.; Olsen, K.; Zirino, A. *Anal. Chem.* **1995**, *67*, 1481-1485.

20. Riso, R. D.; Waeles, M.; Monbet, P.; Chaumery, C. J. *Anal. Chim. Acta* **2000**, *410*, 97-105.

21. Jagner, D.; Sahlin, E.; Axelsson, B.; Ratana-Ohpas, R. *Anal. Chim. Acta* **1993**, 278, 237-243.

22. Riso, R. D.; Le Corre, P.; Chaumery, C. J. *Anal. Chim. Acta.* **1997**, *351*, 83-89.

23. Soares, H. M.V. M.; Vasconcelos, M. T. S. D. *Anal. Chim. Acta* **1995**, *314*, 241-249.

24. Wang, J. *Analytical Electrochemistry*, 2nd Ed; Wiley Interscience: New York, NY, 2000; p 232.

25. Wang, J.; Chen, Q. *Anal. Chim. Acta.* **1995**, *312*, 39-44.

26. Hart, J. P.; Abass, A. K.; Cowell, D. C.; Chappell, A. *Electroanalysis* **1999**, *11*, 406-411.

27. Hitchman, M. L. *Measurement of dissolved oxygen*; Wiley-Interscience: New York, NY, 1978; p 255.

28. Gnaiger, E.; Forstner, H. *Polarographic oxygen sensors: Aquatic and physiological applications*; Springer-Verlag: New York, NY, 1983; p. 370.

29. Revsbech, N. P.; Jørgensen, B. B. *Limnol. Oceanogr.* **1983**, *28*, 749-756.

30. Reimers, C. E.; Fischer, K. M.; Merewether, R.; Smith Jr., K. L.; Jahnke, R. A. *Nature* **1986**, *320*, 741-744.

31. Glud, R. N.; Gundersen, J. K.; Jørgensen, B. B.; Revsbech, N. P.; Schulz, H. D. *Deep-Sea Res.* **1994**, *41*, 1767-1788.

32. Tengberg, A.; De Bovee, F.; Hall, P.; et al. *Prog. Oceanog.* **1995**, *35*, 253-294.

33. Visscher, P. T.; Beukema, J.; van Gemerden, H. *Limnol. Oceanogr.* **1991**, *36*, 1476-1480.

34. Wieland, A.; Kühl, M. *Mar. Ecol. Prog. Ser.* **2000**, *196*, 87-102.

35. Revsbech, N. P.; Nielsen, L. P.; Christensen, P. B.; Sorensen, J. *App. Environ. Microb.* **1988**, *54*, 2245-2249.

12

36. Larsen, L. H.; Kjaer, T.; Revsbech, N. P. *Anal. Chem.* **1997**, *69*, 3527-3531.
37. Damgaard, L. R.; Revsbech, N. P. *Anal. Chem.* **1997**, *69*, 2262-2267.
38. Stüben, D.; Braun, S.; Jeroschewski, P.; Haushahn, P. *App. Geochem.* **1998**, *13*, 379-389.
39. Kühl, M.; Steuckart, C.; Eickert, G.; Jeroschewski, P. *Aquat. Microb. Ecol.* **1998**, *15*, 201-209.
40. Florence, T. M. *Analyst* **1986**, *111*, 489-505.
41. Batley, G. E.; Gardner, D. *Estuar. Coast. Mar. Sci.* **1978**, *7*, 59-70.
42. Taillefert, M.; Luther III, G. W. *Electroanalysis* **2000**, *12*, 401-412.
43. Tercier, M-L.; Buffle, J. *Anal. Chem.* **1996**, *68*, 3670-3678.
44. Wang, J.; Ariel, M. *Anal. Chim. Acta* **1978**, *99*, 89-98.
45. Zirino, A. In *Marine Electrochemistry*; Whitfield, M.; Jagner, D., Eds.; John Wiley & Sons Ltd: Chichester, NY, 1981; pp 421-503.
46. Whitworth, D.; Achterberg, E. P.; Nimmo, M.; Worsfold, P. J. *Anal. Chim. Acta* **1998**, *371*, 235-246.
47. Collado-Sánchez, C.; Pérez-Pena, J.; Gelado-Caballero, M. D.; Herrera-Melian, J. A.; Hernández-Brito, J. J. *Anal. Chim. Acta* **1996**, *320*, 19-30.
48. Korolczuk, M. *Anal. Chim. Acta* **2000**, *414*, 165-171.
49. Wang, J.; Bhada, R. K.; Lu, J.; MacDonald, D. *Anal. Chim. Acta* **1998**, *361*, 85-91.
50. Jacquinot, P.; Müller, B.; Wehrli, B.; Hauser, P. C. *Anal. Chim. Acta* **2001**, *432*, 1-10.
51. Ugo, P.; Moretto, L. M.; Mazzocchin, G. A. *Anal. Chim. Acta* **1995**, *305*, 74-82.
52. Tercier, M.-L; Buffle, J. *Electroanalysis* **1993**, *5*, 187-200.
53. Wang, J.; Greene, B. *Wat. Res.* **1983**, *17*, 1635-1638.
54. De Vitre, R. R.; Buffle, J.; Perret, D.; Baudat, R. *Geochim. Cosmochim. Acta.* **1988**, *52*, 1601-1613.
55. Parthasarathy, N.; Pelletier, M.; Buffle, J. *Anal. Chim. Acta* **1997**, *350*, 183-195.
56. Luther III, G. W.; Reimers, C. E.; Nuzzio, D. B.; Lovalvo, D. *Environ. Sci. Technol.* **1999**, *33*, 4352-4356.
57. Kühl, M.; Revsbech, N. P. In *The Benthic Boundary Layer*; Boudreau, B. P.; Jørgensen, B. B., Eds.; Oxford University Press: New York, NY, 2000; pp 180-211.
58. Stueben, D.; Koelbl, R.; Haushahn, P.; Schaupp, P. *Int. Oceans Syst. Des.* **1998**, *2*, 6-12.
59. Müller, B.; Buis, K.; Stierli, R.; Wehrli, B. *Limnol. Oceanogr.* **1998**, *43*, 1728-1733.
60. Brendel, P. J.; Luther III, G. W. *Environ. Sci. Technol.* **1995**, *29*, 751-761.
61. Rickard, D.; Olroyd, A.; Clamp, A. *Estuaries* **1999**, *22*, 693-701.

62. Taillefert, M.; Bono, A. B.; Luther III, G. W. *Environ. Sci. Technol.* **2000**, *34*, 2169-2177.
63. Luther III, G. W.; Glazer, B. T.; Hohmann, L.; Popp, J. I.; Taillefert, M.; Rozan, T. F.; Brendel, P. J.; Theberge, S. M.; Nuzzio, D. B. *J. Environ. Monitor.* **2001**, *3*, 61-66.
64. Xu, K.; Dexter, S. C.; Luther III, G. W. *Corrosion* **1997**, *300*, 1-18.
65. Tercier-Waeber, M.-L.; Pei, J.; Buffle, J.; Fiaccabrino, G. C.; Koudelka-Hep, M.; Riccardi, G.; Confalonieri, F.; Sina, A.; Graziottin, F. *Electroanalysis* **2000**, *12*, 27-34.
66. Van den Berg, C. M. G.;Kramer, J. R. *Mar. Chem.* **1979**, *106*, 113-120.
67. Ruzic, I. *Anal. Chim. Acta* **1982**, *140*, 99-113.
68. Piotrowicz, S. R.; Springer-Young, M.; Puig, J. A.; Spencer, M. J. *Anal. Chem.* **1982**, *7*, 1367-1371.
69. Müller, F. L. L. *Mar. Chem.* **1996**, *52*, 245-268.
70. Kozelka, P. B.; Bruland, K. W. *Mar. Chem.* **1998**, *60*, 267-282.
71. Capodaglio, G; Turetta, C; Toscano, G; Gambaro, A; Scarponi, G.; Cescon, P. *Int. J. Environ. Anal. Chem.* **1998**, *71*, 195-226.
72. Bruland, K. W.; Rue, E. L.; Donat, J. R.; Skrabal, S. A.; Moffett, J. W. *Anal. Chim. Acta* **2000**, *405*, 99-113.
73. Taillefert, M.; Lienemann, C-P.; Gaillard, J-F.; Perret, D. *Geochim. Cosmochim. Acta* **2000**, *64*, 169-183
74. Scarano, G.; Bramanti, E.; Zirino, A. *Anal. Chim. Acta* **1992**, *264*, 153-162.
75. Müller, F. L. L. *Mar. Chem.* **1999**, *67*, 43-60.
76. Bruland, K.W. *Limnol. Oceanogr.* **1992**, *37*, 1008-1017.
77. Xue, H.; Sigg, L. *Anal. Chim. Acta* **1998**, *363*, 249-259.
78. Xue, H.; Kistler, D.; Sigg, L. *Limnol. Oceanogr.* **1995**, *40*, 1142-1152.
79. Kogut, M. B.; Voelker, B. M. *Environ. Sci. Technol.* **2001**, *35*, 1149-1156.
80. Fisher, E.; Van den Berg, C. M. G. *Anal. Chim. Acta,* **2001**, *432*, 11-20.
81. De Ford, D. D.; Hume, D. N. *J. Am. Chem. Soc.* **1951**, *73*, 5321-5322.
82. Heath, G. A.; Hefter, G. *J. Electroanal. Chem.* **1977**, *84*, 295-302.
83. Correia Dos Santos, M. M.; Simões Gonçalves, M. L. *Electrochim. Acta* **1992**, *37*, 1413-1416.
84. Luther III, G. W.; Rickard, D.; Theberge, S.; Olroyd, A. *Environ. Sci. Technol.* **1996**, *30*, 671-679.
85. Luther III, G. W.; Theberge, T. M.; Rickard, D. *Talanta,* **2000**, *51*, 11-20.
86. Nelson, A.; Mantoura, R. F. C. *J. Electroanal. Chem.* **1984**, *164*, 253-264.
87. Croot, P. L.; Moffett, J. W.; Luther III, G. W. *Mar. Chem.* **1999**, *67*, 219-232.
88. Town, R. M.; Filella, M. *J. Electroanal. Chem.* **2000**, *488*, 1-16.

89. Van den Berg, C. M. G. *Analyst* **1992**, *117*, 589-593.
90. Filella, M.; Town, R. M. *J. Electroanal. Chem.* **2000**, *485*, 21-33.
91. Voelker, B. M.; Kogut, M. B. *Mar. Chem.* **2001**, in press.
92. Buffle, J.; Altmann, R. S.; Filella, M.; Tessier, A. *Geochim. Cosmochim. Acta*, **1990**, *54*, 1535-1553.
93. Town, R.; Filella, M. *Limnol. Oceanogr.* **2000**, *45*, 1341-1357.

In-Situ Measurements in the Water Column

Chapter 2

Submersible Voltammetric Probes for Real-Time Continuous Monitoring of Trace Elements in Natural Aquatic Systems

M.-L. Tercier-Waeber[1], J. Buffle[1], M. Koudelka-Hep[2], and F. Graziottin[3]

[1]CABE, Department of Inorganic and Analytical Chemistry, University of Geneva, 30 Quai E.-Ansermet, 1221 Geneva 4, Switzerland
[2]Institute of Microtechnology, University of Neuchatel, Jacquet-Droz 1, 2007 Neuchatel, Switzerland
[3]Idronaut Srl, Via Monte Amiata 10, 20047 Brugherio (MI), Italy

A summary is given of two state of the art voltammetric analytical systems that allow continuous, real-time monitoring of trace elements (Cu(II), Pb(II), Cd(II), Zn(II) and Mn(II), Fe(II)) in natural aquatic ecosystems. The first system, called the Voltammetric In situ Profiling System (**VIP System**), allowed *in situ* measurements in groundwater and surface water down to a depth of 500 m. The second system, called the Sediment-water Interface Voltammetric In situ Profiling System (**SIVIP System**), has been developed to allow the measurement of real-time, high spatial resolution trace element concentration profiles at the sediment-water interface. Construction of these systems required the development of gel-integrated interconnected or individually addressable microsensor arrays, submersible probes, and the hardware, firmware and software for control of the system components. The main characteristics of the microsensors and probes are summarized and their analytical/environmental features illustrated with examples of laboratory characterization and *in situ* applications.

Introduction

Trace elements may present a severe hazard to the normal functioning of aquatic ecosystems. They are not biodegradable but are involved in biogeochemical cycles and distributed as various physicochemical forms. The proportion of these different forms may vary continuously in space and time. Any variation in the speciation of an element will affect its bioavailability, its rate of transport to the sediment and its overall mobility in the aquatic system *(1)*. Thus, the development of new analytical instrumentation capable of performing *in-situ*, real time monitoring of specific forms of elements in a continuous and reproducible manner in the water column as well as at the sediment-water interface of natural aquatic media is required. This will allow both to get deeper insight into natural processes occurring in these ecosystems and to understand the relationship between anthropogenic releases and their long term impact on man and the environment. The main advantages of *in situ* analytical probes compared to traditional laboratory analysis are: i) rapid detection of pollutant inputs and thus quick appropriate action, ii) minimization of artifacts due to sampling and sample handling, iii) minimization of the overall cost of data collection, iv) accumulation of detailed spatial and temporal data for complete ecosystems, and v) possibility to perform measurements in locations difficult to access (i.e. boreholes, deep lakes and oceans).

Development of such tools is a challenging task for environmental analytical chemists as it requires techniques that combine high sensitivity and reliability, speciation capability, integrity of the samples and unattended operation. A detailed description of *in situ* sensors and probes for water monitoring has been recently published *(2)*. Only a few techniques meet the above requirements ; voltammetric techniques belong to them *(3,4)*. The feasibility and the usefulness of submersible voltammetric probes for *in-situ* trace metal monitoring in the water column have been reported by several authors *(5-7)*. All these systems were prototypes, based primarily on the adaptation of laboratory tools, and limited to short-term (< 1 day) measurements in surface water, i.e. depth < 20 m. Their use for long-term monitoring at greater depth was limited in particular by the following problems : i) insufficient reliability of the voltammetric sensors, ii) the use of conventional- sized electrodes (typically electrode with diameter ≥ 100 μm) which are applicable only in high ionic strength waters (> 10^{-2} M such as sea water) and are sensitive to convection in the test media, iii) the fouling of the sensor surface due to adsorption of natural organic or inorganic matter, iv) liquid junction problems with the reference electrode due to the increased pressure at depth, and/or v) interference from the dissolved oxygen. For measurements at the sediment-water interface, the potentiality of voltammetric techniques has been reported by Luther et al. *(8)*. They used a solid-state Hg-gold amalgam electrode (100 μm diameter) fixed to a micromanipulator to measure vertical profiles of O_2, Mn(II), Fe(II), S(-II) and I(-I) in sediment porewaters with sub-millimeter resolution, using square wave voltammetry (SWV) and linear sweep voltammetry (LSV) techniques. A major limitation of this system is that measurements of a complete profile (typically over a distance ≥ 2 cm) with a resolution of less than a millimeter generally requires several hours. The study of temporal concentration profile variation is thus difficult to achieve due to the long analysis time and to the fact that the sensor has to be repositioned prior to each measurement.

Recently we have developed two more sophisticated voltammetric systems by taking into account all the limitations mentioned above, as well as specific technical requirements such as robustness of the equipment, ease of handling and flexibility, rapidity of data acquisition and transmission, and low energy consumption. The first system, developed to perform *in situ* measurements in surface waters, has been called the Voltammetric In situ Profiling System (**VIP System**) *(9)*. The second system, called the Sediment-water Interface Voltammetric In situ Profiling System (**SIVIP System**) *(10)*, has been developed to enable real-time, high spatial resolution concentration profile measurements of trace elements at the sediment-water interface and other interfaces. The VIP System has been thoroughly tested in the laboratory and has been successfully applied in seawater, lake water and boreholes for *in situ* measurements of Cu(II), Pb(II), Cd(II), Zn(II) at the ppt level and Mn(II), Fe(II) at the ppb level, using either Square Wave Anodic Stripping Voltammetry (SWASV) or Square Wave Cathodic Sweep Voltammetry (SWCSV) *(11-13)*. The SIVIP System is in its infancy and still undergoing laboratory characterization *(14)*. The aim of this chapter is both to summarize the main aspects of these developments, highlighting those which solve the problems mentioned above, and to illustrate the performance and capabilities of these systems for *in situ* trace element monitoring in aquatic systems. More details on *in situ* application of voltammetric techniques in surface waters are given in *(4)*.

Technical and analytical developments of the VIP System

The VIP system is based on advanced microprocessor and telemetry technology and unique gel-integrated microsensors. It has been developed to allow reliable, unattended long term monitoring and profiling of trace elements to a water column depth of 500 meters. The whole system consists of several units: a submersible voltammetric probe, a submersible on-line O_2 removal module, a submersible multiparameter probe (Ocean Seven 316, Idronaut-Milan), a calibration deck unit, a surface deck unit and an IBM compatible PC. A detailed description is given elsewhere *(9)*. The important characteristics of the microsensors and the VIP sub-units are summarized below.

Gel-integrated single or interconnected array microelectrodes

The heart of the VIP voltammetric probe is an Agarose Membrane-covered Mercury-plated Ir-based either single or array microelectrodes (μ-AMMIE or μ-AMMIA respectively) (Figure 1a-b). These microsensors are produced under systematic, well-controlled steps and conditions to insure high reliability and sensitivity of trace metal measurements in complex media *(17-18)*. The single

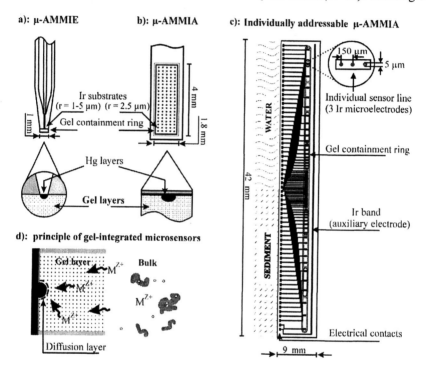

Figure 1. Schematic diagrams of the VIP (a,b) and SIVIP (c) gel-integrated microsensors and their principle (d).(Adapted with permission from references 10, 20. Copyright 2000 by Wiley-VCH, Weinheim, Germany ; and Copyright 1999 by Institute of Physic Publishing, Bristol, UK, respectively).

microelectrode is built by sealing an electroetched Ir wire with a diameter of a few micrometers in a shielded glass capillary, followed by mechanical polishing *(15)*. The microelectrode array is produced by means of thin film technology *(16)*. It consists of 5 x 20 interconnected iridium microdisc electrodes having a diameter of 5 μm each and a centre to centre spacing of 150 μm surrounded by a 300 μm thick Epon SU-8 containment ring for the gel. Both sensors are covered with a 1.5% LGL agarose protective gel membrane through which only low molecular weight ions and molecules can diffuse while colloidal and/or macromolecular materials are excluded minimizing fouling problems (Figure 1d) ; *(17,18))*. Mercury hemispheres are plated and reoxidised electrochemically through the gel, which allows the use of the same agarose protective membrane over a period of typically one month.

General description of the VIP sub-units.

The *submersible voltammetric probe* has been designed in two different models : model 1 with a Delrin housing for *in situ* measurements in surface waters (Figure 2a); model 2 with a Titanium housing for *in situ* measurements of trace elements in groundwater and mining boreholes (Figure 2b). The probe is comprised of several modules: an electronic probe housing (upper part), a pressure compensated flow-through plexiglas voltammetric cell (internal volume = 1.5 ml), a pressure case base incorporating the preamplifier for the voltammetric microsensor, and a submersible peristaltic pump (lower part). This design allows for direct access by the user to the key parts of the probe and thus simplifies the maintenance of the system. The voltammetric cell consists of two parts: an internal flow-through cell and an external cell, held together by means of a cover *(9)*. The working and the counter electrodes are located in the internal flow-through cell. The latter consists of a built-in platinum ring while the former is a gel-integrated either single or array microsensor described above. The compartment between the internal and the external cell is completely filled with 0.1 M NaNO$_3$ in 1.5% LGL agarose gel, which plays several important roles. It acts as a double bridge, with two ceramic porous junctions in contact with the working solution, for the in-house manufactured Ag/AgCl/KCl saturated 3% LGL agarose gel reference electrode located at the bottom of the external cell. It shields both the microsensor and the counter electrode. Most importantly, it acts as a pressure equalizer through a rubber pressure compensator. Pressure compensation of the cell allows *in situ* measurements at great depth and solves liquid junction problem with the reference electrode. The cell is screwed, with o-ring seals, to the cover of the pressure case base. The pressure case base is mechanically assembled to the electronic housing via two titanium rod connectors, through which pass the electrical connections of the three electrodes, the microsensor preamplifier and the submersible pump. The advantages of this configuration are: i) the flow-through voltammetric cell is protected against shocks, ii) maintenance and replacement of the working and reference electrodes are easy since they are simply screwed with o-ring seals, iii) watertightness of all the electrical connections is made possible using standard connectors, and iv) the preamplifier is isolated from the main electronic part, which minimizes an important source of noise. The electronic housing contains all the hardware and firmware necessary to manage: the voltammetric measurements, the interfacing of the multiparameter probe, calibration deck unit and submersible peristaltic pump, and the data acquisition and transfer by telemetry. Data files are stored in an internal non-volatile memory having its own battery, which guarantees high data retention and protection.

The *submersible on-line oxygen removal module* is connected between the sampling pump and the inlet of the flow-through voltammetric cell. It consists of a silicone tubing surrounded by a chemical reducing agent gel. As water is pumped through the silicone tubing, oxygen diffuses through the tubing wall and is consumed on the other side by the reducing agent gel *(13)*. This module is required only for *in situ* trace element monitoring in oxygen-containing freshwaters (see : Environmental application of the VIP System).

The *submersible multiparameter probe* allows to control the exact position of the voltammetric probe model 1 (Figure 2a) at depth and to measure simultaneously the following parameters : temperature, conductivity, salinity, dissolved oxygen, pH and Redox potential. In the case of the probe model 2 (Figure 2b), a temperature and depth sensor have been incorporated into the submersible voltammetric probe.

The *calibration deck unit* enables one to perform in the laboratory, on-shore or on board ship the renewal of the microsensor Hg layer, the calibration of the probe, and the measurements of standards or collected chemically modified natural samples, e.g. acidified raw or filtered samples for the measurements of the colloidal and particulate forms (see : Features and selectivity of gel-integrated microsensors). When the voltammetric probe is ready for deployment, this unit is disconnected.

The *surface deck unit* powers and interfaces, by telemetry, the measuring system with a Personal Computer. This unit allows an autonomy of about 35 hours and can be recharged either in continuous mode using a solar captor or after use. Communication between the Personal Computer and the voltammetric probe is carried out by using the Terminal Emulator under Windows. A user-friendly management software allows the user to control and configure the voltammetric probe operating parameters and functions, such as electrochemical parameters, data acquisition, calibration and maintenance operations. The system can be controlled either by an operator on board or in automatic mode following pre-programmed instructions.

Technical and analytical developments of the SIVIP System

The measurement concept of the SIVIP System is totally different of the VIP System. It is based on an array of individually addressable gel-integrated microsensors, a fast multichannel detection system and telemetry technology.

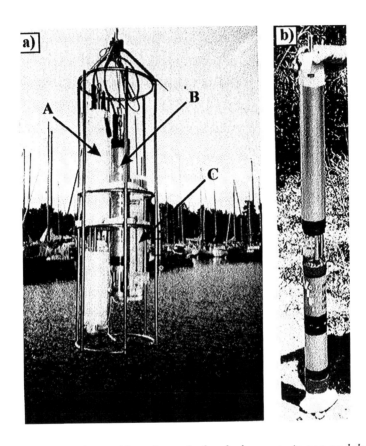

Figure 2. a) VIP submersible unit ready for deployment in sea or lake ; A : voltammetric probe model 1 (dimensions : 86 cm length, 10 cm in diameter ; weight : 8 kg in air, 4 kg in water); B: multiparameter probe, ; C: on-line O_2 removal system. b) VIP voltammetric probe model 2 (dimensions : 100 cm length, 7 cm in diameter ; weight : 6 kg in air, 6 kg in water) for groundwater monitoring.(Figure 2b reproduced with permission from reference 20. Copyright 1999 by Institute of Physic Publishing, Bristol, UK).

This is the first system reported which allows reliable and sensitive simultaneous recording of complete voltammograms over a large number of individually addressable microelectrodes with fast dynamic techniques such as square wave voltammetry (10). The important characteristics of the SIVIP System are summarized below.

The gel-integrated individually addressable microelectrode arrays

The schematic diagram of the individually addressable Agarose Membrane-covered Mercury-plated Ir-based microelectrode arrays (individually addressable μ-AMMIA) is given in Figure 1c. It consists of 64 lines of three Ir microdisk electrodes of 5 μm in diameter and a center-to-center spacing of 150 μm. The spacing distance between sensor lines n° 1 to 23 decreases from 2000 μm to 220 μm and has a constant value of 200 μm between sensor lines n° 23 to 32. The geometry of sensor lines n° 33 to 64 is a mirror image of that for the lines n°1 to 32. This design has been chosen to allow real-time concentration profile measurements over a total distance of 4 cm (i.e. about 2 cm in the water column and 2 cm into the sediments) with a maximum resolution at the interface. It should be noted that this geometry can be readily modified if needed. The device is completed with a 0.25 x 40 mm Ir band that can be used as auxiliary electrode. The microelectrode arrays are prepared on a 4" silicon wafer using thin-film technology (10). The patterned 300 μm EPON Su-8 agarose gel containment ring (18) has been subdivided in 9 compartments, to facilitate the membrane deposition process and minimize adhesion failures that could occur with larger gel membranes. The 9 x 42 mm individual devices are mounted on a 4 layers, 1.5 mm thick, Printed Circuit Board (PCB), wire-bounded, and encapsulated with epoxy resin. As for the VIP microsensor, the sensor lines are covered with a 1.5% LGL agarose gel protective membrane (10), and the Hg layers are electrochemically deposited and reoxidised through the gel (17,18).

General description of the SIVIP sub-units

The *voltammetric probe* is based on a three electrode system: the above individually addressable μ-AMMIA as the working electrode, a home-built Ag/AgCl/KCl sat. in gel reference electrode (9), and an external platinum rod or an on-chip Ir band auxiliary electrode. The voltammetric probe hardware and firmware have been designed to allow simultaneous measurements over the 64 individually addressable microsensor lines using a single potentiostat. This has been achieved by the development of a powerful double-stage multiplexing system. In the first multiplexing stage, the individual sensor line responses are first pre-amplified with a fixed gain (i.e. each individual line has its own preamplifier), then multiplexed into 8 groups of 8 sensor lines. The second multiplexing stage allows to switch simultaneously between the 8 successive sensor lines of each group of 8 every 5 μs (i.e. the maximum time interval for one sampling measurement over the 64 lines is 35 μs). Note that the preamplifiers of the first multiplexing stage not only improve the signal to noise ratio but even more importantly insure that the sensor line potentials are always properly defined at all times, i.e. no sensor lines are left in open circuit when not addressed by the multiplexer. The 64 lines are thus polarized simultaneously by the potentiostat while they are multiplexed, simultaneously by groups of 8, during the measurement phase. Data acquisition is controlled by the firmware. The data flow to a FIFO memory during the sampling time at the end of each square wave pulse and are stored in a temporary memory during the beginning of the next square wave pulse. At the end of the potential ramp scanning, the data stored in the temporary memory are processed and the final complete measurement file (i.e. data file of the 64 voltammograms) stored in a non-volatile memory which has its own battery.

The entire electronics module of the voltammetric probe may be fitted into the lower part of the cylindrical titanium housing of an *in-house microprofiler* (Idronaut, Milan) designed to withstand pressure up to 600 bars. The three electrodes are fitted via titanium connectors in the lower cover of the microprofiler body. The microprofiler controls, via a drive screw activated by a special DC microprocessor-controlled motor, the positioning of the microsensor at the sediment-water interface with a resolution of 200 μm.

A *surface deck unit*, similar to that of the VIP System, powers and interfaces, by telemetry, the measuring system. Communication between the Personal Computer and the SIVIP System is carried out by using the Hyperterminal Emulator under Windows. The management software has a structure, and thus features, similar to that of the VIP System, with some additional functions and commands to manage the microprofiler.

Features and selectivity of gel-integrated microsensors

Measurements with the gel-integrated microsensors are performed in two successive steps: a) equilibration of the agarose gel with the test solution (typically 5 min for a gel thickness of 300 μm) and b) voltammetric analysis inside the gel *(17,18)*. These microsensors have key features required for *in situ*

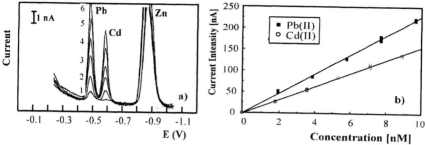

Figure 3. Examples of Pb(II) and Cd(II) measurements at the nanomolar level. a) SWASV voltammograms for direct measurements (curve 1) and successive standard additions of 1.1 nM of both metals (curves 2-6) in a 0.2 μm filtered Arve River (Switzerland) sample ; μ-AMMIE. b) Triplicate calibration curves of both metals in synthetic freshwater ; μ-AMMIA..(Figure 3a reproduced with permission from reference 10. Copyright 2000 by Wiley-VCH, Weinheim, Germany)

measurements, in particular, for both VIP and SIVIP microsensors: 1) high sensitivity and reliability (Figure 3a-b ; *(17,18)*) ; 2) organic and inorganic colloidal and particulate material are efficiently excluded from the agarose gel *(17,18)* and do not interfere with voltammetric measurement ; 3) well-controlled molecular diffusion of metal species occurs in the gel, i.e. ill-controlled hydrodynamic currents of the test water do not influence the voltammetric signal *(17)*; 4) current intensities are a function of the diffusion coefficients of the analytes inside the gel and not in the test media, this is particularly important to allow correct interpretation of *in situ* measurements in porewaters (see : Characterization of the SIVIP System) ; 5) the external medium is not modified by the voltammetric measurements (i.e. the voltammetric diffusion layer is small compared with the gel thickness (Figure 1d)), this is particularly relevant for in situ measurements in sediment where the fluxes at the electrode might influence the porewater concentration profiles as it may be the case for technique of diffusive gradients in thin films (DGT) *(19)*; 6) effects of temperature variation on the voltammetric current intensity can be readily corrected *(20)* (note that peak current variation of 3 to 8% $°C^{-1}$ depending of the analyte were observed which is of significant concern for concentration profiling in water columns where temperature may vary from typically 4 to 25°C); 7) signals are independent of the pressure in the range 1 to 600 bars (i.e. these sensors can be used for monitoring down to 6000 m depth) *(18)* ; 8) microsized-electrodes (i.e. size ≤ 10 μm) have low *iR* drops and reduced double-layer capacitance, i.e. direct measurement can be performed in low ionic stength freshwaters; 9) current intensities measured at micro-sized electrodes are controlled by spherical diffusion and reach a nonzero steady-state value at constant potential, i.e. stirring is not required during the preconcentration step of stripping techniques which is absolutely required to allow anodic stripping voltammetric measurements inside a gel as well as in sediment porewaters; 10) signals measured at micro-sized electrodes are proportional to the diffusion coefficient values of the target compounds and negligible for species larger than a few nm (Figure 4a ; *(4,11)*); and for the SIVIP microsensor : 11) real-time, high resolution whole concentration profiles are obtained at each measurement without repositioning the sensor *(14)*.

Points 1) to 7) are unique key features of the gel-integrated microsensors which solve the problems i) to iii) mentioned in the introduction. They are required to enable rigorous interpretation of voltammetric data obtained from direct measurements in complex media as well as reliable operation of chemical sensors in complex media for a long period of time (typically standard deviation of max. 10% were observed for continuous trace metal measurements, at the nanomolar level, over two weeks using the same Hg layers *(13)*).

Points 8) to 10) are features of voltammetric techniques coupled to microelectrodes. Point 10) is a key feature for trace metal speciation studies. In particular, three environmentally relevant types of metal species, based on their size, can be determined with simple, minimum sample handling (Figure 4a ; *(4)*): i) the dynamic species (defined as free ions and small labile complexes with size smaller than few nm) obtained selectively by direct voltammetric measurements in unmodified samples, ii) the colloidal species, including dissolved inert

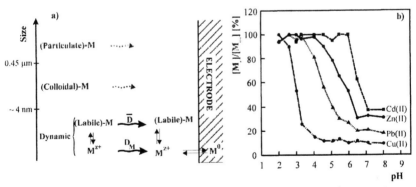

Figure 4. a) Scheme of diffusion of metal species towards voltammetric microelectrodes. b) Change with pH of trace metal SWASV peak currents measured in 0.2 μm filtered Arve River sample using a μ-AMMIA. Total concentrations at pH 2 : Zn(II) = 28.8 nmol L⁻¹ ; Cd(II) = 0.27 nmol L⁻¹ ; Pb(II) = 3.9 nmol L⁻¹ ; Cu(II) = 16.1 nmol L⁻¹.(Figure. 4b reproduced from reference 21. Copyright 2000 ACS Publications).

complexes, (size ≤ 0.45 µm) obtained by difference between metal concentrations measured in acidified (pH 2), filtered samples and dynamic species concentration, iii) the total dissolvable metals adsorbed on particles, defined as the particulate species (size > 0.45 µm), obtained by difference in metal concentrations between raw and acidified (pH 2), filtered samples. Distinction between these three different forms is important for the understanding of the role and the fate of vital or harmful trace elements. In particular, the dynamic fraction is the most important one for bioavailability and ecotoxicity interpretation. It is also the fraction which is the most difficult to measure without analytical artifact (due to sample degradation and risk of contamination) and thus requires direct *in situ* measurements. The colloidal and the particulate forms play different important roles in metal cycling and residence time, e.g. fast sedimentation followed by accumulation in sediments and possible remobilization of the particulate species, and slow coagulation/sedimentation of the colloidal species. Even more information can be gained by voltammetry by acidifying the sample gradually and recording the change in peak current with pH (Figure 4b) and time *(21)*. The latter allows the determination of the kinetics of trace metal desorption in a given aquatic system. The former allows determination of both the fraction of dynamic and dissolvable adsorbed metal ions at the natural pH as well as the average binding energy of metal on the natural particles or colloids from the inflection point $pH_{1/2}$ *(22)* of the pH titration curve. Such measurements are very difficult to perform using classical separation techniques (e.g. ultracentrifugation, ultrafiltration) due to contamination problems.

Finally, feature 11), linked to simultaneous measurements over individually addressable sensor lines, allows the measurement of detailed spatial and temporal concentration profiles with minimum perturbation of the sediment.

Environmental applications of the VIP System

Mn(II) in anoxic lakes

Stability and reliability of in situ voltammetric measurements : Mn(II) concentration profiles were measured within the anoxic hypolimnion of the shallow eutrophic Lake Bret (Switzerland), with the objective of testing the stability, accuracy and reliability of the measurements performed with the VIP System *(11)*. For this purpose, *in-situ* VIP measurements with single and array microelectrodes were compared to voltammetric measurements performed on-site (on a platform at the lake surface), using a microsensor array with and without protective gel. The samples for surface measurements were pumped from exactly the same depth as the VIP position and transferred into an air-tight cell, thermostated at 20°C. Comparison was also made with laboratory ICP-AES analysis of acidified, filtered and ultracentrifuged samples (Figure 5a).

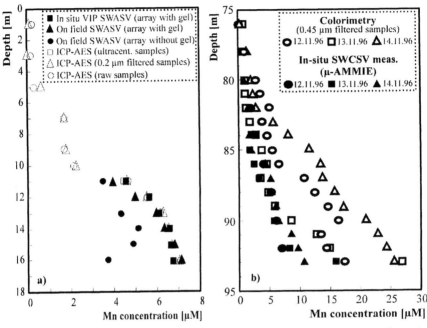

Figure 5. Concentration profiles of Mn(II) measured using various analytical techniques in the anoxic hypolimnion of a) Lake Bret, Switzerland (August 20, 1997) and b) Lake Lugano, Switzerland (November, 1996).(Reproduced from reference 11. Copyright 1998 ACS Publications).

In situ measurements were done without changing the Hg layer during the three days of operation, including calibrations of the probe the day before and the day after each deployment. Calibrations were performed by standard additions into 0.2 μm filtered lake water, using the same operating conditions as for field measurements. The major findings of these studies were as follows:

- Calibration curves were reproducible before and after deployment, and from one deployment to another. Standard deviations of the slopes of calibration curves were in the range of 5.2 to 6.9% (95% probability) for 7 calibrations of each of the following operation modes in various deployments: SWASV on μ-AMMIE, SWCSV on μ-AMMIE and SWASV on μ-AMMIA *(11)*. These results demonstrate the good reproducibility of the Hg layer formation, its stability and the absence of memory effects during deployment. (Note that thanks to this reliability, a complete calibration before each deployment is not required and measurements of one or two standard solutions are sufficient).

- Comparison between voltammetric on-site measurements with and without gel (Figure 5a) clearly shows the efficiency and the importance of the protective agarose gel. Concentrations that are biased to systematically too low values were obtained for the unprotected microsensor. A detailed study showed that the peak current attenuations observed with unprotected sensors were due to adsorption on Hg of lake born iron(III) hydroxide colloids *(11)*. It must be emphasised that, apart from the peak current lowering, no other perturbations in the voltammetric signals, e.g. changes in peak potential or shape, were observed. Note that similar behavior has been also observed for trace metal measurements in the presence of peat fulvic acid and river inorganic particles *(17)*. Thus measurements in the absence of a protective gel would lead to significant underestimation of Mn(II) concentrations, in variable proportions depending on Fe(III) hydroxide concentration in the water, i.e. to *wrong values and shapes* of the concentration-depth profiles in the lake, and the difference observed between *in situ* voltammetry and classical techniques might be wrongly attributed to the presence of colloidal Mn species even though such species are not present (see below).

- The good agreement (Figure 5a) between the concentration values obtained from *in-situ* measurements after temperature effect correction, and both on-site measurements, performed at a constant temperature of 20°C, and ICP-AES analysis in ultracentrifuged samples demonstrate that valid information can be obtained by *in-situ* measurements. In particular, the *in situ* tests show that pressure has no effect on the results, as laboratory tests had already shown *(18)* and that the temperature variation effect on *in situ* voltammetric peaks can be corrected using a temperature calibration performed in laboratory *(20)*.

Intercomparison of methods and speciation data : Validation of *in situ* voltammetric measurements in complicated media by intercomparison with other methods is not easy due to the different sensitivities of the various methods to the different species of the test analyte *(4)*. In particular, for metal ion analysis, voltammetry is mostly sensitive to the dynamic metal species (size ≤ a few nm), while the other detection techniques which can be used for comparison, like AAS or ICP-MS, have little or no sensitivity to metal speciation. Comparisons must therefore include size fractionation, e.g. by filtration and ultrafiltration on low porosity membranes or ultracentrifugation.

Intercomparisons of methods were performed for the analysis of Mn(II), which is less sensitive to analytical artifacts and in particular to contamination during size fractionation. *In situ* voltammetric measurements were made in the anoxic water of Lake Bret, Switzerland (depth 11-18 m; Figure 5a) and Lake

Lugano, Switzerland (80-95m; Figure 5b). Laboratory measurements of acidified (pH=2) samples of raw water, filtered water (0.2 or 0.45 µm pore size membrane) and ultracentrifuged water (30000 rpm for 15 h, corresponding to a size limit of ~ 5 nm) were done by AAS, ICP-AES or colorimetry *(11)*. In Lake Bret (Figure 5a), all the methods gave the same results, showing that the whole of Mn(II) is voltammetrically detectable, (probably most of it is in the Mn^{++} form), and that *in situ* voltammetry provides valid results compared to classical laboratory techniques. In Lake Lugano (Figure 5b), all Mn(II) profiles measured *in-situ* with the VIP System were similar over the measured period while colorimetric measurements, performed in 0.45 µm filtered samples, showed higher values and larger variations in concentration from one day to another. Other available information *(11)* indicates that the differences were due to the resuspension of colloidal Mn (1 nm < Mn < 450 nm), measurable by colorimetry but not by voltammetry, from the sediments of Lake Lugano. These two cases exemplify the additional information which can be gained from *in situ* voltammetry by combining it with spectrometry of both filtered and raw water. The results obtained in Lake Lugano (Figure 5b) also illustrate the importance of unambiguous discrimination between colloidal and ionic forms, and the importance of continuous measurements over an extended period of time versus random sampling to distinguish between seasonal and temporal variability.

Trace metal monitoring in sea water and freshwater.

Sensitivity limit in oxygenated water : O_2 is an electroactive compound in oxic water and may produce two types of interferences : a) production of a signal 4 to 6 orders of magnitude larger that the signal produced by the trace metals of interest, and b) a pH increase at the electrode surface due to the reduction of O_2, in particular during the preconcentration step of stripping techniques. Application of the VIP System for *in situ* measurements in oxygenated sea water (Venice Lagoon - Italy and Gullmar Fjord - Sweden) has demonstrated that a pH increase is not encountered in sea water, due to the high buffering capacity of this medium. Interference a) can be eliminated by using SWASV with a relatively high frequency coupled to background current subtraction *(9)*, confirming the results of previous *in situ* measurements using a macro-sized mercury film electrode *(5)*. Based on a signal-to-noise ratio of 2, the detection limits for Cu(II), Pb(II), Cd(II) and Zn(II) in oxic sea water were found to be 0.1, 0.05, 0.05 and 0.3 nmol.L^{-1} respectively.

In freshwater, interference b) may lead to drastic deformations or even suppression of the trace metal voltammetric signal, due to the formation of hydroxide, carbonate or sulfide precipitates during the stripping step. A submersible on-line oxygen removal system was developed to solve this problem

(Figure 2a ; *(13))*. Efficiency of this module coupled to the VIP System in oxygenated freshwater has been tested and demonstrated in Lakes Leman *(13)*, Luzern and Alpnach (Swizerland). *In situ* VIP measurements and sampling were performed in surface water (depth : 2-5 m) at various stations. Typical examples of the results obtained in the last two lakes for the *in situ* VIP monitoring and the total metal concentrations measured by ICP-MS are given in Figure 6. Detection limits were found to be similar to those in sea water. The ratios of dynamic to total metal concentrations were typically in the range of 5 to 40%. Similar proportions were observed in the Arve river (Figure 4b; *(17, 21))* and in Venice Lagoon *(9)*. Standard deviations, determined from three replicate measurements performed at stations 1, 3, 6 and 8, were found to be $\leq 10\%$ for the three metals. It shows that, as in sea water *(9)*, metal concentrations at the ppt level (10^{-11}-10^{-10} mol L^{-1}) can be measured *in situ* with a good reproducibility. This allows the detection of fast, dynamic changes, as observed from unattended autonomous VIP *in situ* measurements performed in Gullmar Fjord *(9)*, which is important for water quality control monitoring and the study of the fate of trace metal (see below).

Potentiality for trace metal bioavailability studies : The previous examples demonstrated the reliability and sensitivity of the VIP voltammetric probe for in situ measurements of the dynamic fraction of trace elements (i.e. the fraction the most easily bioavailable and difficult to determine using classical techniques). Thanks to these advantages and the ability to perform autonomous, long-term monitoring, the VIP System should be an efficient tool to study both the influence of varying physicochemical conditions on the proportion of the trace metal dynamic species, and the interaction of the dynamic fraction of metal with aquatic biota. The latter is illustrated by the results obtained for *in situ* trace metal profiling in Lake Leman reported in Figure 7. In particular, the ratio of dynamic to total concentrations of Pb(II) ($[Pb]_d/[Pb]_{tot}$) was similar over the entire water column. For Cu(II) and Zn(II) (data not shown for this latter), however, the dynamic fraction and the dynamic to total ratio were found to be much smaller in the upper water layer (typically the first 15 m) than in the bottom layer. The available annual data on the biophysicochemical conditions of Lake Leman show that high primary productivity (i.e., production rates = 22 to 166 mg C/ m^3 day ; Chlorophyll a concentrations = 1.5 to 7.5 mg/m^3) is also observed over the first 15 m of the water column at this period of the year *(23)*. This suggests that an important fraction of the dynamic Cu(II) and Zn(II) species is either assimilated by the phytoplankton or complexed by their exudates *(24)*, or even both, as non-labile species in the upper layer of the water column. The role of biota is supported by the fact that Pb(II), which is known to be not easily assimilated, does not show the same trends as Cu(II) and Zn(II). These data show that extended studies of the variation of the dynamic fraction of the different elements as a function of seasonal primary productivity should enable the collection of useful information on the bioavailability of trace metals.

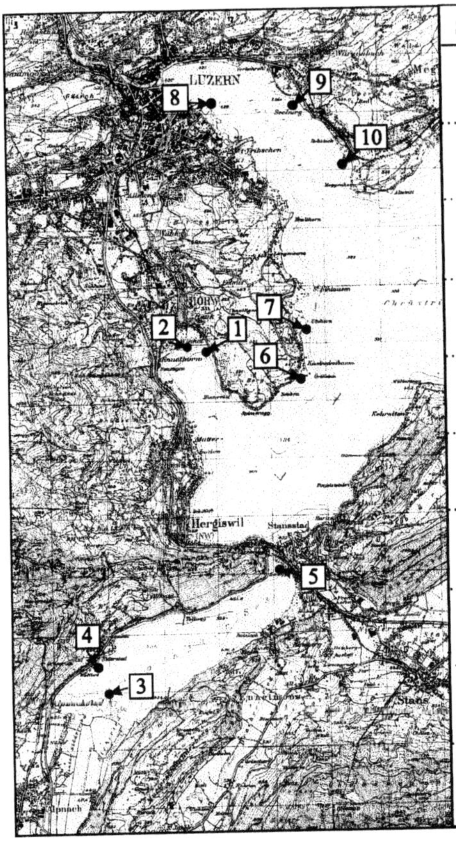

Station	Metal	C_m [nM]	C_{tot} [nM]	C_m/C_{tot} [%]
1	Cu	1.50	12.96	11.6
	Pb	0.65	3.98	16.3
	Cd	0.02	0.06	33.3
2	Cu	1.47	12.65	11.6
	Pb	0.40	2.25	17.8
	Cd	0.02	0.05	40.0
3	Cu	0.94	19.79	4.7
	Pb	0.42	1.55	27.1
	Cd	BDL	0.04	----
4	Cu	0.84	18.57	4.5
	Pb	0.30	0.89	33.7
	Cd	BDL	0.03	----
5	Cu	1.54	19.58	7.9
	Pb	0.24	0.87	27.6
	Cd	BDL	0.04	----
6	Cu	2.00	12.41	16.1
	Pb	0.11	0.61	18.0
	Cd	0.02	0.04	38.0
7	Cu	1.17	10.34	11.3
	Pb	0.09	0.34	26.5
	Cd	0.01	0.04	25.0
8	Cu	1.87	12.95	14.4
	Pb	0.11	0.60	18.3
	Cd	0.02	0.06	33.3
9	Cu	1.13	10.22	11.0
	Pb	0.10	0.35	28.5
	Cd	0.01	0.03	33.3
10	Cu	0.96	10.92	8.8
	Pb	0.11	0.46	23.9
	Cd	0.01	0.04	25.0

Figure 6. Trace metal coastal monitoring in Lakes Luzern and Alpnach (Switzerland, June 2000). C_d: concentrations of the dynamic fraction measured in situ using the VIP unit of Figure 2a ; C_{tot}: total metal concentrations measured in pH 2 acidified samples using ICP-MS. BDL= below detection limit.

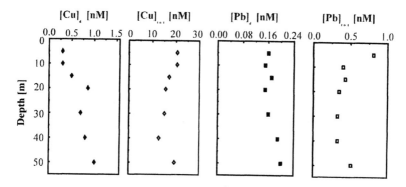

Figure 7. Examples of Cu and Pb concentration profiles measured in Lake Leman (Switzerland, June 1999). $[M]_d$: conc. of the dynamic fraction measured in situ with the VIP voltammetric probe connected to the O_2 removal system ; $[M]_{tot}$: total metal conc. measured in pH 2 acidified samples using ICP-MS.(Adapted from reference 13. Copyright 1998 ACS Publications).

Characterization of the SIVIP System

Reliability and sensitivity of trace metal measurements by SWASV.

The capability of the system to allow simultaneous trace metal measurements over the 64 individual lines of the gel-integrated microsensor arrays by means of voltammetric techniques with fast scan rates was investigated and demonstrated by performing Pb(II) and Cd(II) measurements in standard solutions using SWASV with a pulse amplitude of 25 mV, a step potential of 4 mV and frequencies of 100 and 200 Hz (i.e. scan rates of 400 and 800 mV/s) respectively. The main results obtained were: i) reliable measurements of complete trace metal voltammograms over the 64 sensor lines can be achieved ; ii) influence of the sensor geometry on the signals obtained at the different sensor lines follows the behavior expected from the theory and can be readily corrected ; iii) standard deviations of maximum 10% are obtained for both the average normalized current intensity determined from the 64 lines and the slopes of the calibration curves determined at each sensor line ; and iv) a detection limit of 2 nM is obtained for both metals using a preconcentration time of 5 minutes. A detailed discussion of the results is given in *(10)*.

Profiling capability.

The profiling capability of the SIVIP System has been tested by performing real-time concentration gradient measurements of Tl(I), Pb(II) and Cd(II) as a function of time at well controlled liquid-liquid and liquid-solid interfaces *(14)*. Before each set of concentration profile measurements, proper functioning of the individually addressable μ-AMMIA was tested in a standard solution. After profile measurements, systematic calibrations were performed by standard addition of the target elements to a N_2 degassed 0.1 M NaNO$_3$ + 0.01 M acetate buffer (pH 4.65) solution. The main results obtained for Pb(II) measurements at the liquid-solid interface are summarized below.

Pb(II) diffusion at liquid-solid interface : The procedure used to study Pb(II) diffusion at the liquid-solid interface was as follows: 50 g. of silica was first equilibrated with 500 ml of 0.1 M NaNO$_3$ + 0.01 M acetate buffer solution (pH 4.65) under magnetic stirring for two days in the measurement cell. The stirring was then stopped and, after complete sedimentation of the silica, the individually addressable μ-AMMIA was positioned at the liquid-solid interface in a such way that the half of the sensor was in the solid phase. A given concentration of Pb(II) was added to the overlying liquid layer and nitrogen gas was used to gently bubble the solution for homogenization. This process was carefully controlled to avoid disturbance at the interface. Typical examples of whole concentration profiles, measured in real time, and variation of the concentration profiles as a function of time are shown in Figure 8a.

The shape of the concentration profiles in the liquid and the solid phases are different as the flux, and in particular the effective diffusion coefficient, in the solid phase is a function of the physical characteristics of the medium, i.e. the porosity and the geometric tortuosity of the silica layer. The flux may also depend upon the degree of complexation/adsorption of the metal ion by the solid phase, as was the case for Pb(II) in our system (see eq. 3 below ; for detail see *(14)*). Two analytical expressions of the concentration profiles are thus used in such systems *(25)*, i.e. one for the liquid phase, eq. (1), and one for the solid phase, eq. (2) :

$$C_w(x,t) = \frac{C_0}{1+\sqrt{D_s^{eff}/D_w}}\left[1+\sqrt{D_s^{eff}/D_w}\ erfc\frac{x}{2\sqrt{D_w t}}\right] \tag{1}$$

$$C_s(x,t) = \frac{C_0}{1+\sqrt{D_s^{eff}/D_w}}\ erfc\frac{x}{2\sqrt{D_s^{eff} t}} \tag{2}$$

with, under working conditions of our system :

$$D_s^{eff} \cong \phi\theta^{-2} D_w \frac{\left[Pb^{2+}\right]}{\left[Pb\right]_t} \cong \frac{D_w}{F} \frac{\left[Pb^{2+}\right]}{\left[Pb\right]_t} \qquad (3)$$

where C_0 = initial concentration of the analyte in the top solution phase ; $C_w(x,t)$ and $C_s(x,t)$ = concentrations of the analyte, in the liquid and the solid phases respectively, at depth x (x > 0 in the liquid phase and x < 0 in the solid phase) and time t ; D_w and D_s^{eff} = effective diffusion coefficients of Pb in the liquid and

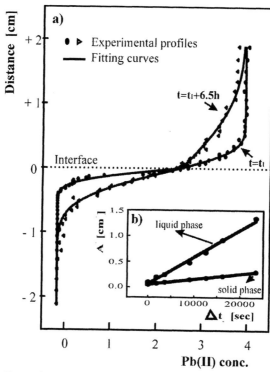

Figure 8. a) : Typical Pb(II) concentration gradients measured by SWASV at a liquid-solid interface using the SIVIP System. Inset b): Linear relationship obtained between A^2 and Δt_n in both phases.

the solid phases respectively ; ϕ = porosity of the solid phase (defined as the volume of the porewater to the total volume of the solid phase and equal to 0.85±0.04 in our system) ; θ = geometric tortuosity of the porous media and F =

formation resistivity factor (defined as the ratio of the resistivity of the solid phase to the liquid phase). The experimental concentration profiles were fitted using the eqs. 1, 2 rewritten as :

$$C_w(x,t) = \frac{C_0}{1+B}\left[1 + B \cdot erfc\frac{x}{A_w}\right] \tag{1a}$$

$$C_s(x,t) = \frac{C_0}{1+B}erfc\frac{x}{A_s} \tag{2a}$$

where $B = \sqrt{D_s^{eff}/D_w}$; $A_w = 2\sqrt{D_w(t_1 + \Delta t_n)}$ for the liquid and $A_s = 2\sqrt{D_s^{eff}(t_1 + \Delta t_n)}$ for the solid phases ; t_1 = measurement time of the first concentration profile, and Δt_n = time interval between the measurements of the first and the nth concentration profile. As expected for a well defined interface, similar B values were obtained in both media (i.e. (0.519 ± 0.10) and (0.520 ± 0.025) for the liquid and solid phase respectively) and a linear response was found between both A_w^2 and Δt_n and A_s^2 and Δt_n (Figure 8b). From the slope of the former, the effective diffusion coefficients of Pb(II) were found to be : $D_w = (1.31 \pm 0.06) \times 10^{-5}$ cm^2/s and $D_s^{eff} = (0.35 \pm 0.05) \times 10^{-5}$ cm^2/s, and from the slope of the latter : $D_w = (0.98 \pm 0.10) \times 10^{-5}$ cm^2/s and $D_s^{eff} = (0.26 \pm 0.05) \times 10^{-5}$ cm^2/s. The value of D_w calculated from the concentration profile measured in the solid phase (i.e. where convection is negligible) is very close to the Pb(II) diffusion coefficient value of 0.95×10^{-5} cm^2/s given in the literature (26). A good agreement was also obtained between the ratio of [Pb^{2+}]/[Pb]$_t$ computed from the data of Figure 8b and complexation measurements of lead with silica, i.e. 0.367 and 0.375 respectively (14).

All these results confirm the reliability and validity of the overall concentration gradients measured with the SIVIP System. They demonstrated also that, despite the fact that the effective diffusion coefficient in the solid phase is about one-third that in the liquid phase, reliable metal concentrations in the solid phase can be determined from the slopes of the calibration curves performed in liquid phase. This is linked to an important feature of gel-integrated microsensors, i.e. measurements are performed inside the gel and thus current intensities measured are a function of the diffusion coefficients of the analytes in the gel and not in the different media. It is worth stressing that this feature is required to allow correct conversion of current intensities measured in porewaters into concentration gradients and thus reliable interpretation of trace elements fluxes at the sediment-water interface.

Conclusion.

It has been shown that the combination of voltammetric principles with recent breakthroughs in electronics and microtechnology allows the development of sophisticated, compact and reliable submersible analytical devices for *in situ* trace element measurements in natural aquatic systems. For the VIP System, further developments are underway to allow *in situ* measurements of other trace metal species in addition to the dynamic fractions (i.e. *in situ* speciation), *in situ* monitoring of other relevant elements and control of the submersible system from a land station. For the SIVIP System, systematic laboratory studies in sediment cores are underway to optimize the analytical conditions, the geometry of the microsensor arrays and /or of the Printed circuit Board (if needed), and the proper control of the positioning of the voltammetric gel-integrated microsensor arrays at the sediment-water interface using the microprofiler as well as to validate the results obtained in real media. This is required before the field deployment of the SIVIP system as *in situ* trace element voltammetric measurements in sediment porewaters is not straightforward, particularly in the presence of sulfide (for details see *(4)*). It must be noted that, due to the complexity of environmental media, a thorough characterization in the laboratory of any novel system similar to those presented here is absolutely required before field deployment. Indeed, a system reliability close to 100 % and a detailed understanding of processes occurring in the bulk solution and at the sensor surface are prerequisite to enable reliable conversion of signals into concentrations and thus the collection of valid environmental data *(4)*.

Acknowledgements

The authors thank : Fabio Confalonieri, Giuliano Riccardi and Antonio Sina (Idronaut) who have developed the software, the hardware of both systems and the submersible pump of the VIP System respectively. Claude Bernard, François Bujard and Serge Rodak (CABE) who have build the VIP flow-through voltammetric cell, the on-line oxygen removal system and all the instrumentation used for constructing the voltammetric single microelectrode. G.C. Fiaccabrino (IMT) who has build the interconnected and individually addressable Ir microelectrode arrays and developed the first multiplexing stage of the SIVIP probe. Jianhong Pei (CABE) for performing measurements presented in Figures 4b and 8.

This work was supported by the European Commission (Marine Science and Technology- MAST III program, contract no. MAS3-CT95-0033) and the Swiss Federal Office for Education and Science (OFES).

38

References

1. *Metal Speciation and Bioavailability in Aquatic System* ; Tessier, A.; Turner, D.R., Eds.; IUPAC Series of Analytical and Physical Chemistry of Environmental Systems; John Wiley & Sons: Chichester, 1995; Vol. 3.

2. *In situ Monitoring of Aquatic Systems*: Chemical Analysis and Speciation ; Buffle, J.; Horwai, G., Eds.; IUPAC Series of Analytical and Physical Chemistry of Environmental Systems; John Wiley & Sons: Chichester, 2000, Vol. 6.

3. Tercier, M.-L.; Buffle, J. *Electroanalysis* **1993**, *5*, 187-200.

4. Buffle, J.; Tercier-Waeber, M.-L. In *In situ Monitoring of Aquatic Systems*: *Chemical Analysis and Speciation*; Buffle, J. ; Horwai, G., Eds.; IUPAC Series of Analytical and Physical Chemistry of Environmental Systems; John Wiley & Sons: Chichester, 2000, Vol. 6, Ch. 9, pp. 279-405 .

5. Tercier, M.-L.; Buffle, J.; Zirino, A. ; De Vitre, R.R.   *Anal. Chim. Acta* **1990** , *237*, 429-457.

6. Wang, J. ; Fostner, N. ; Armalis, S. ; Larson, D. ; Zirino, A.; Olsen, K. *Anal. Chim. Acta.* **1995**, *310*, 223-231.

7. Herdan, J.; Feeney, R.; Kounaves, S.P.; Flannery, A.F.; Storment, C.W.; Kovacs, G.T.A.; *Environ. Sci. Technol.* **1998**, *32*, 131-136.

8. Luther III, G.W.; Taillefert, M.; Bono, A.; Brendel, P.J.; Sundby, B.; Reimers, C.E.; Nuzzio, D.B.; Lovalvo, D. *Mineral. Mag.* **1998** , *62A*, 921-922.

9. Tercier, M.-L.; Buffle, J.; Graziottin, F. *Electroanalysis.* **1998**, *10*, 355-363.

10. Tercier-Waeber, M.-L.; Pei, J.; Buffle, J.; Fiaccabrino, G.C.; Koudelka-Hep, M.; Riccardi, G.; Confalonieri, F.; Sina, A.; Graziottin, F. *Electroanalysis.* **1999**, *12*, 27-34.

11. Tercier-Waeber, M.-L.; Belmont-Hébert, C.; Buffle, J. *Env. Sci. Technol.* **1998**, *32*, 1515-1521.

12. Tercier-Waeber, M.-L.; Buffle, J.; Graziottin, F.; Koudelka-Hep, M.; *Sea Technol.* **1999**, *40*, 74-79.

13. Tercier-Waeber, M.-L.; Buffle, J. *Env. Sci. Tech.* **2000**, *34*, 4018-4024.

14. Pei, J.; Tercier-Waeber, M.-L.; Buffle, J.; Fiaccabrino, G.C.; Koudelka-Hep, M. *Anal. Chem.* (in press).

15. Tercier, M.-L.; Parthasarathy, N.; Buffle, J. *Electroanalysis.* **1995**, *7*, 55-63.

16. Belmont, C.; Tercier, M.-L.; Buffle, J.; Fiaccabrino, G.C.; Koudelka-Hep, M. *Anal. Chim. Acta.* **1996**, *329*, 203-214.

17. Tercier, M.-L. ; Buffle, J. *Anal. Chem.* **1996**, *68*, 3670-3678.

18. Belmont-Hébert, C.; Tercier, M.-L.; Buffle, J.; Fiaccabrino, G.C.; Koudelka-Hep, M. *Anal. Chem.* **1998**, *70*, 2949-2956.
19. Davison, W.; Fones, G.; Harper, M.; Teasdale, P.; Zhang, H. In *In situ Monitoring of Aquatic Systems : Chemical Analysis and Speciation*; Buffle, J., Horwai, G., Eds.; IUPAC Series of Analytical and Physical Chemistry of Environmental Systems; John Wiley & Sons: Chichester, 2000, Vol. 6., Ch. 11, pp. 495-569.
20. Tercier-Waeber, M.-L.; Buffle, J.; Confalonieri, F.; Riccardi, G.; Sina, A.; Graziottin F.; Fiaccabrino, G.C.; Koudelka-Hep, M. *Meas. Sci. Technol.* **1999**, *10,* 1202-1213.
21. Pei, J.; Tercier-Waeber, M.-L.; Buffle, J. *Anal. Chem.* **2000**, *72*, 161-171.
22. *Aquatic Chemistry* ; Stumm, W. ; Morgan, J.J., Eds. ; Wiley & Sons : New-York, 1995.
23. Pelletier, J.P.; Leboulanger, C.; Moille, J.P.; Chifflet, P. In *Commission Internationale Pour La Protection Des Eaux Du Léman (CIPEL ;* Rapport de Campagne 1998) ; CIPEL : Lausanne, CH, 1999 ; pp 61-68.
24. Moffett, J.W.; Brand, L.E. *Limnol. Oceanogr.* **1996**, *41*, 388-395.
25. Crank J. *The Mathematics of Diffusion*; Oxford University Press : New-York, 1975.
26. David, R.L. Handbook of Chemistry and Physics, 75th Edition ; CRC Press: Boca Raton, 1994, cat. 5, p. 90.

Chapter 3

In Situ Voltammetry at Deep-Sea Hydrothermal Vents

Donald B. Nuzzio[1,*], Martial Taillefert[2], S. Craig Cary[2], Anna Louise Reysenbach[3], and George W. Luther, III[2,*]

[1]Analytical Instrument Systems, Inc., P.O. Box 458, Flemington, NJ 08822
[2]College of Marine Studies, University of Delaware,
700 Pilottown, Road, Lewes, DE 19958
[3]Department of Biology, Portland State University, Portland, OR 97201

There is a need to build instrumentation and sensors that can measure *in situ* chemical changes in dynamic environments. Hydrothermal vents are arguably the most dynamic aqueous systems on earth. The orifice of a vent approaches 360 °C and spews vast quantities of dissolved hydrogen sulfide and iron into ambient seawater at 2 °C. These chemical species fuel incredible deep-sea (micro)biological communities, which may be a model for life on other planets. Here we describe an *in situ* submersible analyzer and electrodes for the measurement of aqueous chemical species found near hydrothermal vents. A standard three-electrode arrangement is controlled by a voltammetric analyzer that is deployed from the deep-sea submersible, *Alvin*. Real time measurements for a variety of redox species under flow conditions were made with a 100 μm Au/Hg solid-state working electrode at a depth of 2500 m. The solid-state working electrode was used to detect dissolved O_2, S(-II), Fe(II) and FeS_{aq} molecular clusters. Our *in situ* data show that significant changes can occur in chemical speciation and analyte concentration when waters are sampled and then measured aboard ship.

Introduction

Voltammetric (micro)electrode techniques have been used in a variety of geochemistry and marine chemistry applications. Recently, *in situ* voltammetric measurements have received increasing attention *(1)*. Both direct *(2-4)* and on-line *(5-11,* flow cell) type arrangements have been used for trace metal and major chemical species determinations. However, voltammetric instrumentation has only been deployed in shallow waters to date. There is a need to make measurements in the deep-sea, and this is particularly true of deep-sea hydrothermal vents that have an incredible microbial and macrofaunal community fueling itself via chemosynthesis *(12,13)*. Changes in temperature can be dramatic in this environment and electrodes must be able to respond precisely in waters of lower pH, high pressure, high temperature and high water flow rates. When high temperature waters mix with low temperature waters, their chemistries can differ dramatically . Because organisms live in these dynamic and extreme conditions, it is critical to understand how that chemistry drives or influences biology. In this paper we describe *in situ* electrochemical instrumentation and the initial deployment of it at 9 °N East Pacific Rise (EPR; at a 2500 meter water depth; 250 atm of pressure) and at Guaymas Basin, Gulf of California, 2000 meter water depth (200 atm of pressure). A companion paper in this volume *(11)* demonstrates that current-concentration curves are affected by water flow rates but not by pressure. At high flow rates and reasonable scan rates (~ 1 V s^{-1}), the current-concentration curves become independent of water flow rate for the target redox species.

Deep-sea hydrothermal vents can have high concentrations of iron and sulfide. In this paper we demonstrate the use of voltammetry to measure Fe and S species. Using a solid-state gold amalgam (Au/Hg) working electrode, we show the simultaneous detection and quantification of several sulfur and iron dissolved species. In previous work, dissolved chemical species (O_2, H_2O_2, $S_2O_3^{2-}$, S_x^{2-}, HS$^-$, I$^-$, Fe(II), Mn(II), organically complexed Fe(III), and FeS clusters) have been simultaneously measured *in situ* in sediments and natural waters *(6,14,15)*.

Experimental Methods

Chemicals and solutions

Chemicals used in laboratory experiments were analytical grade from Fisher Scientific Co. Milli-Q quality (Millipore) was used for all reagents. Laboratory measurements were carried out in a 0.55 M NaCl solution or in seawater. Mn(II), Fe(II) and S(-II) standards were prepared from $MnCl_2 \cdot 4\ H_2O$, ferrous ammonium sulfate, and $Na_2S \cdot 9H_2O$. The mercury plating solution was prepared as 0.1 N $Hg(NO_3)_2$ in 0.05N HNO_3.

Electrodes

Gold amalgam PEEK™ electrodes were made as described by Luther et al *(6)* by fixing 100 μm-diameter Au wire soldered to the conductor wire of a BNC cable within a body of 0.125"-diameter PEEK™ tubing, which is commercially available as standard HPLC high-pressure tubing. The metal is fixed within the tubing with West System 105 epoxy resin and 206 hardener. A portion of the black outer coat and braid of the BNC wire are removed to expose the teflon shield and Cu conductor wire so that the Au wire soldered onto the Cu conductor can be inserted into the PEEK™. The epoxy is injected into the PEEK™ tubing which contains the gold wire that was previously soldered to the conductor wire of the BNC cable. Then the teflon is inserted into the PEEK™ tubing until the black coating of the BNC wire fits against the PEEK™ tubing, and the assembly is held with epoxy, which has a moderate setting time (~1 hr), and does not drain out the lower open side. On setting, the epoxy seals the tip and the lower end can be refilled with epoxy if necessary. Then the top end is coated with Scotchkote (3M) electrical coating and Scotchfil (3M) electrical insulation putty. PEEK™ and high-purity epoxy fill permit the determination of metal concentrations without risk of contamination, and at temperatures as high as 150 °C. Pt counter and solid Ag/AgCl reference electrodes were made similarly but 500 μm diameter wire was used for each. These PEEK™ electrodes could be used as is or mated with standard HPLC fittings from Upchurch, Inc for insertion into a flow cell *(11)*.

Once constructed the working electrode (Au) surface was sanded, polished and plated with Hg by reducing Hg(II) from a 0.1 N Hg / 0.05 N HNO_3 solution, for 4 minutes at a potential of -0.1 V, while purging with N_2. The mercury/gold amalgam interface was conditioned using a 90-second -9 V polarization procedure in a 1 N NaOH solution (16). The electrode was then run in linear sweep mode from -0.05 to -1.8 V versus a Saturated Calomel Electrode (SCE) or Ag/AgCl electrode several times in oxygenated seawater to obtain a reproducible O_2 signal.

For DSV*Alvin* work, four Au/Hg electrodes can be controlled by the analyzer. The reference electrode was Ag/AgCl and the counter electrode was Pt wire, both of which were mounted on the basket of DSV *Alvin* so that they would not enter sulfidic waters *(15)*. For hydrothermal vent work, the Ag/AgCl reference was silver wire, which was oxidized in seawater at +9 V for 10 sec to form a AgCl coating. This electrode was used as a solid-state electrode in the seawater medium (I=0.7) so that no pressure effects on filling solutions would hinder electrode performance. Peak potentials of the analytes measured *in situ* and aboard ship were the same and similar to those for a saturated calomel electrode (SCE). All laboratory and shipboard analyses were carried out using an Analytical Instrument Systems (AIS) DLK-100A potentiostat controlled by a microcomputer using software provided by the manufacturer. A DLK-SUB-1 was used for all *in situ* work (see below).

Field Experiments

For hydrothermal vent work, two working electrodes as well as a thermocouple sensor and tubing that lead into a flow cell inlet and a discrete syringe sampler system were placed in a sensor or wand package *(15)* that can be held by a manipulator (arm) of *Alvin*. The wand was held over the vent orifice and areas along the length of vent chimneys. These latter areas are termed diffuse flow because water temperatures can range from 8 to 125 °C and do not emanate from the vent orifice. A flow cell *(11,15)* was fixed in the submersible's basket that was bathed by waters at 2 °C. A submersible electrochemical analyzer (DLK-SUB I) from Analytical Instrument Systems, Inc. was used for data collection (see below).

The electrochemical package was deployed during cruises to 9 °N East Pacific Rise (May, 1999) and Guaymas Basin, Gulf of California (January 2000). Separate discrete samples were taken with a gas tight syringe sampler *(11,16)* from the same waters for comparison with the flow cell measurements. The discrete samples were measured aboard ship by voltammetry *(11)*. We typically made three to five replicate measurements per sample with the flow cell system. Electrodes were calibrated at different temperature *(17)* as well as for flow *(11)*.

Voltammetry

The three-electrode configuration (working, reference and counter) was used to determine the concentration of the species present in natural waters. Linear sweep voltammetry (LSV), cyclic voltammetry (CV) and square wave voltammetry (SQW) were used for analyses. The following conditions were generally applied during the LSV and CV scans: scan rate = 200, 500 or 1000 mV s^{-1}, scan range = -0.1 to -1.75 V, equilibration time = 5 s. Square wave voltammograms were conducted under the same conditions with a pulse height of 24 mV, 1 mV scan increment and 50 mV s^{-1} scan rate. To prevent memory effects, caused by the accumulation of sulfide and metal species on the mercury surface, conditioning steps were applied to the working electrode as per Brendel and Luther (17). To reoxidize metals (Mn, Fe) that are reduced at the amalgam, a potential of -0.1 V was applied over 10-30 seconds before each scan. When sulfide was present, conditioning at −0.9 V for 10 s was employed since the metals and sulfide are not electroactive at that potential (14, 17). All LSV and CV scans shown in the figures below are actual data; i.e., no software smoothing routines are used to enhance the quality.

Results & Discussion

Instrumentation

A schematic of the analyzer, electrodes and sensor package system is provided in Figure 1. The DLK-SUB-1 is an electrochemical analyzer built by Analytical Instrument Systems, Inc. The instrument is designed to perform all of the standard voltammetric analyses that would be available in a shipboard or land-based laboratory. The following standard voltammetric techniques are possible using the DLK-SUB-1: linear sweep, cyclic, normal pulse, differential pulse, square wave. Chronoamperometry as well as stripping techniques for monitoring trace levels of analytes can also be performed. This versatility provides great flexibility to the researcher for *in situ* experiments.

The instrument utilizes the 24 V DC power available from DSV *Alvin* for its power source. All waterproof connections were made with connectors from Impulse, Inc. and the aluminum housing (1 meter length; 20 cm outside diameter) of the instrument was rated for operation at full ocean depth (~6000 m; 600 atm). The instrument is a complete stand-alone package capable of being deployed for long periods of time from *Alvin*, or remotely operated vehicles (ROV) with tethers up to 1500 meters without the need for signal amplification.

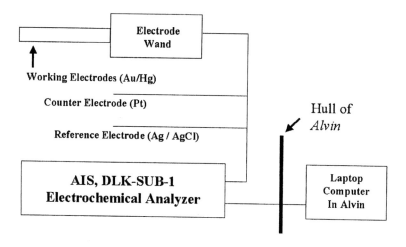

Figure 1. Schematic of the analyzer, electrodes and cable communication through the hull of Alvin.

The voltammetry hardware is linked to an IBM compatible computer inside the housing. The internal computer communicates with another computer through the hull of *Alvin* via a 15-m RS 232 cable and is controlled by an operator, who can reprogram waveforms to respond to the radically different environments found at vents. A separate 1-meter cable is used to make connections with the working, counter, and reference electrodes and the pressure housing. This cable has four inputs for working electrodes (that can be selected one at a time via a multiplexer), one input for the counter electrode and another for the reference electrode. Another input for grounding the reference electrode from the submersible insured signal integrity. The electrode wand is constructed of Delrin and has a stainless steel handle so that the manipulator (arm) of *Alvin* can hold and deploy the electrodes without breaking them.

Hydrothermal vent measurements

The major aqueous species, which are found near hydrothermal vents, react at the Au/Hg electrode according to the following electrode reactions *(17,18)*.

$$O_2 + 2 H^+ + 2 e^- \rightarrow H_2O_2 \qquad\qquad -0.30\ V \qquad (1a)$$
$$H_2O_2 + 2 H^+ + 2 e^- \rightarrow H_2O_2 \qquad\qquad -1.30\ V \qquad (1b)$$
$$HS^- (H_2S) + Hg \rightarrow HgS + H^+ + 2 e^- \qquad ads.\ onto\ Hg <-0.60\ V (2a)$$
$$HgS + H^+ + 2 e^- \leftrightarrow HS^- + Hg \qquad \sim -0.60\ V \qquad (2b)$$
$$FeS_{aq} + 2 e^- + H^+ \rightarrow Fe(Hg) + HS^- \qquad -1.1\ V \qquad (3)$$
$$Fe^{2+} + Hg + 2 e^- \leftrightarrow Fe(Hg) \qquad\qquad -1.43\ V \qquad (4)$$

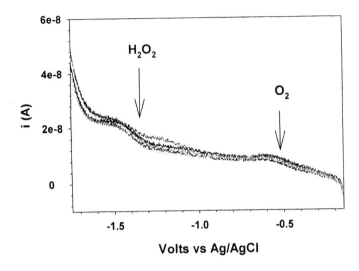

Figure 2. Five LSV scans (200mV s⁻¹) of ambient deep-seawater at 2500 m depth and 2 °C. O₂ is the only major species present.

The electrochemical package was first tested in cold waters away from high temperature vents and high flow rates. Thus, no significant change in signal could be expected. Figure 2 shows five LSV scans of O_2 in ambient 2 °C seawater near the vent site at 2500 meter water depth. The electrochemical analyzer and electrode package shows excellent reproducibility. The O_2 concentration is 46.6 ± 3.3 μM consistent with previous results *(19)*.

The package was also tested 0.5 meter above a vent orifice where the temperature was 25°C. At the orifice the temperature was 360 °C as measured by a second thermocouple without electrodes attached. The electrode measurements were thus performed at a safe distance from the orifice. Black smoke from iron monosulfide and pyrite precipitation was observed emanating from the vent orifice. Figure 3A displays signals for only free H_2S, FeS_{aq} molecular clusters and total S(-2) [S_{AVS} = Σ FeS_{aq} + H_2S]. O_2 is not detected. The reproducibility of these measurements indicates that the vent had constant chemical output at this time as hot iron and sulfide rich vent waters mix with cold ambient seawater. There was no double peak detected near the free sulfide peak *(14,20)*, which indicates that polysulfides were not being formed quickly under these conditions. This is not surprising because the pH is approximately 5 (measured aboard ship) at this location above the vent orifice. Sulfide oxidation is quite slow at pH < 6 *(21)*. We note that one scan (4th in the sequence) showed a cutoff of the S_{AVS} signal that is due to a fluctuation in power from the submersible. Figure 3B shows the concentration versus time plot from five consecutive CV scans. Although a standard for FeS_{aq} clusters is not available, the difference between free H_2S and total S(-2) [S_{AVS}] is an indicator of the FeS_{aq} concentration *(11)*.

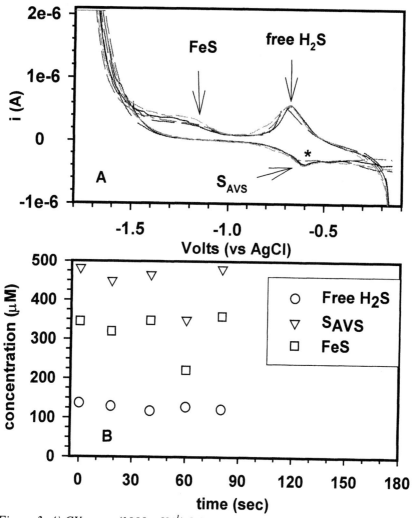

Figure 3. A) CV scans (1000 mVs⁻¹) 0.5 meter above a vent orifice at 25 °C. The asterisk indicates that noise from the submersible cutoff the S$_{AVS}$ signal; B) time course of the data.

In contrast Figure 4A shows a CV measurement in the tube of the polychaete, *Alvinella pompejana*. During measurement at a scan rate of 200 mV s^{-1}, the temperature was 80 ± 20 °C. This dramatic fluctuation in temperature makes it impossible to measure concentrations, but the voltammogram clearly indicates that only FeS_{aq}, Fe(II) and total S(-2) are measurable. Thus, free H_2S was bound with Fe^{2+} as FeS_{aq} molecular clusters. A sample was also drawn into the flow cell (at 2 °C), and analyzed with another electrode under no flow. Figure 4B shows that only a signal for total S(-2) is present in the flow cell. These data indicate that the chemical speciation (probably due to precipitation of FeS) was changed as the waters were cooled. The current also changed two orders of magnitude from Figure 4A to 4B because of the dramatic temperature change. We attribute the noise in Figure 4B to electrical noise from the submersible. The flow cell data corresponds to a concentration of 250 µM. A discrete sample was also taken and analyzed later aboard ship by voltammetry (Figure 4C) using purge and trap methods *(11,22)*. The concentration was found to be 139 µM and indicates that further precipitation of FeS and FeS_2 occurred or that H_2S was liberated from solution. Figure 4C shows a small free H_2S signal indicating that some dissociation of FeS occurred when the sample was brought onboard ship where the temperature was 20 °C. In addition, the smaller S_{AVS} signal indicates that sulfide was lost on sampling and transport to the lab. These data clearly show that chemical speciation and analyte concentration can change with sampling and storage and that real time data are essential in understanding what chemistry an organism actually experiences. For *Alvinella pompejana* the Fe^{2+} binds sulfide as FeS in its habitat, and detoxifies the free sulfide for the organism *(15)*.

Additional reasons for the changes shown in Figure 4A to 4C are that the syringes for discrete samples are not subsampled for measurements for up to six hours after collection allowing for reaction or loss of sulfide if the gas tight syringes malfunction. The samples from these syringes are placed into a voltammetric cell at one atmosphere pressure for measurement, which could lead to a change in the equilibrium for FeS_{aq} (eq. 5) from that at 2500 meter water depth (250 atmospheres). Also, it takes approximately two minutes to fill the syringes during sampling. During that time the temperature can fluctuate dramatically with significant chemical changes, and the manipulator (arm) of *Alvin* has been observed to move during sampling.

$$Fe^{2+} + H_2S \leftrightarrow FeS_{aq} + 2 H^+ \qquad (5)$$

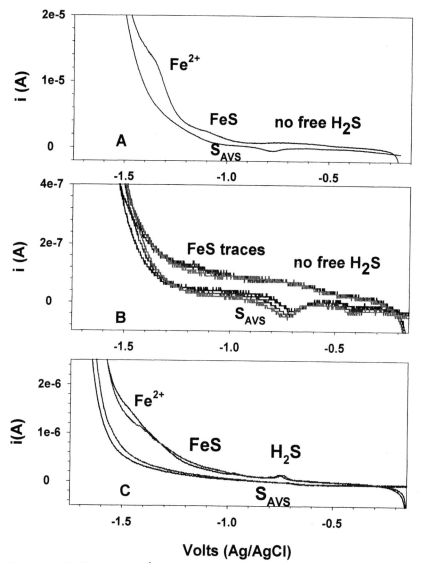

Figure 4. A) CV (200 mVs⁻¹) scan taken in the tube of the polychaete at 80 ±20 °C; B: CV scans (1000 mVs⁻¹) taken of a sample pumped into a flow cell at 2 °C; C) CV scans (200 mVs⁻¹) taken of a subsample aboard ship at 20 °C.

Conclusions

Real time, *in situ* measurements in oceanic hydrothermal vent environments were accomplished with a submersible electrochemical analyzer and PEEK™ Au/Hg electrodes for dissolved O_2, Fe^{2+}, FeS_{aq}, free H_2S and total sulfide ($S_{AVS} = \Sigma FeS_{aq} + H_2S$). The electrochemical instrumentation is robust to work in extreme environments and is commercially available. The electrodes are designed to operate at temperatures up to 150 °C. Our observations demonstrate that real time *in situ* voltammetric analysis provides a more precise measure of chemical speciation and a measure of the chemical concentration of analytes that emanate from vents. In contrast, oxidation and precipitation reactions during sample retrieval and handling of discrete samples can result in an underestimate of these concentrations and in the actual chemical speciation.

Acknowledgments

The authors would like to thank their own lab groups, R. Lutz, T. Shank, the DSV *Alvin* pilots and the crew of the R/V *Atlantis* for assistance. This study was partially funded by grants from the Office of Sea Grant / NOAA (NA16RG0162-03), NSF (OCE-9714302 and SBIR-9760571) and NASA (NAS13-0013).

References

1. Taillefert, M.; Luther, III, G. W.; Nuzzio, D. B. *Electroanal.* **2000**, *12*, 401-412.
2. Tercier, M-L.; Buffle, J.; Zirino, A.; De Vitre, R. R. *Anal. Chim Acta* **1990**, *237*, 429-437.
3. Tercier, M-L.; Buffle, J.; Graziottin, F. *Electroanal.* **1998** *10*, 355-363.
4. Tercier, M-L.; Belmont-Hebert, C.; Buffle, J. *Environ. Sci. Technol.*, **1998**, *32*, 1515-1521.
5. Herdan, J.; Feeney, R., Kounaves, S. P.; Flannery, A. F.; Storment, C. W.; Kovacs, G. T. A.; Darling, R. B. *Environ. Sci. Technol.*, **1998**, *32*, 131-136.
6. Luther, III, G. W.; Reimers, C. E.; Nuzzio, D. B.; Lovalvo, D. *Environ. Sci. Technol.* **1999**, *33*, 4352-4356.
7. De Vitre, R. R.; Buffle, J.; Perret, D.; Baudat, R. *Geochim. Cosmochim. Acta.* **1988**, *52*, 1601-1613.
8. Lieberman, S. H.; Zirino, A. *Anal. Chem.* **1974**, *46*, 20-23.
9. Martinotti, W.; Queirazza, G.; Realini, F.; Ciceri, G. *Anal. Chim. Acta.* **1992**, *261*, 323-334.

10. Newton, M. P.; van den Berg, C. M. G. *Anal Chim Acta* **1987**, *199*, 59-76.

11. Luther, III, G. W.; Bono, A.; Taillefert, M; Cary, S. C. In *Electrochemical Methods for the Environmental Analyses of Trace Element Biogeochemistry*, Taillefert, M.; Rozan, T., Eds. American Chemical Society Symposium Series; American Chemical Society: Washington, D. C., **2001**, this volume, Chapter 4.

12. Cavanaugh, C. *et al. Science* **1981**, *213*, 340-342.

13. Felbeck, H. *Science* **1981**, *213*, 336-338.

14. Luther, III, G. W.; Glazer, B. T.; Hohman, L.; Popp, J. I.; Taillefert, M.; Rozan, T. F.; Brendel, P. J.; Theberge, S. M.; Nuzzio, D. B. *J. Environ. Monit.* **2001**, *3*, 61-66.

15. Luther, III, G. W.; Rozan, T. F.; Taillefert, M.; Nuzzio, D. B; Di Meo, C.; Shank, T. M.; Lutz, R. A.; Cary, S. C. *Nature* **2001**, *401*, 813-816.

16. Di Meo, C. A.; Wakefield, J. R.; Cary, S. C. *Deep-Sea Res. I* **1999**, *46*, 1279-1287.

17. Brendel, P.; Luther, III, G. W. *Environ. Sci. Technol.* **1995**, *29*, 751-761.

18. Theberge, S. M.; Luther, III, G. W. *Aq. Geochem.* **1997**, *3*, 191-211.

19. Johnson, K. S.; Beehler, C. L.; Sakamoto-Arnold, C. M.; Childress, J. J. *Science* **1986**, *231*, 1139-1141.

20. Rozan, T. F.; Theberge, S. M.; Luther, III, G. W. *Anal. Chim. Acta* **2000**, *415*, 175-184.

21. Millero, F. J.; Hubinger, S., Fernandez, M.; Garnett, S. E. *Environ. Sci. Technol.*, **1987**, *21*, 439-443.

22. Luther, III, G. W.; Ferdelman, T. G.; Kostka, J. E.; Tsamakis, E. J.; Church, T. M. *Biogeochem.* **1991**, *14*, 57-88.

On-Line Measurements
in the Water Column

Chapter 4

A Continuous Flow Electrochemical Cell for Analysis of Chemical Species and Ions at High Pressure: Laboratory, Shipboard, and Hydrothermal Vent Results

George W. Luther, III, Andrew B. Bono, Martial Taillefert, and S. Craig Cary

College of Marine Studies, University of Delaware, 700 Pilottown Road, Lewes, DE 19958

An inexpensive and rugged voltammetric flow cell has been developed to test pressure dependence and flow rate on current vs. concentration standard curves and for the simultaneous measurement of electroactive species in field samples. Standard curves show that current is independent of pressure but varies with the square root of the flow rate. The electrochemical system consists of a pump to deliver water into a voltammetric flow cell with a standard three-electrode arrangement that is controlled by a voltammetric analyzer. The flow cell can be used onboard ship or *in situ* at any depth including from the deep-sea submersible, *Alvin*. Field samples were analyzed for a variety of redox species under controlled flow conditions with a solid state working electrode at a depth of 2500 m. In this study, a 100 μm Au/Hg amalgam solid-state working electrode is used to detect dissolved O_2, S(-II), Fe(II) and Mn(II). We describe the flow cell and its capabilities.

Introduction

Voltammetric microelectrode techniques are becoming increasingly popular as a means to quantify electroactive species in natural environments. Since amperometric microelectrodes were first employed in a natural setting to study the biogeochemistry of oxygen in sediments *(1)*, significant improvements have been made in electrode and cell design, as well as data acquisition and processing.

Solid state microelectrodes have several important advantages over larger, more traditional electrodes. Smaller electrode surfaces can operate at low currents, which permits use in solutions with weak ionic strength. Furthermore, because of the decreased IR drop, the working and reference electrodes do not necessarily have to be kept immediately adjacent to one another *(2)*. The smaller surface area results in a decrease in the charging current relative to the faradaic current and thus allows faster scan rates and increased instrumental sensitivities to reversible and semi-reversible redox reactions *(2)*. Recently, the development of voltammetric techniques, which utilize a solid-state gold amalgam (Au/Hg) working electrode, has permitted the simultaneous quantification of several dissolved species both in laboratory and natural settings. Dissolved chemical species (O_2, H_2O_2, $S_2O_3^{2-}$, S_x^{2-}, HS^-, I^-, Fe(II), Mn(II), organically complexed Fe(III), and FeS clusters) have been simultaneously measured in sediment pore waters to spatial resolutions unattainable by traditional subsampling techniques *(3-8)*. Most have been performed in cores brought aboard ship but recent work shows that direct *in situ* sediment and water column work is possible *(5, 9-12)*.

The incorporation of microelectrodes into continuous flow detectors is well established, and electrochemical flow-cells designed for bench-top analyses that work under atmospheric pressure conditions have been commercially available for some time. Continuous flow systems improve over traditional electrochemical cells by permitting minute quantities of a large number of samples to be rapidly evaluated under identical conditions. Furthermore, they reduce errors due to contamination and handling, and provide opportunities for computerized autosampling, pre-analysis conditioning, and data collection *(13-14)*. Additionally, in the case of electrochemical determinations in natural settings, flow cells offer advantages over traditional sampling and analysis techniques since samples are processed immediately and thereby minimize the effects of degassing, oxidation, and microbial metabolism associated with sample storage.

The majority of electrochemical flow-through systems have been developed for use in the pharmaceutical industry to quantify organic compounds *(15-18)*. Several researchers have, however, applied voltammetric flow cell apparatus to natural systems *(19-23)*. The predominant focus in geochemical work has been the determination of trace metals (Mn, Cu, Pb, Cd, Zn, Co, Ni, U).

Consequently, systems have not been designed to evaluate redox-sensitive species such as O_2, Mn(II), Fe(II), $S_2O_3^{2-}$, I$^-$, and S(-II); although, De Vitre *et al.* *(19)* and Tercier *et al.* *(23)* examined the distribution of some redox-sensitive species (Mn(II), Fe(II), S(-II)) in Lake Bret (Switzerland) using a hanging mercury drop electrode in a flow cell design.

Electrochemical flow-cell systems for geochemical analyses tend to be specialized. Most researchers have selected hanging drop mercury electrodes (HDME) to avoid potential memory problems associated with mercury film electrodes *(19, 20, 22)*. The bulky and elaborate HDME working electrodes, however, cannot work at high pressure and preclude *in situ* deployment of the apparatus. Large solid state electrodes were first used in a flow cell design by Lieberman and Zirino *(20)*, who used mercury film on a graphite tube, and more recently by Tercier et al. *(23)* with a mercury film on a 3 mm glassy carbon electrode. Since then, an array of mercury coated Ir electrodes (Ir/Hg) has been used for trace metal analyses *(10)*. In addition to permitting *in situ* deployment, the solid state electrode simplifies the flow cell design, endures higher flow rates, and permits the miniaturization of the working electrode with the benefits that accompany the use of microelectrodes. Most flow cell designs have also involved significant pre-analysis handling of the sample, including filtration, purging or degassing, and various pre-concentration and chelation steps to determine low levels of trace metals *(19-22)*. Completely closed flow cells free from sample pre-treatment, like the design employed by Tercier et al. *(23)*, prevent alteration of the sample prior to analysis and are necessary for operating *in situ* under high pressure conditions.

In this paper, we describe a portable, closed-system voltammetric cell made from the durable polymer, polyethyletherketone (PEEK™). Using the Au/Hg solid-state working electrode permits simultaneous *in situ* analysis of a wide range of aqueous dissolved species in both field and laboratory settings at pressures at least up to 250 atmospheres. The overall goal of this work was to determine the effect of flow and pressure to calibrate the working electrode for hydrothermal vent conditions. First, laboratory testing of the design established the effects of flow and pressure on the response of solid state Au/Hg microelectrodes. Shipboard field tests of the flow cell system were performed on waters from the Chesapeake Bay. *In situ* deployment from the submersible DSV *Alvin* enabled comparison to established sampling and analysis techniques and evaluation of the utility of the apparatus as an *in situ* tool for geochemical study. This flow cell shows significant promise in a wide range of possible future applications including determination of respiration rates of single larvae (Glazer and Luther, unpublished) because the total cell volume is about 0.7 mL.

Experimental Methods

Chemicals and solutions

All chemicals used in laboratory experiments were analytical grade, and all water was Milli-Q quality (Millipore). All laboratory measurements were carried out in a 0.55 M NaCl solution (Fisher). Mn(II), and S(-II) standards were prepared from $MnCl_2 \cdot 4\ H_2O$ (Fisher), and $Na_2S \cdot 9H_2O$ (Fisher). The mercury plating solution was prepared as 0.1 N $Hg(NO_3)_2$ in 0.05N HNO_3. The glassware and flow cells were cleaned and stored in a 10% (v/v) HCl solution.

Flow cell System

The flow-cell is constructed entirely of PEEK™ (polyethyletherketone) and employs standard HPLC fittings (0.125 inch), which permit the rapid removal or exchange of electrodes and sample tubing, and the use of the cell at elevated temperatures and pressures. The flow cell is approximately 10 cm in length and 1 inch in diameter; a hole of 0.0625 inch (1.59 mm) diameter is drilled along its length with ports at each end for sample input / output (Figure 1). Three holes are drilled perpendicular to the lengthwise hole for a three solid-state electrode system; the total cost per cell is less than $500 including parts and labor. The flow cell is a completely closed system and permits analysis of aqueous samples without manipulation or exposure to the atmosphere. The reference is a 500 μm Ag/AgCl or Pt electrode, the counter is a 500 μm Pt electrode and the working electrode is a 100 μm Au/Hg electrode.

Gold amalgam PEEK™ electrodes were made as described by Luther et al *(9)* by fixing 100 μm-diameter Au wire soldered to the conductor wire of BNC cable within a body of 0.125"-diameter PEEK™ tubing, which is commercially available as standard HPLC high-pressure tubing. The metal is fixed within the tubing with the West System 105 epoxy resin and 206 hardener. A portion of the black outer coat and braid of the BNC wire are removed to expose the teflon shield and Cu conductor wire so that the Au wire soldered onto the Cu conductor can be inserted into the PEEK™. The epoxy is injected into the PEEK™ tubing which contains the gold wire that was previously soldered to the conductor wire of the BNC cable. Then the teflon is inserted into the PEEK™ tubing until the black coating of the BNC wire fits against the PEEK™ tubing, and the assembly is held so that epoxy, which has a moderate setting time (~1 hr), does not drain out the lower open side. On setting, the epoxy seals the tip and the top end can be refilled with epoxy if necessary. Then the top end is coated with Scotchkote

(3M) electrical coating and Scotchfil (3M) electrical insulation putty. PEEK™ and high-purity epoxy fill permits the determination of metal concentrations without risk of contamination, and at temperatures as high as 150 °C. Pt counter and solid Ag/AgCl reference electrodes were made similarly but 500 μm diameter wire was used for each. These PEEK™ electrodes are mated with standard HPLC fittings from Upchurch, Inc for insertion into the flow cell.

Once constructed the working electrode (Au) surface was sanded, polished and plated with Hg by reducing Hg(II) from a 0.1 N Hg / 0.05 N HNO_3 solution, for 4 minutes at a potential of -0.1 V, while purging with N_2. The mercury/gold amalgam interface was conditioned using a 90-second -9 V polarization procedure in a 1 N NaOH solution (3). The electrode was then run in linear sweep mode from -0.05 to -1.8 V versus a Saturated Calomel Electrode (SCE) or Ag/AgCl electrode several times in oxygenated seawater to obtain a reproducible O_2 signal.

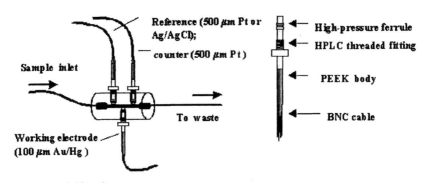

Figure 1. A schematic of the flow cell and electrode designs. The input and output tubing also utilize the high-pressure HPLC threaded fittings and ferrules. The flow cell is 100 mm in length and 25.4 mm in diameter.

For DSV *Alvin* work, up to three working electrodes could be placed into the flow cell. The reference electrode was Ag/AgCl and the counter electrode was Pt wire, both of which were mounted on the basket of DSV *Alvin* so that they would not enter sulfidic waters (5, 11). For hydrothermal vent work, the Ag/AgCl reference was silver wire, which was oxidized in seawater at +9 V for 10 sec to form a AgCl coating. This electrode was used as a solid state electrode in the seawater medium (I=0.7) so that no pressure effects on filling solutions would hinder electrode performance. Comparison of peak potentials for the analytes measured *in situ* and aboard ship was the same and similar to those for a saturated calomel electrode (SCE). The tubing leading into the flow cell's inlet was placed in a sensor package (11), which was held over the vent orifice and areas along the length of vent chimneys. These latter areas are termed diffuse flow because water temperatures can range from 8 to 125 °C and do not emanate from the vent orifice; the cell sits in a basket at 2 °C. A submersible electrochemical analyzer (DLK-sub I) from Analytical Instrument Systems, Inc. was used for data collection *(11)*.

All laboratory and shipboard analyses were carried out using an Analytical Instrument Systems (AIS) DLK-100 potentiostat controlled by a microcomputer using software provided by the manufacturer. Chesapeake Bay field sampling was performed using a Rabbit-Plus peristaltic pump (Rainin Instrument Co., Inc.) with a flow rate of 12 mL min^{-1} through Teflon® tubing. For DSV *Alvin* work, the submersible analyzer was used for data collection. A General Oceanics T5 submersible pump was used to fill the cell; the cell could be used under flow conditions but was typically used under diffusion control conditions when the pump was turned off since hydrothermal vent waters may undergo rapid change *(11)*.

Laboratory Experiments

The effect of solution flow past the electrode surface was determined by connecting the flow cell in line with a Scientific Systems Inc. Model 200 HPLC pump, attached to a Model 210 Guardian and LP-21 Lo Pulse dampener. Experiments with this system permitted the measurement of current versus flow only. Addition of a 15 cm C-18 (3 μm) HPLC column to the exit port of the flow cell permitted measurement of electrode response with increasing internal pressure within the cell by varying the flow rate, up to a maximum internal pressure of 200 atmospheres. In these experiments, a 150 μM Mn(II) solution was pumped from the solvent reservoir through the pump and flow cell (only flow rate varies), or through the pump, flow cell, and column (both flow rate and pressure vary).

Field Experiments

Field work was performed in the Chesapeake Bay during the summer anoxic season on 22 August 1996. Values for the concentration of oxygen were determined by three methods: (i) an *in situ* O_2 sensor contained within the conductivity-temperature-depth (CTD) package, (ii) analysis of waters sampled by bottle cast using Winkler titrations, and (iii) direct on-deck sampling of waters using the voltammetric flow cell. Agreement was favorable as previously published *(9)*. Waters pumped into the flow cell from the anoxic layers of the Chesapeake were compared against samples from the bottle casts. The bottles were sampled using the syringe method described in Luther *et al (24)* and were analyzed via hanging drop mercury electrode (HDME) stripping voltammetry for S(-II), Fe(II), and Mn(II). For hydrothermal vent work, the flow cell was mounted onto the basket in front of the deep-sea submersible *Alvin*, during a cruise to Guaymas Basin, Gulf of California in January 2000. Separate discrete samples were taken with a gas tight syringe sampler from the same waters for comparison with the flow cell measurements. The discrete samples were measured aboard ship by voltammetry *(11)*. We typically made three to five replicate measurements per depth with the flow cell system. Electrodes were calibrated at different temperature *(3)* as well as for flow in this paper.

Voltammetry

The three-electrode configuration was used to determine the concentration of the species present in natural waters. Linear sweep voltammetry (LSV), cyclic voltammetry (CV) and square wave voltammetry (SWV) were typically used for analyses. CV and SWV were used to test for reversibility. SWV is the method of choice for low level detection for a reversible signal; LSV and CV were used for fast scans and for higher concentrations of analytes. The following conditions were generally applied during the LSV scans: scan rate = 200 to 1000 mV/s, scan range = -0.1 to -1.7 V, equilibration time = 5 s. Square wave voltammograms were conducted under the same conditions with a pulse height of 24 mV, 1 mV scan increment and 200 mV/s scan rate. To prevent memory effects, caused by the accumulation of sulfide and metal species on the mercury surface, conditioning steps were applied to the working electrode as per Brendel and Luther *(3)*. To re-oxidize metals (Mn, Fe) that are reduced at the amalgam, a potential of -0.1 V was applied over 30 seconds before each scan. When sulfide was present, conditioning at -0.8 V for 10 s was employed since the metals and sulfide are not electroactive at that potential *(3, 5)*.

Results & Discussion

Effect of flow and pressure

The effect of flow on an electroactive species at the electrode surface is described by the Levich equation (eq. 1):

$$I = knFCD^{2/3} r^{3/2} U^{1/2} v^{1/6} \quad (1)$$

where I represents current under mass-transport (not diffusion or zero flow) controlled conditions, k is a constant coefficient, n is the number of electrons transferred, F is the Faraday constant, C is the concentration of the electroactive species, D is the diffusion constant, r is the radius of a disk-shaped but stationary electrode, U is the rate of flow through the cell, and v is the kinematic viscosity.

In Figures 2A,B the effect of flow on the current response (SWV) of the electrode at constant atmospheric pressure to a 150 μM Mn(II) solution corresponds closely to eq. 1, in close agreement with previous work *(13, 14 ,20, 25, 26)*. The tendency of current to plateau at increased flow rates suggests that the conditions at the 100 μm electrode surface are no longer controlled by the rate of mass transport of the analyte to the electrode surface. The relationship in Figure 2B confirms a linear relationship ($r^2 = 0.972$) of I with $U^{1/2}$ before current levels off at flow rates above 2 ml/min, which corresponds to 1.68 cm/s based on the diameter of the flow cell's hole. Similar results were found using CV scans (data not shown) and for O_2 and H_2S/HS^- solutions.

We observed the change in current with flow-induced pressure using a 150 μM Mn(II) solution, using the same apparatus as the flow experiment but with the addition of an HPLC column after the cell to provide internal back-pressure. The HPLC fittings were tested to pressures as high as 200 atmospheres, corresponding to mid-ocean ridge depth (2000 m). The behavior of the electrodes under induced pressure indicates that the increase in current is comparable to that of flow without induced pressure (Figure 2C). When the current response of the electrode is plotted against both flow and flow-induced pressure, the increase in current fits the Levich relationship for flow ($I \propto U^{1/2}$). The strong correlation between the two plots confirms that the response of the electrodes, while dependent on flow, is independent of pressure. The relationship between pressure and the current response of a solid-state electrode has been reported in only one other work, which used an Ir/Hg electrode in a pressure chamber *(10)*.

Current change with flow rate and scan rate change

The second part of this work involved calibrating electrodes for in situ use at hydrothermal vents *(11)*, where the diffuse water flow can have high flow rates. Ideally, one would like to calibrate an electrode onboard ship using the pilot ion method *(27)* at no flow and use this sensitivity to relate it to the in situ conditions. Calibrations for S(-II) were performed because hydrothermal fluids are mainly composed of dissolved sulfide.

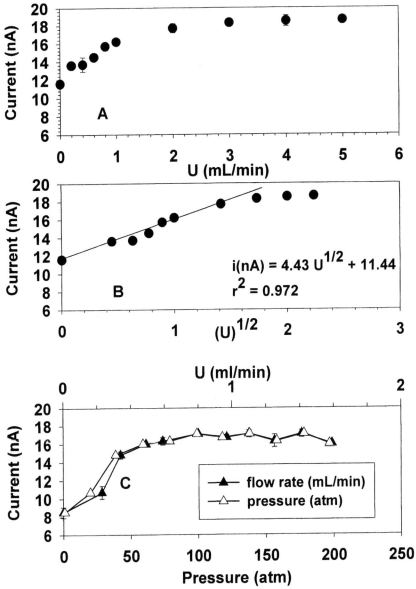

Figure 2. (A) the effect of flow at constant atmospheric pressure on the response of the working electrode to a 150 µM Mn(II) solution (SWV); (B) the relationship between current and (flow rate)$^{1/2}$ at constant atmospheric pressure; (C) the effect of flow and pressure on the response of working electrode under conditions of flow-induced pressure.

The electrodes were first calibrated at zero flow with Mn(II) and then calibrated for S(-II) at different flow rates with the flow cell. A sensitivity ratio, R, was calculated via the expression in eq. 2:

$$R = \frac{S^{f}_{S(-II)}}{S^{f=0}_{Mn(II)}} \qquad (2)$$

where $S^{f=0}_{Mn(II)}$ is the sensitivity of Mn(II) at flow $f = 0$ and $S^{f}_{S(-II)}$ is the sensitivity of S(-II) at flow f. Figure 3 presents the effect of the flow rate on the sensitivity ratio R by cathodic cyclic voltammetry (CSV, scan from -0.1 to -1.8 V with return to -0.1 V) at different flow rates. Calibrations for sulfides were determined by considering both the positive wave (Figure 3a) and the negative wave (Figure 3b). The positive current wave is due to the formation of HgS and HgS_x films at the electrode at positive potentials (eq. 3, formally an electrochemical oxidation of the Hg which pre-concentrates sulfide on the Hg) and the reduction of the films to reform Hg on scanning negatively *(6;* forward reaction of eq. 4*)*. The negative wave is due to the reformation of the HgS and HgS_x films at the electrode on scanning positively (reverse reaction of eq. 4). This electrochemical reoxidation of Hg by sulfide is formally analogous to the reduction wave of Mn(II) at a Hg electrode.

$$HS^- + Hg \rightarrow HgS + H^+ + 2\,e^- \qquad \text{reaction more positive than} -0.6\text{ V} \qquad (3)$$

$$HgS + H^+ + 2\,e^- \leftrightarrow HS^- + Hg \qquad \text{at} -0.6\text{ V} \qquad (4)$$

In a complex aqueous system containing several sulfur species such as HS^-, S(0), S_x^{2-}, and $FeS_{(aq)}$ at the pH of seawater, the positive wave (cathodic) will provide information on the sum of H_2S/HS^-, S(0), and S_x^{2-} only *(6)*, while the negative wave (anodic) will account for total S(-II) which includes H_2S/HS^- and FeS_{aq}. Thus, a comparison of these two waves can be used to quantify FeS_{aq}. The results show for the anodic wave (Figure 3A, no pre-concentration) that the sensitivity ratio R increases with the flow rate at a scan rate of 200 mV s^{-1}, but then reaches a plateau about 2-3 ml min^{-1} analogous to figure 2 for Mn(II). At 500 and 1000 mV s^{-1}, there is <25 % change in R over these flow rates. These

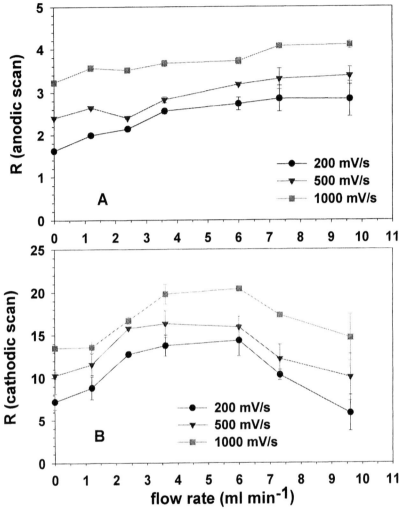

Figure. 3. S(-II) / Mn(II) sensitivity ratios (R) by cathodic cyclic voltammetry as a function of the flow rate in the flow cell at 22 °C. The ratios R were related to the sensitivity of Mn(II) with no flow and calculated from a) the positive wave and b) the negative wave. They are shown for three different scan rates of 200, 500, and 1000 mV s⁻¹.

results can be used to predict the concentration of total dissolved sulfide in situ at high flow regimes. The high water flow rates at hydrothermal vents exceeds the flow rate where the sensitivity ratio R (and current) plateaus. We recommend the use of high scan rates under flow conditions because the R factor (current) is hardly affected.

For the cathodic wave (Figure 3B; pre-concentration) R is larger than the current for the anodic wave because of the pre-concentration effect (see also Fig. 5C below in SWV). There is also an increase in R from diffusion control conditions at low flow rates with a plateau occurring near 2 ml min^{-1}. Because sulfide is reacting with the Hg film to form HgS during the scan, there is a decrease in R at high flow rates due to saturation of the Hg film with sulfide. This saturation effect is noticeable at about 600 μM free sulfide for the 100 μm Au/Hg electrode when scanning in this cathodic direction. Both Figure 3A and B demonstrate that R increases with scan rate, which is consistent with the well known dependence of I on (scan rate)$^{1/2}$.

Chesapeake Bay data

The waters of the Chesapeake Bay were sampled during the summer season when deep basin waters overlying the sediments become anoxic *(28-30)*. The upper Bay is about 30 meters deep and the lower Bay is 20 meters deep due to the Potomac River depositing sediments to the lower Bay. This allows for entrainment of high salinity water during tidal motion and the formation of stagnant bottom waters in the upper Bay. These waters become sulfidic when oxygen is depleted and sulfate is used as the terminal electron acceptor for organic matter decomposition in the sediments and water column. The salinity and temperature profiles from conductivity-temperature-depth (CTD) data presented in Figure 4 indicate two small thermo-clines at 8 and 13 meters, while a marked halo-cline at 13 meters separates the surficial brackish (S $\sim$ 7) from deeper, more saline waters (S $\sim$ 16‰). Voltammetric scans are presented in Figures 5A-C that are representative of the oxic (0.5 m; LSV), suboxic (no detectable O_2 and H_2S at 16 m; SWV) and sulfidic (19.5 m; SWV) waters, respectively. Figure 5C shows that the cathodic scan plot has a higher current than that for the anodic scan plot because of the pre-concentration effect on starting the scan at positive potentials. This effect allows for low detection limits of free sulfide [on the order of < 0.1 μM] for this Au/Hg electrode.

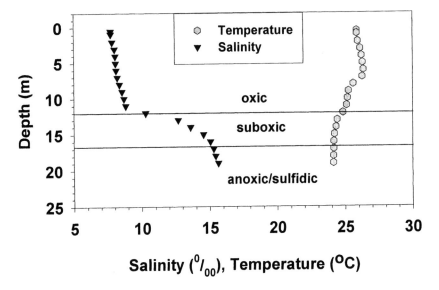

Figure 4. Salinity and temperature profiles at the sampling site in the Chesapeake Bay on August 22, 1996.

Data from one profile (0 - 19 m) through the oxic, suboxic and into the anoxic/sulfidic zone are shown in Table I. There is marked similarity between oxygen concentrations determined via the in situ CTD system, which uses a Clark-style amperometric O_2 sensor, as well as by Winkler-titration from a bottle cast, and the sampling by direct introduction into the voltammetric flow-cell using a peristaltic pump. Precision for all methods is less than 2%. Values for the concentration of O_2 are consistent with values obtained by the CTD array as well as by Winkler titrations, with apparent under-estimations for the Winkler values at 9 and 11 meter depths. The lower Winkler values resulted from scavenging of oxygen from the sample as the Niskin bottles sat for about an hour between sample collection and the addition of Winkler reagents for O_2. The need to sample and add the reagents immediately for O_2 using the Winkler method is well known. The low concentrations reported from the CTD O_2 sensor in the anoxic/sulfidic zone reflect the measurement of current at one fixed potential rather than an I-E curve so that a true zero or residual current as shown in Figure 5A cannot be determined. Detection limits for the Winkler titration and the Au/Hg electrodes are 5 and 3 μM, respectively.

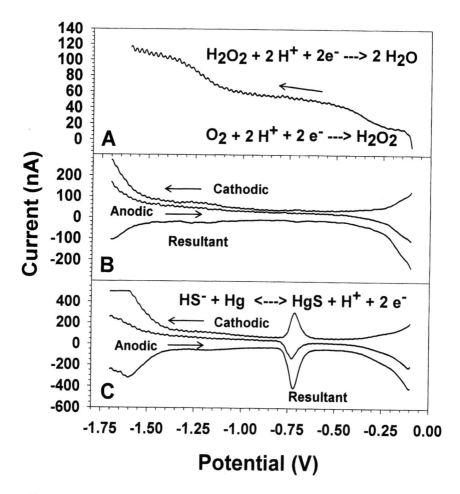

Figure 5. Representative voltammograms of Chesapeake Bay water sampled using the flow cell (flow rate of 10 mL/ min). In LSV (A), reduction reactions are plotted as positive currents, whereas in SWV (B,C), the resultant current is plotted as negative. (A) surface oxic waters (0.5 m); (B) suboxic, waters (16 m) (C) anoxic, sulfidic waters (19.5 m).

Table I. Au/Hg electrode measurements are compared with traditional sampling and analytical measurements. The low level O_2 - CTD measurements reflect that a current is measured at one potential and true "zero" is unknown. All values are in micro-molar units. ND = not detected; NA = not analyzed.

depth (m)	O_2 electrode	O_2 CTD	O_2 Winkler	H_2S electrode	H_2S HDME
0.5	279.0	277.9	278.8	ND	ND
9	191.0	210.5	152.2	ND	ND
11	169.0	167.6	89.5	ND	ND
16	ND	1.43	ND	0.20	0.09
18	ND	1.16	ND	8.40	8.96
19	ND	1.12	ND	13.5	NA

Sulfide concentrations determined by the flow cell were similar to those obtained by HDME. Sulfide was first detected at trace levels (0.2 µM) at 16 meters and increased with depth to values as high as 14 µM at 19 meters. No overlap of H_2S and O_2 was in evidence in the profiles observed in the Chesapeake Bay. Other researchers (28-30), who used separate techniques for measurement of each analyte, have noted this lack of overlap. Dissolved Mn(II) and Fe(II) were not detected with the electrodes in the flow cell suggesting that, in these waters, there is no overlap between O_2 and these dissolved metals. This lack of overlap has also been observed in the Black Sea water column (31) and the porewaters from the Scotian shelf / slope region (9).

Hydrothermal vent measurements

The flow cell was used in Guaymas Basin at a 2000 meter water depth in January 2000 from DSV *Alvin*. Data collected from the flow cell and from discrete samples [in diffuse flow waters ranging in temperature from 2-100°C] are compared in Figure 6. The discrete samples were also treated with basic zinc acetate to precipitate all forms of acid volatile sulfide [FeS, H_2S, HS⁻, S_x^{2-}] as ZnS. These treated samples were later acidified with 3M HCl to release and trap H_2S in 1 M NaOH for subsequent analysis by voltammetry (32) and the methylene blue method *(33)*, which is most commonly used by oceanographers. Both lab-based methods gave identical results.

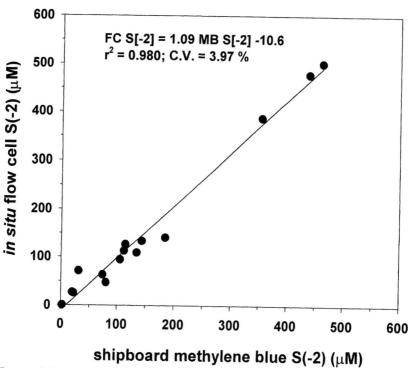

Figure 6. Comparison of total sulfide [S(-2)]measurements made with the in situ flow cell and with discrete samples aboard ship.

Figure 6 shows that *in situ* flow cell data can agree well with the shipboard data. The slope is slightly larger than one and the intercept is negative indicating that some small amounts of sulfide were lost as the discrete samples were transported from the bottom to the surface. There are other data that do not agree and are not plotted because speciation changes on sampling *(34)*. In those instances, the *in situ* flow cell data are typically much higher than the discrete samples. This is due to several possible problems with collecting the discrete samples under extreme environment conditions. First, when iron is present in the samples FeS precipitates in the syringes (this is particularly true of samples taken at 9°N East Pacific Rise, 2500 meter water depth). A twofold or more loss of total sulfide may occur as hot waters at 250 atm pressure are brought to 1 atm pressure and cool to ambient seawater temperature of 2°C in the syringes.

Secondly, the syringes are not subsampled for discrete measurements for up to six hours after collection allowing for reaction or loss of sulfide if the gas tight syringes malfunction. Thirdly, it takes approximately two minutes to fill the syringes during sampling. During that time the temperature can fluctuate dramatically with significant chemical changes (*11*), and the manipulator (arm) of *Alvin* has been observed to move during sampling.

Conclusions

A portable, rugged and inexpensive voltammetric flow cell has been designed and constructed to make precise and accurate measurements of several dissolved electroactive species such as O_2, S(-II), Fe(II), and Mn(II) simultaneously in marine and freshwater systems. The use of the cell permits (sub)micromolar to millimolar level analyses to be performed quickly and without any sample manipulation or exposure to atmospheric conditions.

The cell design was tested under conditions of both flow and high internal pressure. The increase in current with flow rate closely followed the relationship, $I \propto U^{1/2}$, predicted by the Levich equation. At flow rates higher than 2 ml/min for a 100 µm diameter Au/Hg working electrode, current changes relatively little as mass transport ceases to be rate limiting. Current did not demonstrate a dependence on internal pressure. Because the response of the working electrode is independent of pressure, the unit or the electrodes alone may be deployed over considerable depths without special calibration requirements.

The cell was used during a field study in the Chesapeake Bay and continuous measurements of oxygen and sulfide in the water column were performed. Data from the flow cell corresponded to values obtained using more traditional sampling techniques, and did not suffer from sampling artifacts that can significantly alter these measurements and which we observed in bottle samples that were not sampled and treated promptly. The unit was able to detect the presence or absence of a broad suite of redox species simultaneously and demonstrated that oxygen and sulfide did not coexist at the same depths within the waters of the Chesapeake Bay.

Real time measurements at hydrothermal vent environments were accomplished with this cell. PEEK™ is durable under high pressure and in seawater. Our observations indicate that real time analyses give a better understanding of the chemistry that the vent organisms are actually experiencing. Discrete samples underestimate in many instances what the chemistry of the system actually is.

The performance of the cell indicates that it is suitable for benchtop high-pressure work as well as deployment in high-pressure environments such as the deep ocean. The portability of the cell and its use in conjunction with a fully portable, DC-powered voltammetric analyzer facilitate its use in field applications that require remote sampling. Finally, the PEEK™ electrodes themselves are robust and have been successfully deployed from a remotely operated vehicle (ROV) directly into near-shore sediments *(9)* to the waters of the deep sea *(5, 11)* as in this work.

Acknowledgments

The authors would like to thank D. Hicks for help with designing and fabricating the flow cell, D. Nuzzio for electrochemical assistance, the DSV *Alvin* pilots, the crew of the R/V *Cape Henlopen*, and the crew of the R/V *Atlantis*. This study was funded by the Office of Sea Grant / NOAA (NA16RG0162-03). The ship-time for the Chesapeake Bay work was provided by NSF (OCE-9314349). Hydrothermal vent research was funded by NSF (OCE-9714302).

References

1. Revsbech, N. P.; Sorensen, J.; Blackburn, T. H.; Lomholt, J. P. *Limnol. Oceanogr.* **1980**, *25*, 403-411.
2. Wightman, R. M. *Anal. Chem.*, **1980** *53*, 1125A-1134A.
3. Brendel, P.; Luther, III, G. W., *Environ. Sci. Technol.* **1995**, *29*, 751-761.
4. Luther, III, G. W.; Sundby, B.; Lewis, B. L.; Brendel, P. J.; Silverberg, N. *Geochim. Cosmochim. Acta.* **1997**,*61*, 4043-4052.
5. Luther, III, G. W.; Glazer, B. T.; Hohman, L.; Popp, J. I.; Taillefert, M.; Rozan, T. F.; Brendel, P. J.; Theberge, S. M.; Nuzzio, D. B. 2001. *J. Environ. Moni*toring, **2001**, 3, 61-66.
6. Rozan, T. F.; Theberge, S. M.; Luther, III, G. W. *Anal. Chim. Acta* **2000**, *415*, 175-184.
7. Taillefert, M.; Bono, A. B.; Luther, III, G. W. 2000 *Environ. Sci. Technol.34*, 2169-2177.
8. Theberge, S. M.; Luther, III, G. W. *Aquatic Geochem.* **1997**, *3*, 191-211.
9. Luther, III, G. W.; Reimers, C. E.; Nuzzio, D. B.; Lovalvo, D. *Environ. Sci. Technol.* **1999**, *33*, 4352-4356.
10. Belmont-Hébert, C.; Tercier, M. L.; Buffle, J.; Fiaccabrino, G. C.; de Rooij, N. F.; Koudelka-Hep, M. *Anal. Chem.* **1998**, *70*, 2949-2956.
11. Luther, III, G. W.; Rozan, T. F.; Taillefert, M.; Nuzzio, D. B; Di Meo, C.; Shank, T. M.; Lutz, R. A.; Cary, S. C. *Nature* **2001**, *410*, 813-816.

72

12. Herdan, J.; Feeney, R., Kounaves, S. P.; Flannery, A. F.; Storment, C. W.; Kovacs, G. T. A.; Darling, R. B. *Environ. Sci. Technol.*, **1998**, *32*, 131-135.
13. Mahoney, L.; O'Dea, J.; Osteryoung, J., *Anal. Chim. Acta* **1993**, *281*, 25-33.
14. Björefors, F.; Nyholm, L. *Anal. Chim. Acta* **1996**, *325*, 11-24.
15. Canete, F.; Rios, A.; Luque de Castro, M. D.; Valcarcel, M. *Anal. Chim. Acta*, **1988**, *214*, 375-384.
16. Chaney, Jr. E. N.; Baldwin, R. P. *Anal. Chim. Acta* **1985**, *176*, 105-112.
17. Kounaves, S.; Young, J. *Anal. Chem.* **1989**, *61*, 1469-1472.
18. Wang, J.; Frelha, B. *Anal Chem.* **1983**, *55*, 1285-1288.
19. De Vitre, R. R.; Buffle, J.; Perret, D.; Baudat, R. *Geochim Cosmochim Acta* **1988**, *52*, 1601-1613.
20. Lieberman, S. H.; Zirino, A. *Anal. Chem.* **1974**, *46*, 20-23.
21. Martinotti, W.; Queirazza, G.; Realini, F.; Ciceri, G. *Anal. Chim. Acta* **1992**, *261*, 323-334.
22. Newton, M. P.; van den Berg, C. M. G. *Anal. Chim. Acta.* **1987**, *199*, 59-76.
23. Tercier, M-L.; Buffle, J.; Zirino, A.; De Vitre, R. R. *Anal. Chim. Acta.* **1990**, *237*, 429-437.
24. Luther, III, G. W.; Church, T. M.; Powell, D. *Deep-Sea Res.* **1991**, *38(S2)*, S1121-S1137.
25. Amez del Pozo, J.; Costa-García, A.; Tuñón-Blanco, P. *Anal. Chim. Acta* **1993**, *289*, 169-176.
26. Dalangin, R. R.; Gunasingham, H. *Anal. Chim. Acta* **1994**, *291*, 81-87.
27. Meites, L. Polarographic Techniques (2nd ed.) **1965**, Wiley Interscience Publishers, NY.
28. Luther, III, G. W.; Ferdelman, T.; Tsamakis, E. *Estuaries* **1988**, *11*, 281-285.
29. Millero, F. J. *Est. Coast. Shelf Sci.*, **1991**, *33*, 521.
30. Officer, C. B.; Biggs, R. B.; Taft, J. L.; Cronin, L. E.; Tyler, M. A.; Boynton, W. R. *Science* **1984**, *223*, 22-26.
31. Murray, J. W.; Codispoti, L. A.; Friederich, G. E. In *Aquatic Chemistry, Interfacial and Interspecies Processes*; Huang, C.P. O'Melia, C. R.; Morgan, J. J., Eds.; Advances in Chemistry Series; American Chemical Society: New York (NY), **1995**; Vol. 44, pp 157-176.
32. Luther, III, G. W.; Ferdelman, T. G.; Kostka, J. E.; Tsamakis, E. J.; Church, T. M. *Biogeochem.* **1991**, *14*, 57-88.
33. Cline, J. E. *Limnol. Oceanogr.* **1969**, *14*, 454-458.
34. Nuzzio, D. B.; Taillefert, M.; Cary, S.C.; Reysenbach; A. L.; Luther, III, G. W. In *Electrochemical Methods for the Environmental Analyses of Trace Element Biogeochemistry*, Taillefert, M.; Rozan, T., Eds. American Chemical Society Symposium Series; American Chemical Society: Washington, D. C., **2001**, this volume, Chapter 3.

Chapter 5

Field Application of an Automated Voltammetric System for High-Resolution Studies of Trace Metal Distributions in Dynamic Estuarine and Coastal Waters

Eric P. Achterberg, Charlotte B. Braungardt, and Kate A. Howell

Department of Environmental Sciences, Plymouth Environmental Research Centre, University of Plymouth, Plymouth PL4 8AA, United Kingdom

The increasing environmental pressures on estuarine and coastal waters call for improved monitoring techniques of chemical constituents to aid management decisions. Automated stripping voltammetry is a suitable technique for continuous, near real-time monitoring of trace metals in marine systems. This contribution describes the application of voltammetric monitoring techniques in estuarine and coastal waters of the UK and Spain. The high spatial and temporal resolution of the data obtained, allows a thorough interpretation of the trace metal sources and behaviour. Future trends in this field research include submersible sensors which can be remotely deployed for a period of several weeks.

Introduction

Estuaries and coastal waters are dynamic environments with often important temporal and spatial variability in physical, chemical and biological parameters. Many coastal systems are strongly influenced by anthropogenic activities, with 40 percent of the world's population living within 100 km of a coastline. Population increases, changes in agricultural practices and spread of aquaculture, in addition to further industrialisation and natural resources exploitation are enhancing the environmental pressures on coastal environments. Coastal systems are of ecological, economical and recreational importance. They are biologically productive, and form nursery grounds for a high proportion of commercial fish and shellfish species. The need for sustainable management of coastal zones, has led the European Union to implement integrated coastal zone management practices (*1*) and draft a comprehensive and far reaching Water Framework Directive. In the United States, the Estuaries and Clean Water Act of 2000 (S. 835) and the Oceans Act of 2000 (S. 2327) (http://www.house.gov) have improved legislation concerning the marine environment. In order to achieve the objectives of coastal management plans, it is essential that naturally occurring biogeochemical processes are understood, particularly as processes in coastal waters often have complex cause-effect relationships. For this reason, our ability to monitor coastal waters forms an important management tool.

The traditional monitoring approach for marine waters involves the collection of discrete samples using a survey vessel. Discrete surface water samples can be obtained using a pump with a bottom-weighted hose. At greater depths, sampling is carried out with the use of specialised samplers (e.g. Niskin or Go-Flo bottles) which are attached to a hydrowire or rosette frame and deployed with the use of a winch. The samples are often filtered on-board ship and analysed in a land-based laboratory. Commonly used laboratory techniques for dissolved trace metal analysis in seawater include Graphite Furnace Atomic Absorption Spectrometry (GFAAS) and Inductively Coupled Plasma Mass Spectrometry (ICP-MS) (after liquid-liquid or solid-phase extraction for trace metal preconcentration and matrix removal), chronopotentiometry, colorimetry, chemiluminecence and stripping voltammetry. This approach of laboratory based analyses of discrete samples is time-consuming and therefore expensive. Only a limited number of samples can be collected using discrete sampling techniques and as a result important changes in water quality may not be noticed.

Trace metal concentrations in estuarine and coastal waters are often low. Inadequate sampling and sample treatment techniques have for a long time posed a high risk of contamination during collection and analysis of seawater samples. Sample contamination may arise from components of the sampling gear and from sample handling. Advances in equipment design and sampling protocols have improved contamination-free sample collection (*2*). Furthermore, improved understanding of post-sampling contamination has resulted in the introduction of ultra-trace working practices (*3*). These include the cleaning of sample bottles, filters and filtration equipment with acid prior to use, and the handling of samples in a clean environment (class-100 laminar flow hood in a clean room). These precautions against sample contamination are essential in order to obtain high quality trace metal data, but also importantly reduce the number of samples per day that can be processed by a worker.

In the marine field, the use of continuous near real-time and in-situ automated monitoring techniques is becoming more widespread (*4,5*). A major advantage of this approach is the reduced contamination risk due to a minimisation of sample handling. In addition, the computer controlled nature of the automated analytical techniques enhances the quality of the data. Furthermore, the approach results in a higher sample throughput and hence an increased amount of environmental data. The application of high resolution monitoring approaches in estuarine and coastal waters is of particular importance due to the dynamic nature of these systems, and will therefore allow small-scale processes to be investigated (*6*).

Electrochemical techniques for marine waters

Electrochemical techniques are commonly used for water quality assessments. The measurements of pH (hydrogen ion selective glass electrode; potentiometry) and oxygen (Clark electrode; amperometry) are often part of monitoring exercises. A sensitive and versatile technique for the determination of trace metals in seawater is stripping voltammetry. Voltammetric techniques are based upon the measurement of a current response as a function of the potential applied to an electrochemical cell. Stripping voltammetry combines a preconcentration step with a stripping step (*7,8*), thereby enhancing the sensitivity and selectivity. With the low concentrations of trace metals in seawater (typically $< 10^{-8}$ M), the preconcentration step is important to reduce the limit of detection of the methods. During the preconcentration step, the trace metal of interest is collected onto or in a working electrode and during the stripping step the collected metal is oxidised or reduced back into solution. Anodic stripping voltammetry (ASV) and adsorptive cathodic stripping voltammetry (AdCSV) are the most suitable stripping voltammetric techniques for the determination of trace metals in seawater. ASV typically involves the addition of a small amount of acid (nitric acid or hydrochloric acid) or pH buffer (e.g. acetate) to the sample prior to analysis, whereas AdCSV requires the addition of an AdCSV ligand in addition to a pH buffer. The AdCSV ligand complexes the trace metal of interest and for most AdCSV methods, this metal-ligand complex has electroactive properties (*8*). In ASV metals are reduced during the preconcentration step, and their oxidation current is determined during the potential scan towards more positive potentials. In the majority of the AdCSV methods, the metal-ligand complex is adsorbed onto the working electrode during the preconcentration step, and the reduction current of the metal is determined during the potential scan towards more negative potentials. Important advantages of the stripping voltammetric techniques include the extremely low detection limits (10^{-10} – 10^{-12} M), their multi-elemental and speciation capabilities, and their suitability for on-line, ship-board and in-situ applications (7,8,9). ASV has been successfully applied for trace measurements of Cu, Cd, Pb and Zn in seawater (10). ASV is suitable for the determination of other elements (e.g. In, Tl), however their typical seawater concentrations are too low, or the analysis is subject to interferences. AdCSV allows for the determination of more than 20 metals in natural waters (11,8,12).

The instrumentation used for stripping voltammetry is relatively low-cost, and typically consists of a voltammetric analyser, a three-electrode-cell (working, reference and counter electrodes), and a computer for automated measurements and data acquisition. For automated on-line measurements, inert valves and peristaltic pumps for sample and reagent transport are required. Hanging mercury drop electrodes (HMDE) and mercury film electrodes (MFE) are the most commonly used working electrodes for trace metal measurements in seawater using stripping voltammetry. The HMDE is a reliable working electrode, both in the laboratory and field. With the HMDE, a new electrode surface is produced with the formation of each new drop, which eliminates analyte memory effects between scans, and also is important for unattended trace metal monitoring activities. The HMDE is the preferred working electrode for trace metal measurements in seawater using AdCSV, but can also be used for ASV. The MFE is the preferred working electrode for dissolved metal analysis using ASV. The MFE may be formed by in-situ plating of Hg on glassy carbon (13), or by preliminary deposition. The advantages of the MFE are the robustness and the high sensitivity due to the high surface to volume ratio (13). Glassy carbon is the most commonly used material for MFE, but the use of gold (14), iridium (15,16), graphite pencil (17) and carbon fibre (18) have also been reported.

The manifold of an automated monitoring system using a HMDE (Metrohm, VA Stand 663) and a voltammetric analyser (μAutolab, Ecochemie) for analysis of trace metals in field situations is presented in Figure 1. Such systems have been deployed on-board ships and from the banks of estuaries (19,6,20,21). The analysis is performed in an automated batch-mode, whereby aliquots of ca. 10 ml are analysed at a rate of one complete fully calibrated measurement every ca. 10-20 min (22). The collection of water is conducted using a strong polyethylene (PE) or polyvinyl chloride (PVC) hose and a peristaltic pump (Watson Marlow).

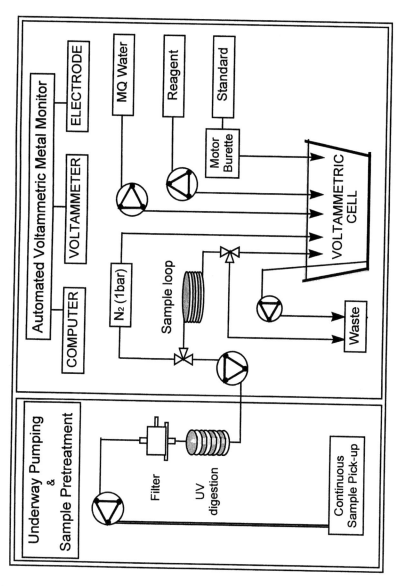

Figure 1. Manifold of an automated voltammetric system for ship-board analysis of trace metals (Reproduced with permission from reference 8. Copyright 1999 Elsevier).

Water is typically pumped at 1-2 l min^{-1}, which results in a minimum residence time of the sample in the hose and a constant conditioning of the hose surface with fresh sample. For the analysis of the dissolved metal fraction, suspended particulate matter is removed from the water using on-line filtration. In turbid estuarine and coastal waters, a tangential filtration device (with membrane filters, 0.4 or 0.2 μm) is most appropriate because of its self-cleansing ability (6). However, in clearer marine waters the use of filtration capsules (e.g. Sartobran P, Sartorius; 0.45 μm prefilter and 0.2 μm final filter) is practical, as they can be used for large volumes of water (and hence long periods of time (several days)). On-line UV-digestion is applied in order to destroy surfactants which may interfere with the voltammetric measurement, and to break down metal complexing organic compounds and hence allow the determination of total dissolved trace metals in seawater (23). The efficiency of UV-digestion in releasing metals (incl. Ni and Cu) from organic complexes is >99% (23). The UV-digestion unit contains a medium pressure mercury vapour lamp (400 W, Photochemical Reactors) surrounded by a quartz glass coil (i.d. 1.0 mm, length ca. 3.5 m) (24). The filtered and UV-treated sample is pumped into a sample loop (9.90 ml, Teflon®) using a peristaltic pump, which form part of an automated sample and reagent transport system. This system also rinses the voltammetric cell between samples and delivers reagents to the cell. The sample loop is enclosed by two inert three-way valves (Teflon®, Cole Parmer). The sample loop is emptied into the voltammetric cell using nitrogen gas. Teflon® tubing is used throughout the monitor, with the exception of the peristaltic pump tubing (Santoprene®). A high precision syringe pump (Cavro) is used for metal standard addition to the sample for quantification purposes. Figure 2 shows a flow diagram of a software-controlled analytical cycle during automated analysis. A typical trace metal analysis involves the addition of reagent to the sample by pump, a purging step using oxygen-free nitrogen gas (4 min), a sequence of 3 deposition periods, each followed by voltammetric scans (typically square wave), an internal standard addition with a subsequent sequence of 3 deposition periods followed by voltammetric scans. Dedicated software is used for data acquisition, peak evaluation and data storage. The software is self-decisive, and is able to reject sub-quality scans with a standard deviation above a pre-set value (typically 6%) and will initiate additional measurements. The software is also able to initiate additional standard additions, in case the increase in peak height as a consequence of the first standard addition was insufficient (i.e. less than 100%). The calibration of each sample using standard additions results in high quality data. This approach is important in estuarine and coastal waters, as it takes into account the changes in sample matrix (e.g. as a result of salinity variations), which may result in changes in the sensitivity of the measurement.

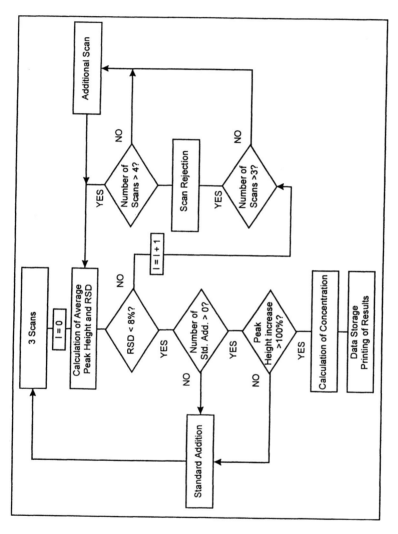

Figure 2. Flow diagram of soft-ware controlled analytical cycle during automated trace metal analysis with voltammetric monitor. The cycle starts with 3 voltammetric scans and ends with data storage and printing of results.

Trace metal monitoring in estuarine environments

Estuaries are highly dynamic and reactive zones, where fluvial discharges mix with seawater and dissolved elements interact with organic material and particles in the water column. Important factors influencing trace metal behaviour in estuarine waters include freshwater inputs, pH, redox conditions, tidal mixing and resuspension of sediments, colloid formation and coagulation, sorption, biological cycling and organic complexation (25,26). The dynamic nature of estuarine waters requires trace metal monitoring activities at a resolution similar to the processes affecting trace metal distributions (20).

Study at single geographic point

The instrumental set-up employed for the automated determination of dissolved trace metals at a geographical point in an estuary is shown in Figure 3. The instrumentation is operated from a van, and powered by a portable generator. A braided PVC hose is used to collect the water samples from the estuarine channel. The hose is submerged to a depth of ca. 0.5 m using a float and an anchor.

An important feature of a tidal cycle study at a single geographical point in an estuary is that the trace metal concentrations are affected only by water column processes (end-member mixing, adsorption, desorption, precipitation, coagulation etc.) and hence this approach allows a thorough interpretation of biogeochemical processes. In contrast, in an axial transect study, metal concentrations along a transect are affected by processes occurring in the water column in addition to metal inputs (run-off and discharges), and consequently the data is often more difficult to interpret.

Trace metal behaviour in the Tamar estuary

Figure 4 shows the results of automated measurements of dissolved Ni in the Tamar estuary (southwest England) during a tidal cycle (July 1997). The Tamar is situated in a mineral rich region, and has a number of abandoned metal mines in its catchment area. The Tamar receives important trace metal loads (Cu, Zn, Pb, As, U, Sn, Ni), as a consequence of surface run-off (27,28). The Ni analyses were performed using AdCSV, with dimethyl glyoxime as the added AdCSV ligand and HEPES as the pH buffer (final concentrations 0.2 mM and 10 mM (pH 7.8), respectively) (21). In order to circumvent problems with slow kinetics of Ni-DMG formation following addition of standard for sample calibration, DMG (0.2 mM) (instead of acid) was added to the Ni standard to ensure rapid

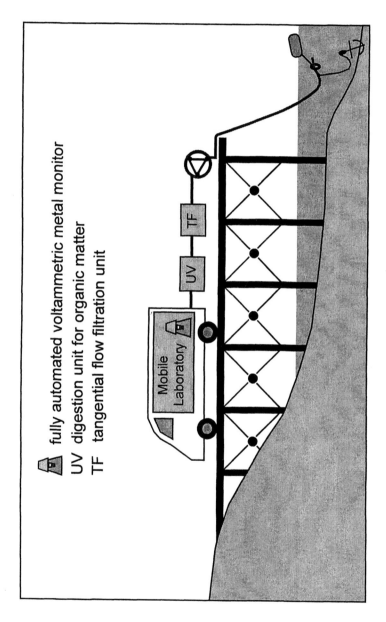

Figure 3. Set-up for estuarine trace metal monitoring with continuous sample collection and automated AdCSV trace metal analysis (Reproduced with permission from reference 8. Copyright 1999 Elsevier).

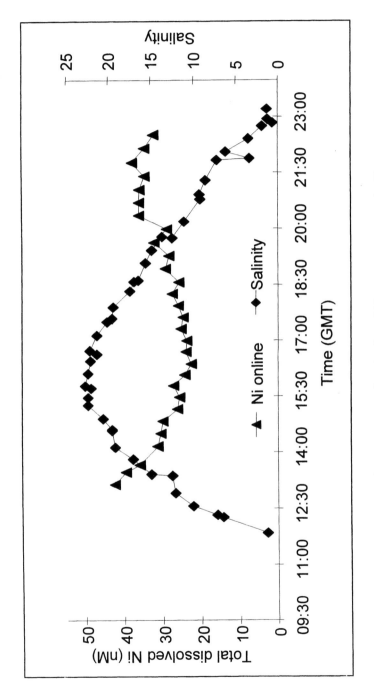

Figure 4. On-line continuous dissolved total Ni measurements in the Tamar estuary during a tidal cycle (July 1997). Discrete salinity data were obtained from conductivity measurements (Reproduced with permission from reference 21. Copyright 1998 Elsevier).

establishment of Ni equilibrium (< 30 s) and the provide stability of the metal standard. The limit of detection and working range for the AdCSV Ni method are presented in Table I, whereas the results of method validation using a certified reference material are presented in Table II.

Table I: Limits of detection and linear ranges of Ni, Cu and Zn analysis using square wave AdCSV for field measurements in the Tamar (Ni), surface waters of the Gulf of Cadiz (Cu, Zn; nM range) and in the mid-Odiel estuary (Cu, Zn; μM range). Values in brackets refer to the simultaneously analysed metal (i.e. Zn).

	Ni	Cu (Zn) nM range	Cu (Zn) μM range
Limit of detection (LOD)	0.1 nM	0.48 (0.81) nM	-
Linear range	100 nM	150 (300) nM	4 (7) μM
R^2 for linear range	0.99	0.99	0.99

Table II: Analysis of UV-irradiated seawater reference material (CASS-3) by AdCSV in batches of 10 ml. Confidence intervals refer to ± 2 SD of the sample mean.

CASS-3	n	ACSV result (nM)	certified (nM)
Ni	6	6.48 ± 0.40	6.58 ± 1.06
Cu	4	8.17 ± 1.05	8.14 ± 0.98
Zn	8	17.7 ± 2.58	19.0 ± 3.82

The advantages of the continuous, near real-time measurements are evident from Figure 4, with ca. 4 fully calibrated dissolved Ni analyses per hour. The dissolved Ni concentrations ranged between 21 and 43 nM. The tidal cycle was undertaken during a period with low river flow (ca. 6 m^3 s^{-1}; long-term mean flow 34 m^3 s^{-1} (29)), a tidal range of 3 m (mean tidal range: 3.5 m) and consequently a large variation in salinity was observed (24 salinity units) at our sampling point. The highest Ni concentrations were observed in waters with a low salinity (e.g. [Ni] = 38 nM at salinity 7), and low Ni concentrations were observed during periods with enhanced salinity (e.g. [Ni] = 21 nM at salinity 23). A direct inverse relation between salinity and dissolved Ni was observed (Figure 5), and indicates that dilution of Ni enriched freshwater with cleaner seawater was an important process regulating Ni behaviour in our estuarine

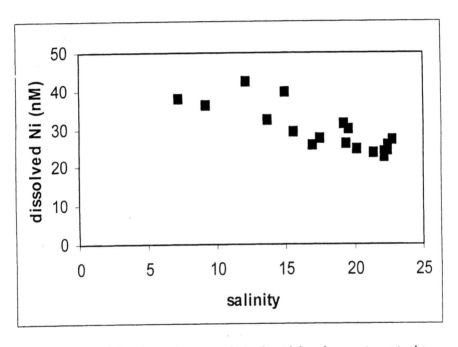

Figure 5. Total dissolved Ni versus salinity for tidal cycle experiment in the Tamar (July 1997).

system. A conservative behaviour for dissolved Ni has previously been observed for the higher salinity regions of the Severn (*30*) and Forth (*31*), in line with our findings. However, important removal of dissolved Ni by suspended particulate matter has been reported for the low salinity region (not covered during our tidal cycle) of the Tamar (*32,33*).

Trace metal behaviour in the Huelva estuary

The Huelva estuary is situated in southwest Spain (Figure 6), and receives metal-rich waters from the rivers Tinto and Odiel, which rise in an important mining district (Iberian Pyrite Belt; Cu, Zn, Fe, Au mines) and flow through heavily industrialised zones. Oxidation of iron pyrite in the orebodies and spoil heaps at the mining sites results in the formation of sulphuric acid and dissolved Fe, with additional release of metals. An automated voltammetric monitor was applied in this system in June 1997. The estimated freshwater discharge for the Tinto and Odiel in June 1997 was 0.08 and 0.3 m^3 s^{-1}, respectively (long-term mean water flows 3 and 15 m^3 s^{-1}). In Figure 7, the variations in dissolved Cu, Zn, salinity and pH are presented over a tidal cycle in the mid-estuary. Dissolved Cu and Zn were determined using AdCSV, with 8-hydroxyquinoline as AdCSV ligand and HEPES as pH buffer (final concentrations 15 μM and 10 mM, respectively), following 100 times dilution of the samples. The dilution was required as a result of the extreme conditions with respect to the metal concentration and sample matrix encountered at this site. The limit of detection and working range of the AdCSV Cu and Zn methods can be found in Table I. The quality of the trace metal data was validated through the analysis of certified reference material (see Table II), and furthermore corroborated by land-based ICP-MS analysis of hourly sampled discrete water samples, which showed close agreement with the continuous, near real-time determined trace metal concentrations (*6*). A strong temporal variability in metal concentrations was observed in this estuarine system as a result of enhanced metal inputs and tidal water movements. The mean tidal range in the Huelva estuary is 2.5 m. The metal concentrations at the single geographic point in the Huelva estuary were high, with Cu (Zn) concentrations ranging between ca. 2 (40) and 120 (200) μM. In line with observations in the Tamar estuary for Ni, the lowest Cu and Zn concentrations occurred during the periods with high tide (high salinity), and the highest concentrations during periods with low tide (low salinity). Figure 7 shows that the variations in pH were the mirror image of the metal variations. A plot of Cu and Zn versus conductivity shows an inverse relationship, indicating a conservative behaviour of these metals in the Huelva estuary (Figure 8). This conservative behaviour was also observed for these elements during axial transects in the Huelva (*34*). Although such a behaviour has been observed in

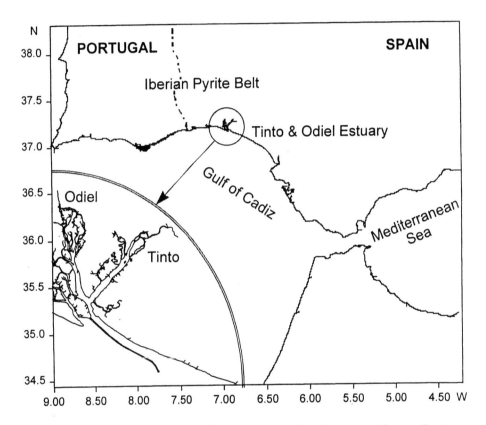

Figure 6. Location of the Gulf of Cadiz and the sulphide bearing Iberian Pyrite Belt in the southwest of the Iberian Peninsular. The inset shows the confluence of the Tinto and Odiel estuaries to form the Huelva estuary.

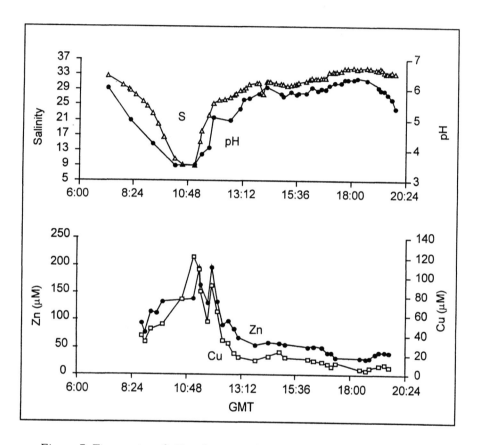

Figure 7. Time series of pH, salinity, total dissolved Cu and Zn over a tidal cycle at a fixed position in the Huelva estuary (June 1997).

88

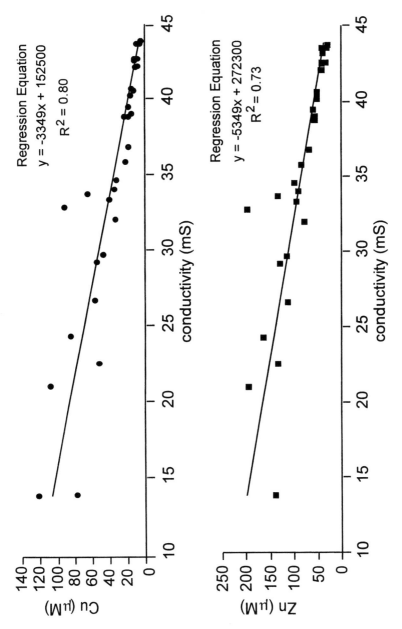

Figure 8. Total dissolved Cu and Zn versus conductivity for tidal cycle experiment in the Huelva estuary (June 1997).

other estuarine systems (e.g. Severn (*30*)), Cu and Zn are often reported to show a non-conservative behaviour in estuaries, with typically removal in the low salinity turbidity maximum zone and a mid-estuarine maximum due to anthropogenic and sedimentary inputs (Cu and Zn in Tamar (*35*) and Scheldt (*36*), and Zn in Forth (*31*)). The main difference which distinguishes the Huelva from other estuarine systems is its acidity, and the conservative Cu and Zn behaviour in the Huelva can be explained by the low pH in this system, which keeps trace metals in solution by proton competition for the available sorption sites on suspended particulate material (*37*).

Trace metal monitoring in coastal waters

The application of automated voltammetric instrumentation on-board ship, in conjunction with a continuous underway sampling approach, has allowed the acquisition of high quality and high resolution coastal trace metal data. Figure 9 shows the set-up for continuous ship-board sampling and analysis. Underway pumping is used as a means of sample collection, and this obviates the need for the vessel to halt for the collection of discrete samples. An effective pumping system involves a peristaltic pump and a braided PVC hose. The hose is hung overboard and attached to a 'fish' (torpedo-like structure, KIPPER-1) which is towed from a strong cable attached to a winch. The fish stays at a constant depth (ca. 2-4 m), even at speeds exceeding 10 knots, due to its design and weight (ca. 40 kg). The fish is made from solid carbon steel with an inlet at the front and a hole through the centre for the sampling hose. The fish is painted with a non-metallic epoxy-based paint.

Distribution and behaviour of trace metals in the Gulf of Cadiz

Figure 10 shows the results of a times series of salinity and dissolved Zn over a full tidal cycle undertaken ca. 3 km off the mouth of the Huelva estuary (see Figure 6), whilst the vessel (*B/O Garcia del Cid*) was anchored (October 1998). The salinity time series shows a limited salinity range (S = 36.33-36.35), indicating that the physical characteristics of the marine waters at the sampling position were not strongly affected by a tide-driven estuarine plume from the Huelva or adjacent Guadiana and Rio Piedras systems. This observation is partly the result of the low freshwater flow entering the Huelva estuary from the feeding rivers (October 1998, estimated flow of Tinto and Odiel: 0.1 and 0.4 m^3 s^{-1}, respectively), and also the result of the residual current structure in the Gulf of Cadiz, with currents moving in an west-east direction and estuarine plumes remaining close to the coast. Ship-board on-line voltammetric measurements of

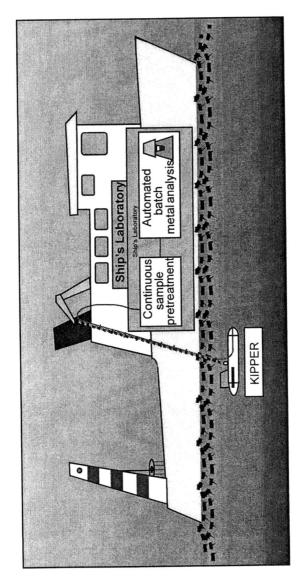

Figure 9. Drawing of ship-board continuous underway sampling and analysis system (Reproduced with permission from reference 8. Copyright 1999 Elsevier).

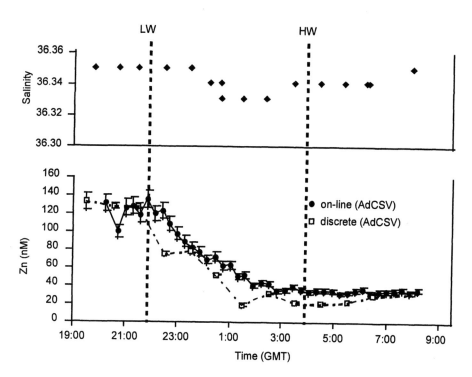

Figure 10. Time series of salinity and total dissolved Zn over a tidal cycle undertaken at a fixed position in the Gulf of Cadiz (October 1998). On-line dissolved Zn measurements were performed on-board ship using automated voltammetry with a pumped seawater supply, and discrete dissolved Zn measurements were performed in the laboratory in Plymouth in samples collected (hourly) using Niskin samplers.

dissolved Zn (8-hydroxy-quinoline as AdCSV ligand), and results of AdCSV Zn analysis in discrete samples (ammonium pyrrolidine dithiocarbamate as added AdCSV ligand) taken in parallel at hourly intervals using modified trace metal clean Niskin bottles, are presented in Figure 10b. The dissolved Zn concentrations from the on-line measurements varied between ca. 30 and 140 nM. The highest dissolved Zn concentrations (100-136 nM) were observed around the time of low tide at the mouth of Huelva estuary. The lowest concentrations (31-38 nM), which are representative of background Zn concentrations in the nearshore waters of the Gulf of Cadiz (*38*), were observed during high tide. The dissolved Zn concentrations in discrete samples agreed well with the results from the on-line analysis. The differences, with slightly lower Zn concentrations in discrete sample, can most likely be explained by differences in sampling depths. The discrete samples were obtained from a depth of ca. 7-8 m, whereas the pumped water intake was positioned at a depth of ca. 3 m. The high number of data points (n = 41) obtained during the on-line analysis (compared with 13 discrete samples), provides a good picture of the temporal trend of dissolved Zn in these coastal waters.

Coastal surveys with continuous underway sampling and near real-time on-line trace metal measurements have been undertaken in the Gulf of Cadiz in the period 1996-1998. The studies were performed using a small vessel (*Cirry Tres,* 9 m length) for surveying of shallow near coastal waters, and a larger vessel (*B/O Garcia del Cid,* 45 m length) for off-shore work. Two fully automated voltammetric systems were used on-board ship. Dissolved Cu and Zn (8 hydroxy-quinoline as AdCSV ligand) were determined simultaneously with one, and dissolved Co and Ni (dimethyl glyoxime as AdCSV ligand) with the other system. Figure 11 shows the sampling grids, with the locations of on-line determined trace metal samples for surveys in April 1998 (Figure 11a; *Cirry Tres*; >60 data points) and October 1998 (Figure 11b; *B/O Garcia del Cid*; 190 Zn, 270 Ni data points). Figure 12a shows the contour plots of salinity, total dissolved Zn and Ni for the April 1998 survey. The contour plots were generated using Surfer (Win 32), and the interpolation between the data points was carried out using Kriging as the gridding method. The results from the on-line measurements were not corrected for tidal excursion, which has to be considered when interpreting the contour plots. The salinity contour plot indicate low salinity surface water patches to the south-east of the Huelva estuary (S<34.2) and to the west of the Huelva estuary (S<34.4) and an additional pocket (S<34.2) was observed at the western fringe of the survey area. The metal concentrations were elevated in the plume coming out of the Huelva estuary, to the south-east of the estuary's mouth (≤300 nM Zn, ≤ 10 nM Ni) and in the north-west corner of the surveyed area (108 nM Zn, 5.1 nM Ni).

During the October 1998 survey (Figure 12b), areas of low salinity (≤36.0) were observed associated with the Guadiana in the west, the Huelva estuary in

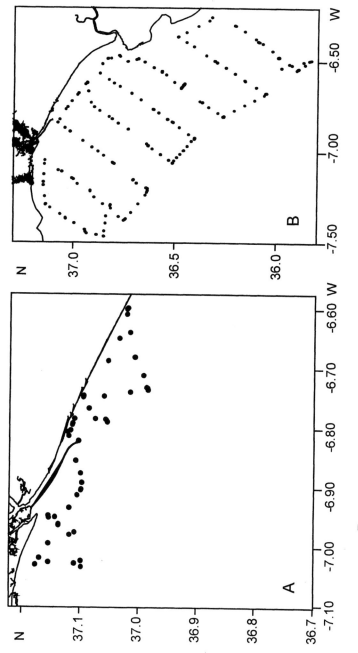

Figure 11. Locations of sampling positions in the Gulf of Cadiz where on-line trace metal measurements were undertaken in April 1998 (Figure 11a) and October 1998 (Figure 11b).

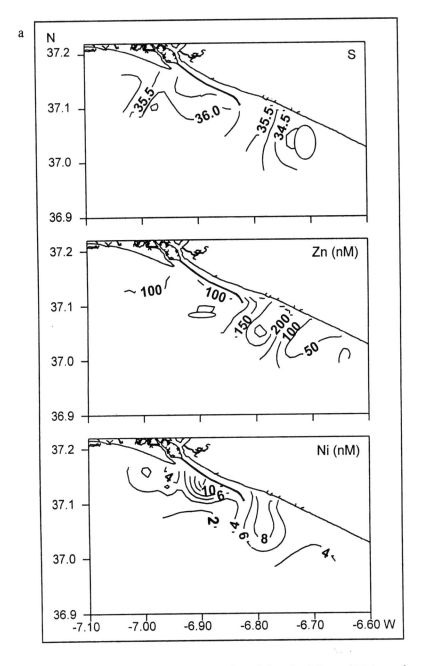

Figure 12. Contour plots of salinity and total dissolved Zn and Ni in surface waters of the Gulf of Cadiz during Cirry Tres *cruise (April 1998; Figure 12a); and* Garcia del Cid *cruise (October 1998; Figure 12b).*

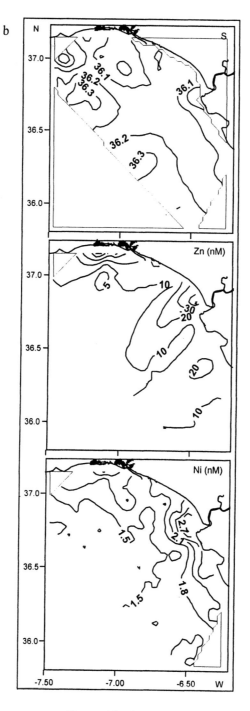

Figure 12. Continued.

the north and the Guadalquivir in the east of the Gulf of Cadiz. In surface waters of the central and offshore areas of the Gulf of Cadiz, the salinity was greater than 36.2, with maxima above 36.6. High concentrations of Zn (18-90 nM) and elevated Ni levels (2.0-6.0 nM) were observed along the shoreline. A steep concentration gradient between the outflow of the Huelva and Guadalquivir estuaries was observed. In an offshore direction, the Zn and Ni concentrations showed a decrease as a result of mixing of metal enhanced near coastal waters with more pristine Atlantic Ocean surface waters. The concentrations at offshore locations in the Gulf of Cadiz were ≤11 nM Zn and ≤2.0 nM Ni. Relative to off shore waters, Ni concentrations in the nearshore waters of the Gulf of Cadiz were not as elevated compared with Zn. This can be explained by the absence of Ni enrichment, and a high abundance of Zn, in the geology of the Iberian Pyrite Belt. The observations obtained using the near-real time ship-board monitoring approach were in good agreement with trace metal values reported in literature (*38,39*). However, the high spatial resolution of the data has allowed a more thorough interpretation of sources and behaviour of trace metals in the Gulf of Cadiz.

Distribution and behaviour of trace metals in English coastal waters

As a result of the wet climatic conditions of the British Isles, freshwater inputs into coastal waters are much more important compared with southern Spain. Major fluxes of dissolved trace elements around the coast of England are associated with inputs of freshwater by rivers, and of seawater by the English Channel and the northern North Sea (*40,41,42*). The strong influence of rivers on English Coastal Waters is exemplified by Figure 13, which shows salinity and dissolved Cu and Zn data for a coastal track undertaken by RRS *Challenger* (October 1994). The cruise started in Barry (south Wales) and finished in Hull (northeast England). Continuous sampling and analyses was undertaken during the cruise, resulting in high resolution data for the English Coastal Waters. The data in Figure 13 have been plotted in the sequence in which they were obtained using the underway sampling approach. Positions of the vessel in the vicinity of estuarine plumes have been indicated by vertical dotted lines. Low salinity values associated with river water inputs into the coastal waters are apparent in Figure 13. The cruise in October 1994 took place during a relatively dry period, with river flows below the mean monthly long-term (1961-1990) flow. Enhanced dissolved Cu and Zn concentrations were observed in the vicinity of rivers and estuaries. Important trace metal signals were observed for the Fal, Thames, Humber, Tees, Tyne and Wash. These systems are anthropogenically perturbed with relatively large mining, industrial and urban waste water discharges. Metal-salinity relationships obtained using the high resolution data showed a non-

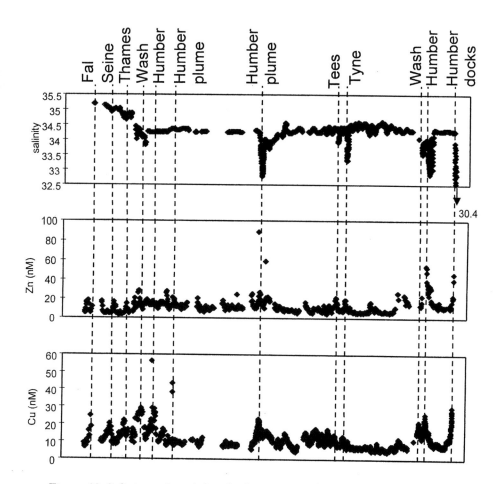

*Figure 13. Salinity and total dissolved Zn and Cu plotted against location of vessel (*RRS Challenger*) in English Coastal Waters during cruise in October 1994. Data are plotted in sequence in which they were obtained using underway sampling (Adapted with permission from reference 20. Copyright 1999 Elsevier).*

conservative behaviour for both Cu and Zn, indicating that processes other than mixing between water masses was influencing their behaviour in English Coastal Waters (20).

Conclusions

The application of ship-board voltammetric instrumentation for high resolution monitoring of trace metals in coastal seas has distinct advantages. The trace metal data obtained in the Gulf of Cadiz revealed a complex mixing pattern of contaminated estuarine with more pristine Atlantic waters. This data will be used in numerical models to evaluate the biogeochemical behaviour of the trace metals investigated during our studies. The enhanced spatial resolution of trace metals observed during our work in the English Coastal Waters allowed us to identify signals from fluvial sources which would possibly be missed using a traditional sampling and analysis approach. In addition, our approach aided the interpretation of trace metal behaviour in coastal waters. The high resolution measurements in coastal waters therefore result in data sets that allow a more thorough deconvolution of the complex temporal and spatial behaviour of trace metals in such waters.

A recent development in trace metal monitoring is the fully submersible voltammetric probe. One such probe, which is currently used in our laboratory, is the voltammetric in-situ profiling (VIP) system (9,43,44). The VIP allows the determination of Cu, Cd, Pb and Zn in seawater using ASV, with a gel integrated mercury coated microelectrode as the working electrode (45). The iridium microelectrode array displays an excellent sensitivity and reproducibility, and the agarose gel prevents fouling of the electrode and allows metal speciation measurements. Although a series of papers has been published on the analytical developments involving the VIP, the application of the probe for in-situ real-time trace metal measurements in estuarine and coastal waters has yet to be fully exploited. The remote deployment of the probes, with satellite communication, will allow the systems to be used in pollution control and water quality management scenarios.

Acknowledgements

The authors would like to thank the European Union (TOROS project, contract ENV4-CT96-0217; IMTEC project, contract (ENVK3-CT-2000-00036), the Natural Environment Research Council (GT 4/98/MS/234) and the University of Plymouth for funding. Dr. Whitworth is acknowledged for the

Tamar data. We thank the crew and scientist on the *Cirry Tres, Garcia del Cid* and *Challenger* for their assistance.

References

1. European Commission. Towards a European Integrated Coastal Zone (ICZM) Strategy. Office for Official Publications of the European Communities, Luxembourg, 1999.
2. Morley, N.H.; Fay, C.W.; Statham, P.J. *Adv. in Underwat. Technol., Ocean Sci. and Offshore Engin.* **1988**, *16*, 283-289.
3. Howard, A.G.; Statham, P.J. Inorganic trace analysis: philosophy and practice; Wiley, 1993.
4. Johnson, K.S.; Coale, K.H.; Jannasch, H.W. *Anal. Chem.* **1992**, *64*, 1065A-1075A.
5. Andrew, K.N.; Blundell, N.J.; Price, D.; Worsfold, P.J. *Anal. Chem.* **1994**, *66*, 916A-922A.
6. Braungardt, C.; Achterberg, E.P.; Nimmo, M. *Anal. Chim. Acta* **1998**, *377*, 205-215.
7. Wang, J.In *Laboratory techniques in electrochemical chemistry*; Kissinger, P.T., Heineman, W.R., Eds.; Marcel Dekker: New York, 1996; pp 719-738.
8. Achterberg, E. P and Braungardt, C. B. *Anal. Chim. Acta* **1999**, *400*, 381-397.
9. Tercier, M.L.; Buffle, J.; Graziottin, F. *Electroanal.* **1998**, *10*, 355-363.
10. Zirino, A.; Leiberman, S.H.; Healy, M.L. In *Marine Electrochemistry*; Berkowitz, B.J., Horne, R., Banus, M., Howard, P.L., Pryor, M.J., Whitnack, G.C., Weiss, H.V., Eds.; Electrochemical Soc.: Princeton, N.J., 1973; pp 319-332.
11. van den Berg, C.M.G. *Analyst* **1989**, *114*, 1527-1530.
12. Paneli, M.G.; Voulgaropoulos, A. *Electroanal.* **1993**, *5*, 355-373.
13. Florence, M. *J. of Electroanal. Chem.* **1970**, *27*, 273-281.
14. Luther III, G.W.; Brendel, P.J.; Lewis, B.L.; Sundby, B.; Lefrancois, L.; Silverberg, N.; Nuzzio, D.B. *Limnol.Oceanogr.* **1998**, *43*, 325-333.
15. de Vitre, R.R.; Tercier, M.-L.; Tsacopoulos, M.; Buffle, J. *Anal. Chim. Acta.* **1991**, *249*, 419-425.
16. Herdan, J.; Feeney, R.; Kounaves, S.P.; Flannery, A.F.; Storment, C.W.; Kovacs, G.T.A.; Darling, R.B. *Environ.Sci.Technol.* **1998**, *32*, 131-136.
17. Bond, A.M.; Mahon, P.J.; Schiewe, J.; VicenteBeckett, V. *Anal. Chim. Acta* **1997**, *345*, 67-74.
18. Bond, A.M.; Czerwinski, W.A.; Llorente, M. *Analyst* **1998**, *123*, 1333-1337.
19. Achterberg, E.P.; van den Berg, C.M.G. *Mar. Poll. Bull.* **1996**, *32*, 471-479.

20. Achterberg, E.P.; Colombo, C.; van den Berg, C.M.G. *Cont.Shelf Res.* **1999**, *19*, 537-558.
21. Whitworth, D.-J.; Achterberg, E.P.; Nimmo, M.; Worsfold, P.J. *Anal. Chim. Acta* **1998**, *371*, 235-246.
22. Achterberg, E.P.; van den Berg, C.M.G. *Anal. Chim. Acta* **1994**, *284*, 463-471.
23. Achterberg, E.P.; van den Berg, C.M.G. *Anal. Chim. Acta* **1994**, *291*, 213-232.
24. Achterberg, E.P.; Braungardt, C.B.; Sandford, R.C.; Worsfold, P.J. *Anal. Chim. Acta* **2001**, in press.
25. Muller, F.L. *Mar. Chem.* **1996**, *52*, 245-268.
26. Stumm, W.; Morgan, J.J. In *Aquatic Chemistry-Chemical equilibria and rates in natural waters;* John Wiley & Sons: New York, 1996; pp 1-1022.
27. Butler, E.I.; Tibbitts, S. *J.Mar.Biol.Assoc.UK.* **1972**, *32*, 681-699.
28. Morris, A.W.; Bale, A.J.; Howland, R.J.M. *Est. Coast. Shelf Sci.* **1982**, *14*, 649-661.
29. Grabemann, I.; Uncles, R.J.; Krause, G.; Stephens, J.A. *Est. Coast. Shelf Sci.* **1997**, *45*, 235-246.
30. Apte, S.C.; Gardner, M.J.; Gunn, J.E.; Vale, J. *Mar. Poll. Bull.* **1990**, *21*, 393-396.
31. Laslett, R.E.; Balls, P.W. *Mar.Chem.* **1995**, *48*, 311-328.
32. Morris, A.W. *Sci. Tot. Env.* **1986**, *49*, 297-304.
33. Morris, A.W.; Bale, A.J.; Howland, R.J.M.; Millward, G.E.; Ackroyd, D.R.; Loring, D.H.; Rantala, R.T.T. *Wat. Sci. Technol.* **1986**, *18*, 111-119.
34. Elbaz-Poulichet, F.; Morley, N.H.; Cruzado, A.; Velasquez, Z.; Green, D.; Achterberg, E.P.; Braungardt, C.B. *Sci. Tot. Env.* **1999**, *227*, 73-83.
35. Ackroyd, D.R.; Bale, A.J.; Howland, R.J.M.; Knox, S.; Millward, G.E.; Morris, A.W. *Estuar.Coast.Mar.Sci.* **1986**, *23*, 621-640.
36. Zwolsman, J.J.G.; Van Eck, B.T.M.; Van der Weijden, C.H. *Geochim.Cosmochim.Acta* **1997**, *61*, 1635-1652.
37. Braungardt, C.B.; Achterberg, E.P.; Elbaz-Poulichet, F.; Morley, N.H.; Cruzado, A.; Velasquez, Z.; Nimmo, M. **2001,** In preparation.
38. van Geen, A.; Boyle, E.A.; Moore, W.S. *Geochim. Cosmochim. Acta* **1991**, *55*, 2173-2191.
39. van Geen, A.; Rosener, P.; Boyle, E. *Nature* **1988**, *331*, 423-426.
40. Burton, J.D.; Althaus, M.; Millward, G.E.; Morris, A.W.; Statham, P.J.; Tappin, A.D.; Turner, A. *Phil.Trans.R.Soc.London* **1993**, *343*, 557-568.
41. Tappin, A.D.; Hydes, D.J.; Burton, J.D.; Statham, P.J. *Continent. Shelf Res.* **1993**, *13*, 941-969.
42. Tappin, A.D.; Millward, G.E.; Statham, P.J.; Burton, J.D.; Morris, A.W. *Est. Coast. Shelf Sci.* **1995**, *41*, 275-323.

43. Belmont-Hébert, C.; Tercier, M.L.; Buffle, J. *Anal.Chem.* **1998**, *70*, 2949-2956.

44. Tercier, M. L.; Buffle, J.; Koudelka-Hep, M.; Graziottin, F. In *Electrochemical Methods for the Environmental Analyses of Trace Element Biogeochemistry*, Taillefert, M.; Rozan, T., Eds. American Chemical Society Symposium Series; American Chemical Society: Washington, D. C., **2001**, this volume, Chapter 2.

45. Tercier-Waeber, M.-L.; Buffle, J.; Confalonieri, F.; Riccardi, G.; Sina, A.; Graziottin, F.; Fiaccabrino, G.C.; Koudelka-Hep, M. *Meas. Sci. Technol.* **1999**, *10*, 1202-1213.

Chapter 6

Permeation Liquid Membrane Coupled to Voltammetric Detection for In Situ Trace Metal Speciation Studies

N. Parthasarathy, P. Salaün, M. Pelletier, and J. Buffle[*]

CABE, Department of Inorganic, Analytical and Applied Chemistry, University of Geneva, Sciences II, 30 Quai E. Ansermet, CH–1211 Geneva 4, Switzerland

Permeation liquid membranes (PLM) are emerging as a versatile analytical tool for in situ speciation measurements of trace metals under natural waters. Their attractive features include the ability to separate and preconcentrate target species in a single step. A PLM comprising didecyl 1,10 diaza crown ether - lauric acid in phenylhexane/toluene (1:1) transports Cu, Pb, Cd and Zn across a microporous membrane from aqueous phase of interest to a stripping solution containing a strong ligand. Preconcentration factors in the range 100 - 3000 can be obtained in 5 - 120 min and a strip solution containing a strong complexing ligand. Detection limits of less than 10 pmol. L^{-1} can then be achieved in real time by voltammetry at a Hg-Ir microelectrode. A theoretical transport model has been developed to interpret metal speciation and validated for Cu complexed to a selection of synthetic and natural ligands.

Introduction

The importance of trace element speciation analysis in environmental samples is widely recognized. In natural water, trace metals exist under various chemical forms such as the free hydrated ions, inorganic and organic complexes and metals associated with colloidal particles (*1-5*). These play various roles in geochemical and biogeochemical cycling of elements, including transport, bioaccumulation, and toxicity. The determination of free metal ion concentration is therefore important for understanding these processes. Trace metals in natural waters, are present at low total concentration (<100 nmol. L^{-1}) and their free concentrations are still lower. The determination of free metal ions or specific species at such low concentration still remains a challenging task to analytical chemists since very few techniques combine speciation determination and high sensitivity. The methods available are the direct methods viz. voltammetry, particularly using microelectrodes, and separation and preconcentration methods such as liquid-liquid extraction, equilibrium dialysis, ultrafiltration (*1,5-7*) with their advantages and disadvantages. Of the methods available, voltammetry, in particular square wave anodic stripping voltammetry using microelectrodes (*8*), is a powerful tool for making in situ speciation measurements. However this technique does not measure the free ion but the sum of mobile and labile complexes. Another, alternative, newly emerging technique to meet this challenge is the use of permeation liquid membrane (PLM), also called supported liquid membrane (SLM), based on solvent extraction principles (*9-16*). The various aspects of this analytical technique are described in ref. (*10*). It consists of a thin microporous, hydrophobic membrane impregnated with a species-selective hydrophobic extractant, C, dissolved in water-immiscible organic solvent separating a source (sample) solution containing the target species and a strip solution containing a complexing ligand stronger than the carrier (Figure 1 A). The metals are transported uphill across the PLM, the transport is diffusion limited and the driving force is the chemical potential gradient of M'. The attractive feature of this technique is the simultaneous separation and preconcentration of the test metal ions and the selectivity achieved with a tailor-made species-specific carrier. While this technique has been widely used in the determination of traces of organic compounds (*16*), very few studies have been conducted for metal analysis or speciation in natural waters (*10,14,17-20*).

We have developed two PLM systems for trace metal speciation in natural waters comprising of a flat sheet or a hollow fiber membrane impregnated with an equimolar mixture of species-selective macrocyclic carrier 1,10 didecyl 1,10 diaza 18 crown 6 ether (22DD) and lauric acid (LAH) dissolved in phenylhexane / toluene mixture (1:1 v/v) (*10,14,20*). 22DD transports Cu(II), Pb(II), Cd(II) and Zn(II) selectively across the membrane at neutral pH. The attractive features of this technique are that it is simple, versatile and selective; it allows separation and preconcentration in one step thus avoiding contamination risks, and, in addition, high preconcentration factors can be achieved. It also allows real time measurements by coupling to analytical instrumental attendant methods, it is readily automated and eliminates matrix effects common to direct metal analysis. More importantly, PLM devices can be used for direct on site, in field and in situ preconcentration of target species in natural waters followed by metal analysis either in the laboratory or in situ. This can be achieved by combining a PLM with sensitive detectors that are suitable for in situ measurements, such as voltammetry. In this chapter the state of the art of PLM and its potential application for in situ speciation studies of Cu and Pb will be reviewed with special emphasis on its coupling with voltammetric flow-through microdetectors.

Theory

Principle and metal transport mechanism through PLM

PLM principles are described in detail in (*10*).

A permeation liquid membrane consists of a macroporous support impregnated with a hydrophobic organic solvent containing a cation selective carrier. The membrane is placed between the sample (source) and receiver (strip) solutions (Figure 1 A). The permeation of metal species through the PLM can be described as the combination of extraction into organic phase followed by back extraction. These two steps occur simultaneously and the transport is a carrier-assisted coupled transport.

The mechanism of metal transport through PLM is shown in Figure 1 B (*1*). The transport of metal ions occurs by: i) diffusion of metal, M, through the stagnant aqueous Nernst layer ii) complexation with the carrier, C, at the source/membrane interface; iii) diffusion of the complex, MC, across the membrane and iv) decomplexation of MC at the strip/membrane interface. The metal transport across this membrane can be driven by proton gradient, co anion gradient or counter cation gradient (10). In these systems, metal species are transported against their concentration gradients. In addition, by using strip solution volumes much lower than that of the sample solution, the target metal species can be preconcentrated. For analytical speciation measurements, perturbation of the test medium should be avoided. Thus, the counter transport of the cation, such as that used with the 22DD-LA-phenylhexane-toluene PLM system, is the most suitable counter transport for these measurements .

With the PLM used in this work, the flux is controlled by diffusion either in the source solution or in the membrane (*10*). Under steady state diffusion conditions and for inert complexes (i.e. complexes which do not dissociate at the source/membrane interface), the overall flux is given by :

$$J = \frac{C_{so}/\alpha_s}{\left(\dfrac{\delta_s}{D_s} + \dfrac{\ell}{D_{MC}K_p[C]}\right)} \tag{1}$$

and for labile complexes :

$$J = \frac{C_{so}}{\left(\dfrac{\delta_s}{D_s} + \dfrac{\ell\alpha_s}{D_{MC}K_p[C]}\right)} \tag{2}$$

where D_s is the diffusion coefficient of free M in the source solution, $\overline{D}_s$ the mean diffusion coefficient of all the M species in the source and D_{MC} is the diffusion coefficient of the metal carrier complexe in the membrane phase. δ_s is the thickness of the stagnant aqueous layer on the source side of the membrane, K_p is the partition coefficient of M at the solution/membrane interface, ℓ is the thickness of the membrane and C is the concentration of the carrier, C_{so} is the concentration of the metal in the source solution. α_s is the degree of complexation of the metal in the source phase and is given by :

$$\alpha_s = \frac{C_{so}}{[M]} = 1 + K_{ML}[L] \tag{3}$$

where [L] is the concentration of complexing ligand in the source solution, K_{ML} is the equilibrium constant of ML and [M] is the free metal ion concentration.

Equation 2 shows that the flux is related to the complexation parameter α_s and is proportional to the free metal ion concentration ([M] = C_{so}/α_s). In the absence of complexing agent α_s = 1, the flux is proportional to the total metal concentration. Thus α_s can be calculated from the ratio of fluxes in the absence and presence of complexing parameters.

Equation 2 can be rewritten as

$$\frac{C_{so}}{J} = \frac{\delta_s}{D_s} + \frac{\ell}{D_{MC}K_p[C]}\alpha_s \tag{4}$$

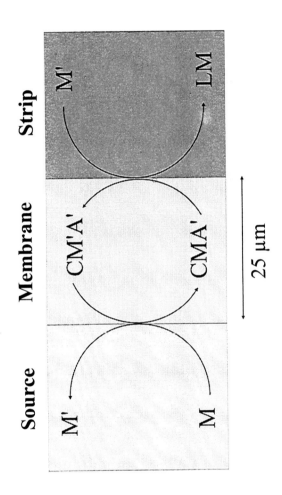

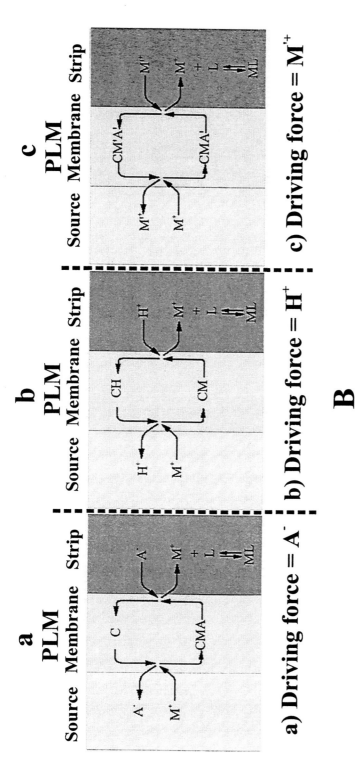

Figure 1 A): Schematic representation of a PLM system. M : test analyte, L : stripping agent, C : carrier, M' : counter anion, A' : co anion; B) Transport mechanism across the PLM.

Equations 1 and 4 show that, if inert complexes are formed, J is proportional to [M] (eq 1). On the other hand, if labile complexes are formed (eq 4) two limiting cases exist:

$$D_{MC}K_p[C]<<\overline{D_s}/\delta_s \text{ (membrane diffusion limited transport)}$$

then J is directly proportional to [M].

$$D_{MC}K_p[C]>>\overline{D_s}/\delta_s \text{ (source diffusion limited transport) then}$$

all the metal complexes will contribute to the overall flux J, after dissociation and J will be proportional to C_{so}.

Therefore Equations 1 and 4 are useful to predict or interpret the permeation behavior of metal species and to determine whether the rate-limiting step for metal transport is the diffusion in the source solution or in the membrane. In addition, flux measurements made under various conditions, particularly by varying δ_s, ℓ, and carrier concentration, may provide insight into the factors governing the permeation of complexes and provide information on whether the flux depends on M or C_{so}.

Devices

Two basic PLM configurations are shown in Figure 2. In the PLM system the source and the strip solutions are separated by a chemically inert microporous support impregnated with the hydrophobic solvent and carrier. The organic liquid is held in the pores of the membrane by capillary forces. The flat sheet PLM system (Figure 2 A) is useful in laboratory studies to optimize operating conditions while the hollow fiber PLM system (HFPLM) (Figure 2 B) is more suitable for field applications, due to its large surface area to volume ratio which yields a larger flux. This system can be adapted for in situ preconcentration of trace elements in natural waters.

Operation and Interpretation

Preconcentration of metals using off-line PLM systems

The advantage of PLM techniques is that metal separation and preconcentration can be accomplished in one step by using smaller volumes of

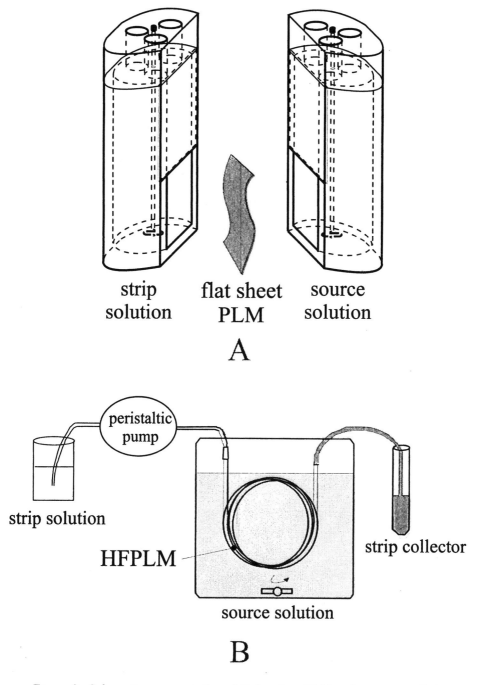

strip
solution

flat sheet
PLM

source
solution

A

peristaltic
pump

strip solution

HFPLM

strip collector

source solution

B

*Figure 2 : Schematic representation of A) flat sheet PLM and transport cell, B)
single hollow fiber PLM (HFPLM).* (Adapted with permission from reference
20. Copyright 1997 Elsevier Science)

strip solution than source solution. This is important when determining metal ions present at trace levels such as those found in most natural waters. In order to attain a large preconcentration factor: $F = C_{st}^t / C_{so}$ (where C_{st}^t and C_{so} are the concentration of the metal in the strip solution at time t and the initial metal concentration in the source solution, respectively), the volume of the strip solution, V_{st}, should be much smaller than the volume, V_{so}, of the source solution. For any value of V_{so} and V_{st}, F increases with time and reaches a limiting value F^{max} given by (10) : $1/F^{max} = V_{st}/V_{so} + \alpha_s/\alpha_{st}$ where α_{st} is the degree of complexation in the strip solution. Using the flat sheet PLM system described in Figure 2 A, the value of $F^{max} = 16$ was obtained for Pb and Cd after 180 min, independent of initial metal concentration. The source and strip solutions volumes were 80 and 5 ml respectively (10,20).

For attaining higher preconcentration factors at shorter times, larger V_{so}/V_{st} should be used. In the hollow fiber PLM (HFPLM) system (Figure 2 B), microliter scale volumes of strip solutions and very large volumes of source solutions can be used. Accurel® pp 5q/2 polypropylene hollow fiber (inner diameter = 600 μm; outer diameter = 800 μm ; pore size = 0.2 μm) was used as the support. A single hollow fiber (HF) module was used in the transport experiments. The detailed description of the experimental system is given in (10,20). Hollow fibers with lengths varying from 14 to 22 cm have internal volumes of 40-70 μL.

The hollow fiber is impregnated with the carrier from the outside (well impregnated membranes are transparent), the excess is removed on the inside and outside with Milli Q® water as described in (20). Strip solution (5.10^{-4} mol.L^{-1} CDTA; pH= 6.56) is filled in the lumen side of the HF by means of a peristaltic pump at a flow rate of 0.1 ml/min with the outside of the fiber dipping in a beaker containing Milli Q water. At time t = 0, the beaker of water is replaced by the source solution containing the target metals (MES buffer pH=6.0). The source solution is stirred at 480 rpm to minimize the thickness of the aqueous stagnant layer and the strip solution is left stagnant. Source and strip solutions are collected at prescribed times and metal concentrations determined by using flame or graphite furnace atomic absorption spectrometry (AAS). Preconcentration factor vs. time curves for Cu, Pb and Cd (Figure 3) show that high preconcentration factors of up to 3000 for Pb and 800 for Cu are obtained in 120 min using strip and source solutions volumes of 68 μl and 250 ml respectively. A preconcentration factor of 80 is obtained for Cd under the same conditions (20). The lower preconcentration factor for Cd is due to its lower partition coefficient. Preconcentration factors are independent of initial metal concentration in the source solution over a wide concentration range (5 μmol. L^{-1} to 1 nmol.L^{-1}). This is very important for the quantitation of trace metals in real

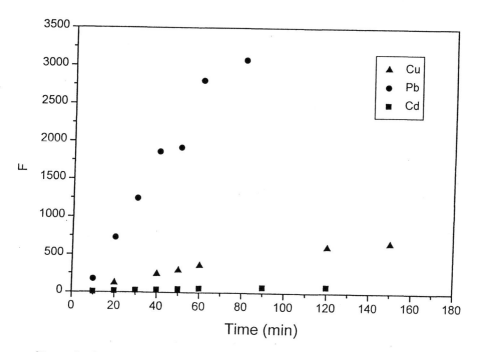

Figure 3 : Preconcentration factor vs. time for Cu, Pb and Cd through hollow fiber PLM ; source solution : (●) Pb (10^{-6} mol.L^{-1}), (▲) Cu ($5\ 10^{-6}$ mol.L^{-1}) and (■) Cd ($5\ 10^{-5}$ mol.L^{-1}) in MES buffer (pH = 6.0), V_{so} = 250 ml; strip solution : CDTA ($5\ 10^{-4}$ mol.L-1), pH 6.56, V_{st} = 68 μl. (Adapted with permission from reference 20. Copyright 1997 Elsevier Science)

world samples. The reproducibility of preconcentration factors, calculated from seven replicate measurements, is within 5% (*20*).

The concentration of analyte in a real world sample can be computed from the F value corresponding to the same preconcentration time in a standard solution. Once the preconcentration factors are determined for a given HF, it can be used for computing the analyte concentration of the sample without recalibration which considerably reduces the analysis time. The membrane is stable for at least a week. The possible causes of instability of the PLM are (*22, 10*) : i) the solubility of the carrier and other components of the organic solution; ii) increased pressure drops across the membrane resulting in the expulsion of organic phase from the membrane and causing a leakage of the source solution into the strip solution; iii) osmotic pressure gradient; iv) formation of emulsion at the source /membrane interface due to lateral shear forces. This last process decreases the mass transfer rate at the sample/membrane interface. In our system, the emulsion formation is probably the cause of membrane instability.

On-line coupling of PLM to micro liter flow-through voltammetric cell

Analysis of concentrated test ions or compounds in the strip solution are usually determined off-line in the laboratory by collecting (sampling) the strip solution at prescribed intervals, followed by trace metal analysis with AAS, ICP-MS, UV-Visible spectrometry or voltammetric techniques (*10, 14, 17*). However for on site or in situ monitoring of trace metals in environmental samples, on-line compact analytical systems are required to allow real time analysis and to minimize sample handling and contamination risks. Voltammetric detectors are particularly well suited for this purpose. Moreover, combining the preconcentration steps of anodic stripping voltammetry (ASV) and PLM makes it a highly sensitive analytical tool. There is only one report in the literature that has coupled a PLM to a 200 µl strip cell with in situ potentiometric stripping analysis at mercury coated reticulated vitreous carbon electrode for trace metal determination in urine (*23*). To increase the preconcentration factor F for trace metal analysis in natural waters, a smaller volume flow-cell (<50 µL) is required, which entails the use of microelectrodes. We have developed two types of microliter volume flow-cells with microelectrodes: one for on-line coupling with HFPLM and another one where PLM and microvoltammetric electrodes are integrated on a chip (AMMIA: agarose gel membrane mercury plated Ir based electrode). To our knowledge there is no other report of such developments in the literature. There are several advantages for coupling PLM and AMMIA (PLM/AMMIA): the electrochemical device is compact compared to the classical hanging dropping electrode; and it can be adapted for field measurements on site, or, in situ real time measurements. In situ measurements cannot be accomplished with PLM/ICP-MS or PLM/GFAAS devices. Furthermore, voltammetry is less expensive than the multiple element ICP-MS. Donnan dialysis method in combination with ASV, such as described by Cox et al (30), have similar advantages except that the solutions on the two sides must have the same ionic strength. One of the main disadvantages of our method is that the presence oxygen might interfere with the voltammetric measurements.

Two main problems have to be solved for designing such a microliter voltammetric cell coupled to PLM (25,26):

i) The cell geometry i.e. the cell must be constructed with hydrodynamic conditions permitting sensitive and reproducible voltammetric measurements. This is a difficult problem to solve in the case of micro flow-through cells. This aspect is described in detail in (28). The choice of material is another important factor in the cell construction. The material should not adsorb metals and should not perturb the sample flow. The latter could alter the electrode signals through electrostatic effects. In addition the material should be solvent resistant for use in conjunction with PLM;

ii) Adsorption on the working voltammetric electrode of traces of PLM carrier or solvent, leached into the strip solution. It has been shown that this may drastically interfere with voltammetric measurements and may sometimes completely suppress the signal if the strip solution remains for a long time (t > 10 min) in the voltammetric cell (25). Methods to remove this effect have been discussed previously (25,29).

HFPLM coupled to microvoltammetric flow-through cell (HFPLM/MFTC)

Based on these considerations, the optimum microliter volume voltammetric cell design coupled to HFPLM is shown in Figure 4 (29). It comprises the HFPLM device, a 50 µl PEEK® (polyether-etherketone) voltammetric flow-through cell (resistant to organic solvent and trace metal free), pumps, and switching valves. A single Hg-Ir microelectrode is used as the working electrode (WE). Plexiglas buffer vessels at the outlet of the pump serve to remove micro air bubbles and the reticulated vitreous carbon removes traces of organics coming from dissolution of the membrane carrier. Conditions, including flow rate and flushing time for washing the cell prior to a new measurement, were optimized for performing square wave anodic stripping voltammetry (SWASV) in deoxygenated strip solutions containing Pb, Cd and Cu (28,29). The strip solution used is 10^{-3} mol. L^{-1} $Na_4P_2O_7$ (pH=6.5) It must be noted that the mercury plating on iridium is performed in an external cell (12 min deposition at -400mV vs. Ag/AgCl/3 mol. L^{-1} KCl//1 mol. L^{-1} $NaNO_3$ reference electrode (RE)) and the Hg-Ir electrode is conditioned by running SWASV prior to inserting it into the flow cell. Calibration curves of Pb, Cd and Cu in 10^{-2} mol. L^{-}

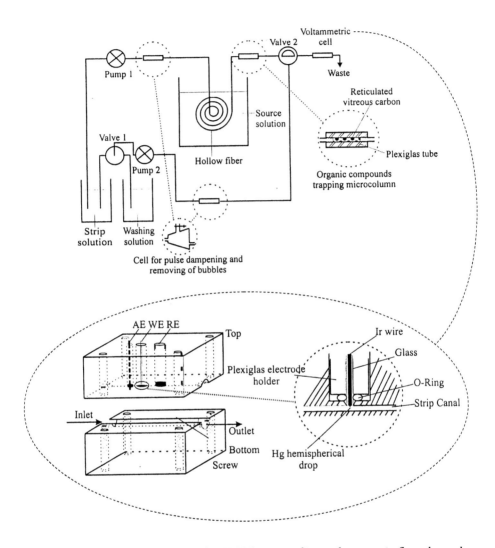

Figure 4 : On-line coupling of HFPLM to microliter voltammetric flow-through cell (HFPLM/MFTC). Pumps 1 and 2 : peristaltic pumps ; valves 1 and 2 : electrically actuated valves. Inset : schematic representation of flow-through voltammetric cell. WE, RE, AE : working (Hg-Ir microelectrode), reference (Ag/AgCl/3 mol.L⁻¹ KCl//1 mol.L⁻¹ NaNO₃) and auxiliary electrode (Pt) respectively.

[1] $Na_4P_2O_7$, which is used as strip solution in HFPLM/MFTC systems, are linear in the studied range (10 nmol. L^{-1} – 1 μmol. L^{-1}). The time needed to push the preconcentrated samples into the cell (50 sec), the time needed to wash the cell after each preconcentration measurement (2 min) and the pre electrolysis time for SWASV measurements (30 sec to 300 sec, depending on the concentration of the metals in the strip solution) were optimized to perform on-line measurements with this system. All these operations were computer controlled. The software program was written in our laboratory and the hardware consists of a home-made potentiostat and amplifier coupled to a PC.

Preconcentration factors F vs. time are determined for Pb and Cu in the same way as in the off-line method described above. Typical voltammograms obtained at various preconcentration times are shown in the inset of Figure 5 for Pb. The plots of F vs. time for Cu and Pb show that preconcentration factors are comparable with those obtained using the off-line detection method (Figure 5). It should be noted that for times longer than 20 min, SWASV currents decrease, probably because the Hg film becomes passive when it is left in an open circuit. More studies are needed to elucidate the cause of this behavior. The reproducibility of the measurements calculated from seven replicate samples was within 5%. Preconcentration times of 20 min enables to get a sensitivity of 10^{-11} mol.L^{-1} for Pb and Cu.

PLM and chip integrated microvoltammetric cell

Coupling PLM and voltammetric detectors can also be realized by placing the working electrode directly in the strip solution just over the PLM. This new configuration allows real time measurements of F vs. time. The analysis is conducted while separation and preconcentration are occurring. Thus, complete curve of F vs. time can be followed in one step, which decreases significantly the overall analysis time. The experimental set-up is shown in Figure 6 and is described below. The flat sheet membrane is pressed between a PEEK® block containing the source solution channel and a microfabricated electrode system containing: the strip channel, the working microelectrode array (WE) and the Ir auxiliary electrode (AE). The strip channel size is fixed by microfabricated structures of Epon SU-8 and a spacer of PTFE which also insures, together with O-rings (not represented on Figure 6), a good seal in the system. The working electrode consists of a square array of 15 (3*5) thin film mercury coated iridium microelectrodes separated from each other by 150 μm. The array is placed in front of the PLM membrane in the strip solution. The active strip volume is less than 1 μL. In order to get a short response time, the distance between the PLM and WE must be as small as possible (500 μm). Adsorption of the organic solvent on the mercury microelectrodes is avoided by incorporating C18

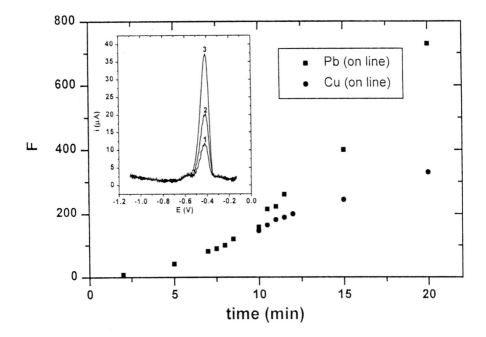

Figure 5 : Preconcentration factor vs. time for (●) Cu and (■) Pb using HFPLM/MFTC . Source : 250 ml of Pb (10 nmol.L⁻¹) and Cu (10 nmol.L⁻¹) in 10⁻² mol.L⁻¹ MES buffer (pH = 6.0). Strip: 10⁻² mol.L⁻¹ Na₄P₂O₇ (pH = 6.5), V_{st} = 60 μl. Inset : Typical voltammograms for preconcentration of Pb using HFPLM/MFTC. 1, 2 and 3 represent HFPLM preconcentration times of 2, 5 and 10 min respectively. N.B. : for HFPLM preconcentration of 10 min or more, deposition time of 2 min is used. SWASV conditions : 1 min preclean at +25 mV, 5 min deposition at -1100 mV, scan potential : -1100 to +25 mV, pulse amplitude : 25 mV, step amplitude : 8 mV and frequency = 100 Hz ; current amplification : 1000.

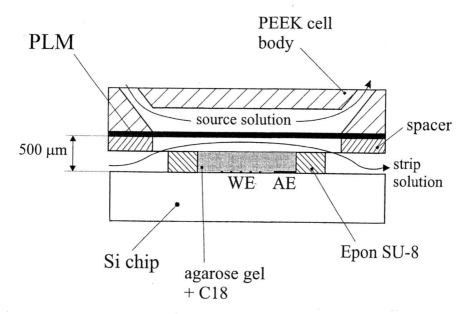

Figure 6 : Schematic representation of flat sheet PLM coupled on-line with Hg-Ir square array microelectrodes (PLM/CIMC).

particles into a 1.5% agarose gel. Metal complexes are not adsorbed on their surface in pyrophosphate media at neutral pH, but traces of PLM carrier or solvent leached into the strip solution adsorb on these particles (25,26).

The iridium auxiliary electrode is also of microsize (0.25 mm^2) and is placed close to the WE array. The reference electrode is a classical Ag/AgCl/3 mol.L^{-1} KCl//1 mol.L^{-1} NaNO$_3$ placed downstream in the strip channel. Plating of mercury on the iridium square array of microdiscs (5 µm diameter) is performed externally through the agarose/C18 gel by electrodeposition of 5 10^{-3} mol.L^{-1} Hg^{2+} in 0.1 mol.L^{-1} HClO$_4$ at -400 mV for a given time. The electrode isthen placed and pressed on the PEEK® block and PLM without damaging the deposited Hg film. The surface tension of Hg and the strong Hg-Ir cohesion forces make the Hg film very stable, even in an open circuit (28). As for the HFPLM/MFTC, pyrophosphate is used as the strip ligand and SWASV is the detection technique. Such technique allows measurements in pH buffered oxygenated pyrophosphate solutions.

Typical preconcentration curves vs. time for Pb and Cd are shown in Figure 7. Corresponding SWASV voltammograms of simultaneous preconcentration of Pb, Cd, Cu and Zn are presented in the inset. The optimum source flow rate is 9.86 mL.min^{-1} corresponding to a linear flow velocity v = 4.53 cm.s^{-1}. Lead, cadmium, copper and zinc are preconcentrated in the strip solution from an oxygenated source solution with [Pb] = 10 nmol. L^{-1}, [Cd] = 100 nmol.L^{-1} and [Cu] = 50 nmol.L^{-1} in 10^{-2} mol.L^{-1} MES at pH 6.0. After the preconcentration step, the strip channel is washed by passing a new strip solution at a flow rate of 10 µL.mn^{-1} (corresponding to v = 1 cm.s^{-1}) for 12 min. The system is then ready for another measurement.

Speciation studies using PLM and applications to natural waters

In order to test whether PLM can be applied to metal speciation studies, the interaction of Cu with several synthetic ligands such as oxalate (OX), Tiron, sulfosalicylic acid (SSAL), 8-hydroxy quinoline sulphonic acid (HQSA), were studied by measuring the Cu fluxes through PLM in the absence and presence of these ligands (10, 16, 17). All the ligands tested form negatively charged complexes with Cu(II) which do not pass through the membrane. In addition, these complexes can be considered inert under the conditions used. Therefore, it is expected that only the free metal ion flux is measured by PLM. From the ratio of the fluxes in presence and absence of ligands, the degree of complexation α_{exp} is computed (eq 3) and compared with the theoretically calculated α_{th} values using the stability constants reported in the literature (10, 17). A plot of α_{th} vs. α_{exp} yields a slope of 1 (Figure 8) confirming that indeed only the free metal concentration is measured under these conditions. These results show that PLM

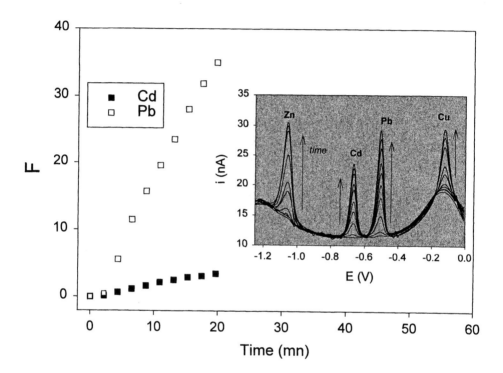

Figure 7 : Preconcentration factor vs. time for Cd (■) and Pb (□) using the PLM/CIMC system. Inset: Corresponding SWASV voltammograms. Conditions: Source : [Pb] = 10 nmol.L⁻¹ [Cd] = 100 nmol.L⁻¹, [Cu] = 50 nmol.L⁻¹ and impurities of Zn in oxygenated solution of MES buffer (10⁻² mol.L⁻¹, pH 6.0), v = 4.53 cm.s⁻¹ ; strip : oxygenated solution of Na₄P₂O₇ (10⁻² mol.L⁻¹, pH = 6.0). SWASV parameters : 50 s precleaning at 0 V, 60 s deposition at -1.25 V, scan potential : -1.25 to 0 V, pulse amplitude : 25 mV, step amplitude : 9 mV, frequency : 50 Hz.

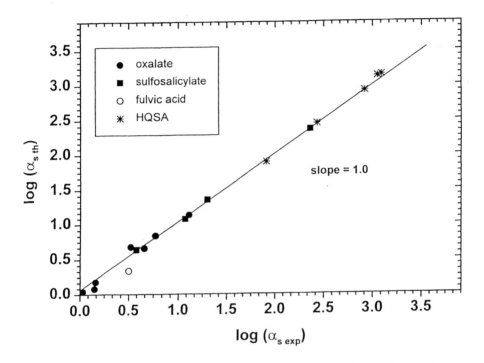

Figure 8 : Comparison of theoretical $\alpha_{s\,th}$ with experimental $\alpha_{s\,exp}$ found using the off-line PLM system. Conditions and symbols used: : source solution : (■) Cu (50 μmol.L⁻¹) and SSAL (50-1000 μmol.L-1) in MES buffer (pH = 6.0); (●) Cu (5-50 μmol.L⁻¹) and OX (200-800 μmol.L⁻¹) in MES buffer (pH = 6.0); (✱) Cu (50 μmol.L⁻¹) and HQSA (40-80 μmol.L⁻¹) in MES (pH = 6.0); (○) Cu (10 μ mol.L⁻¹) and FA (11 μmol.L⁻¹) in MES buffer (pH = 6.0). (Reproduced with permission from reference 10. Copyright 2000 Wiley & Sons Ltd.)

can discriminate between free metal ions and hydrophilic charged complexes. The striking feature of these speciation measurements is that there is a good correlation between the experimental and theoretically predicted free metal concentrations.

Speciation measurements of Cu (II) with fulvic acid, a major organic complexing ligand which forms negatively charged complexes with Cu, were conducted with the PLM system (*17*).

Fluxes were determined with the two PLM systems (flat sheet and HFPLM) in solutions containing:

i) a constant Cu concentration (10 μmol.L^{-1}) and varying the Suwannee river fulvic acid concentration (5-20 mg.L^{-1}) at pH=6.

ii) a constant Suwannee river fulvic acid concentration (20 mg.L^{-1}) and varying the Cu concentration (0.25 μmol.L^{-1} to 6 μmol.L^{-1}).

Fluxes of Cu in the presence of fulvic acid were smaller than those in its absence, in agreement with the fact that Cu-fulvic complexes do not pass through the membrane due to their negative charge. The conditional stability constant of Cu-fulvic acid complex is calculated from the experimentally determined degree of complexation using eq (3). It was found to be log $\beta^* = 5.6 \pm 0.3$ which is in good agreement with values reported in the literature *(31)*. It must be noted that no clogging or fouling of the membrane were observed. PLM is thus not prone to this type of interference which poses problems in other techniques. These results suggest that PLM may be applied for speciation studies in natural waters.

The off-line HFPLM system presented in Figure 2 B was used for preconcentration of trace metals, in particular Pb and Cu, in rivers (Arve and Rhone in Geneva (Switzerland)), lakes (Lake Bret, Lake Luzern (Switzerland)), a polluted stream (Chenevier, Geneva, Switzerland) and seawater (Britanny, France). The samples were preconcentrated for two hours and metal concentrations were determined off-line by ICP-MS. The samples measured on site and in the laboratory were filtered through 0.2 μm. The total metal concentrations were determined by ICP-MS after acidification of the filtered samples (Table I). These results show that in all cases the concentrations of Pb and Cu determined using HFPLM are much lower than the total concentrations. Thus the off-line HFPLM systems can be used to determine the concentrations of the free metal ion, the lipophilic complexes and the neutral metal inorganic complexes which may potentially cross the membrane. On-line determination of Pb and Cu in Arve River and in a polluted water (Chenevier, Geneva, Switzerland) using the coupled HFPLM/MFTC system shows that the results are comparable with off-line analysis within the limits of experimental errors (Table I). Preconcentration times of 20 min and 120 min were respectively used for on-line and off-line determinations. The slightly higher concentrations found in the off-line determinations, if significant, stem from the slowly diffusing neutral Cu or Pb carbonato complex. Future studies should verify this hypothesis.

Table I. Comparison of on-line and off-line determination of Pb and Cu in
natural waters with HFPLM

Sample	Total dissolved (nmol.L^{-1})		Free metal (nmol.L^{-1})			
	Pb	Cu	Pb		Cu	
			Off-line	On-line	Off-line	On-line
Arve	0.92	68.1	0.04 ± 0.005	0.04 ± 0.02	3.2 ± 0.02	2.1 ± 0.05
Chenevier	52.0	234	0.67 ± 0.02	0.58 ± 0.04	1.9 ± 0.01	1.6 ± 0.03

These preliminary results suggest that the application of the on-line detection system will reduce the analysis time and allow real time measurements in real world samples. In that regards, the in situ preconcentration of Pb and Cu in the water column of a lake(Lake Luzern, Switzerland) was also performed with an off-line HFPLM system, and the total and free metal ion concentrations were determined off-line by ICP-MS. The results showed that free concentrations of Pb and Cu were 0.002 nmol. L^{-1} and 1.0 nmol. L^{-1} respectively and the total metal ion concentrations were higher, 18.0 nmol. L^{-1} and 11.4 nmol. L^{-1} respectively, indicating that few picomolar level detection limits can be achieved.

Acknowledgments

We thank S. Rodak for building the potentiostat and amplifier used in this study and for writing the necessary software programs to conduct voltammetric measurements and run the automated PLM-voltammetric systems (HFPLM/MFTC). We thank C. Bernard and F. Bujard for constructing the diffusion cell, flow-through voltammetric cell, buffer vessel, and organic trapping micro-column. The authors also thank Dr. M. Martin for the ICP-MS analyses. We thank AKZO NOBEL for donating hollow fibers and HOECHST for providing flat sheet Celgard® membranes.

We gratefully acknowledge the Swiss National Science Foundation for its financial support.

References

1. Buffle, J. *Complexation Reactions in Aquatic Systems - an analytical approach*, Horwood, Chichester, UK 1988; p 692.

2. Stumm, W. Ed., *Chemical Processes in Lakes*, Wiley, New York 1985; p 435.
3. Ure, A. M.; Davidson, C. M., Eds. *Chemical Speciation in the Environment*, Blakie Academic Press and Professional, Glasgow, UK 1995; p 408.
4. Hill, S. J. *Chem. Soc. Rev.* **1997**, *26*, 291-298.
5. Batley, G. E, Ed. *Trace metal speciation, Analytical methods and problems*, CRC Press 1989; p 360.
6. Tercier, M. L; Buffle, J. *Electroanalysis* **1993**, *5*, 187-200
7. Van den Berg, C. M. G. *Anal. Chim. Acta* **1991**, *250*, 265-276.
8. Tercier-Waeber, M. L.; Belmont., C.; Buffle, J. *Environ. Sci. Technol.* **1998**, *32*, 1515-1521.
9. Danesi, P.R. *Sep. Sci. Technol. 1984/85*, *19*, 857-894.
10. Buffle, J.; Parthasarathy, N.; Djane, N. K.; Mathiasson, L. In *In situ monitoring of aquatic systems: chemical analysis and speciation*; Buffle, J.; Horvai, G., Eds.; Wiley & Sons: New York, NY, 2000, Vol.6, Ch.10.
11. Uto, M.; Yoshida, H.; Suwara, M.; Umezawa, Y. *Anal. Chem.* **1986**, *58*, 1798-1803.
12. Cox, J.A.; Bhatnagar, A. *Talanta* **1990**, *37*, 1037-1041.
13. Papontani, M.; Djane, N. K.; Ndungu, K.; Jonsson J. A.; Mathiason, L. *Analyst* **1995**, *122*, 1073-1477.
14. Parthasarathy, N.; Buffle, J. *Anal. Chim. Acta* **1994**, *284*, 649-659.
15. Bartsch, R. A.; Way J. D. Eds, In *Chemical separations with liquid membranes*; American Chemical Symposium Series, American Chemical Society: Washington, DC, 1996, Vol. 642, p 432.
16. Jonsson, J. A.; Mathiasson, L. *Trends Anal. Chem.* **1999**, *18*, 325-334.
17. Parthasarathy, N.; Buffle, J.; Gasssama, N.; Cuenod, F. *Chem. Analyst* (*Warsaw*) **1999**, *44*, 455-470.
18. Djane, N. K.; Ndung'u, K.; Malcus, F.; Johansson, G.; Mathiasson, L. *Fresenius J. Anal. Chem*, **1997**, *358*, 822-827.
19. Djane, N. K.; Ndung'u, K.; Johansson, G.; Sartz, H.; Tomstrom, T.; Mathiasson, L. *Talanta* **1999**, *48*, 1121-1132.
20. Parthasarathy, N.; Pelletier, M.; Buffle, J. *Anal. Chim. Acta* **1997**, *350*, 183-195.
21. Guyon, F.; Parthasarathy, N.; Buffle, J. *Anal.Chem.* **1999**, *77*, 819-826.
22. Kemperman, A. J. B.; Bargeman, D.; Van Den Boomgaard, T.; Strathmann, H. *Sep. Sci. Technol.* **1996**, *31*, 2733-2762.
23. Malcus, F.; Djane, N. K.; Johansson, G.; Mathiasson, L. *Anal. Chim. Acta.* **1996**, *327*, 295-300.
24. Djane, N. K.; Aramalis, S.; Johansson, G.; Mathiasson, L. *Analyst.* **1996**, *123*, 393-396.

124

25. Keller, O. C.; Buffle, J. *Anal. Chem.* **2000**, *72*, 936-942.
26. Keller, O. C.; Buffle, J. *Anal. Chem.* **2000**, *72*, 943-948.
27. Stulik, K.; Pacakova, V. In *Electroanalytical measurements in flowing liquids*; Ellis Horwood Ltd, John Wiley & Sons: New York, NY, 1987, p 297.
28. Tercier-Waeber, M. L.; Buffle, J. In *In situ monitoring of aquatic systems: chemical analysis and speciation*; Buffle, J.; Horvai, G., Eds.; Wiley & Sons: New York, NY, 2000, Vol.6, Ch. 9.
29. Parthasarathy, N.; Buffle, J. **2001**, *Electroanalysis*, in press.
30. Cox, J.; Twardowsky, Z. *Anal. Chim. Acta*, **1980**, *119*, 39-45.
31. Buffle, J.; Greter, F. L; Heardi, W. *Anal. Chem.* **1977**, *49*, 216-222.

In-Situ Measurements
at the Sediment–Water Interface

Chapter 7

In Situ Measurements in Lake Sediments Using Ion-Selective Electrodes with a Profiling Lander System

Beat Müller, Martin Märki, Christian Dinkel, Ruth Stierli, and Bernhard Wehrli

Swiss Federal Institute of Environmental Science and Technology (EAWAG), Limnological Research Center, CH–6047 Kastanienbaum, Switzerland

This chapter illustrates the capabilities of *in situ* applications of potentiometric chemical sensors at the sediment-water interface of deep lakes. A lander system is described which allows positioning of the sensors with a stepping motor and a video-endoscope. A procedure has been developed to determine the position of the sediment-water interface directly from the standard deviation of acquired data. Preliminary studies illustrate the ability of the device to detect spatial heterogeneity and the seasonality in pore-water profiles. The different ion-selective sensors available for freshwater systems allow us to gain new insights in the nitrogen cycle and calcite dissolution.

Introduction

Porewater profiles in lake sediments reveal important information on the rates and the spatial and seasonal distribution of biogeochemical processes. It has long been recognized that the coarse resolution of conventional dialysis samplers cannot resolve some of the steep gradients occurring at the sediment-water interface in limnic systems (*1, 2*). The low salinity of most continental waters offers the opportunity to study early diagenesis *in situ* with ion-selective electrodes (ISEs) (*3-7*). The first submersible lander system to apply ion-selective electrodes at the sediment-water interface was introduced by Clare Reimers (*8*). Since that time several elaborate profiling landers have been constructed (*9-14*). Several microelectrode profilers for O_2 measurements are commercially available. These systems were developed recognizing the necessity to measure the uptake of O_2 by the sediment to gain information on the mineralization of organic carbon in the world oceans and thus, on the global carbon cycle. The first microsensors applied were amperometric O_2 electrodes. Thereafter, potentiometric microsensors for pH with a glass sensing tip and Severinghouse-type electrodes for CO_2 were developed and applied (*15-17*). Ag/AgS-electrodes sensitive to S^{2-} followed (*18- 20*).

In this chapter we introduce our profiling system, which has been designed for porewater studies in freshwater lakes up to 400 m deep. We report results from several trial studies with the new device. The case studies were chosen to illustrate the range of relevant research questions, which can be addressed with the profiling system. A method to detect the exact position of the sediment-water interface when working with multiple sensors is discussed. Using an array of 12 pH electrodes simultaneously in Lake Zug we then illustrate the reproducibility and the range of spatial heterogeneity encountered in field applications. The consistency of high-resolution porewater profiles is demonstrated with repeated measurements at the deepest site in Lake Alpnach during a full annual cycle. Finally, we outline the potential to analyze processes like denitrification, nitrification and calcite dissolution by combining ISEs for NO_3^-, NH_4^+, H^+ and CO_3^{2-} in a field test in Lake Baldegg. As some of this work is still going on, we have only sketched the progress in the analysis of biogeochemical processes, which is now possible with *in situ* profiling technology.

Lander for Ion-Selective Analysis (LISA)

The submersible lander for *in situ* analyses with chemical sensors at the sediment-water interface of freshwater lakes is shown in Figure 1. Its profiler is centered within an aluminum frame with three legs and large load-bearing surfaces to prevent it from sinking into soft sediments. The lander is 1 m tall and

Figure 1. The Lander for Ion-Selective Analysis (LISA) on Lake Alpnach, Switzerland.

has an overall width of 2.5 m. Its weight is 203 kg on land and 74 kg in water. The feet can be removed for transport. A stepping motor (escap PH632) is mounted vertically in the center of the tripod that controls the position of the sensors. A video endoscope allows one to observe the sediment surface and to a certain extent the position of the sensors. The submersible endoscope optics is a custom development by Treier Endoscopy, Beromünster, Switzerland.

Three pressure cases for water depths up to 400 m are mounted on the tripod. One pressure case contains the power supply and the control unit of the stepping motor. The power supply with 48 Ah lead-acid batteries is designed to allow at least 12 hours of operation. Another case contains the electronic unit for signal conditioning and the data logger (CR10X, Campbell Scientific Ltd.). RS232 connection and an asynchronous short-range modem transfer data between the data logger and a computer on board. The third casing houses a fiber optic waveguide that provides a focused light beam. A Sony CCD camera (SSC-M370) with pressure-safe endoscope optics yields black and white video with a resolution of 570 lines. The on-line video image allows the approximate positioning of the sensors at the sediment water interface. It yields qualitative information on bottom currents, benthic fauna and sediment heterogeneity.

Production of a bow-wave is prevented due to the handling of the lander with a tether and a winch and the geometry of the wide-spread feet. Its position above the sediment was observed simultaneously by a sonic depth finder. Landing speed was less than 10 cm s^{-1}. The profiling commenced after approximately 1 h when all electrodes acquired stable values in the bottom water.

All functions of the stepping motor and data acquisition are controlled on-line from the ship with a notebook via the data logger. Transmission of data in real time allows adjustment of step size, early detection of measurement errors and greatly improves the reliability of the instrument. The cable consists of a 4-wire line cable for data connection and a coaxial cable (RG6 A/W, 75Ω) for the video signal.

The motor moves a sensor rack with a total of 18 electrodes at a step resolution of <10 μm. Electrodes are mounted within an area of 20 x 20 mm (see Figure 3b). The rack is equipped with a pressure compensating system to avoid damages of the probes. The effect of water pressure in the membrane potential was found to be negligible (21). In addition to two O_2 microelectrodes the rack can be equipped with miniaturized solid-state Ag/AgS-electrodes and/or liquid membrane electrodes for pH, NH_4^+, NO_3^-, Ca^{2+}, and CO_3^{2-}. Preparation and performance of the potentiometric sensors with a tip diameter of 600 μm is described in (7). A Ag/AgCl electrode (Metrohm) is used as reference. To protect the data acquisition system from electromagnetic interference all control signals for the motor are separated from the electronic ground by an optocoupler. To prevent electronic noise on the high impedance signal of the ISEs the distance between the electrode and the impedance converter is kept as short as possible. The converter is positioned just above the electrodes within the holder rack. For each channel an instrumentation amplifier (Burr Brown, INA116) subtracts the electrode signal from the Ag/AgCl reference electrode by using a Platinum wire as common ground (see INA 116 datasheet, Burr Brown). The low impedance output voltage is amplified by an adjustable gain. The electronic part for the signal conditioning of O_2 microelectrodes is accommodated in the same pressure case as the data logger. The amperometric signal of the O_2 electrodes is converted into a voltage signal and amplified with a gain.

Following each deployment of the LISA a sediment core was sampled with a gravity corer. The pH was measured in the overlying water with a field pH meter with temperature compensation. Water samples from the core were taken for Winkler titration, and alkalinity. Ca^{2+}, Mg^{2+}, K^+, Na^+, NH_4^+, NO_3^-, Cl^-, SO_4^{2-}, and phosphate were analyzed with standard methods. These values were used to validate the *in situ* measurements with ISEs in the water above the sediment. Precision of the electrostatic measurements is generally much better than their accuracy, i.e. it is reliable to measure concentration changes accurately but quite difficult to determine the absolute concentration of the ions. In other words: the slopes of the calibration curves are much better reproducible than their intercept which is affected easily by all kinds of stress exerted on the membrane. Thus, following the procedures of oceanographers (*17, 22, 23*) bottom water measurements determined with standard methods were used as reference points to adjust electrode profiles to absolute concentrations.

Determining the diffusive boundary layer thickness

To date there has been uncertainty in the determination of the magnitude of the diffusive boundary layer (DBL). The upper boundary in profiles of parameters such as O_2 is readily apparent from the sharp bend in the vertical concentration profile. Transition from the aqueous zone of mere molecular diffusion to the zone of consumption or production within the sediment can only be detected by a small change of the concentration gradient.

So far, scientists have tried to determine the contact of their sensors with the sediment surface visually and thus many electrode landers - like our own - are equipped with video or endoscope cameras. Due to the microscopic size of the sensor tips and the heterogeneous pattern of the grains or flakes at the sediment surface such remote optical observations are essential for a coarse positioning of the sensors. In addition, the topographic patterns and the heterogeneity of the sediment surface can be documented. However, the endoscope did not stand the test of a precise determination of the sediment-water interface.

The information about the sediment surface is contained in the acquired data. Gundersen and Jørgensen (*24*) observed a systematic decrease of fluctuations in their O_2 measurements with decreasing distance from the surface in flow chamber experiments. Hence, we analyzed the standard deviation of multiple measurements and confirmed a systematic decrease in the fluctuations towards the sediment surface. This can be rationalized with the present ideas of the structure of the water overlying the sediment.

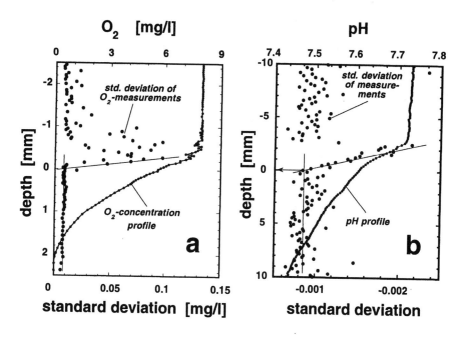

Figure 2. Determination of the sediment surface using standard deviations of measurements. The continuous lines depict the concentration profiles calculated from the averages of all measurements at one depth step. Dots represent the corresponding standard deviations. **a.** O_2 measurements from Lake Alpnach (May 16, 2000), 35 m depth. Standard deviations from 200 measurements at each depth. **b.** pH measurements from Lake Zug (Sept 5, 2000), 40 m depth. Standard deviations from 30 measurements at each depth.

The bottom water is mixed by turbulent eddies. The concentration gradient within the DBL is linear since no production or consumption occurs. Within the DBL the vertical mixing of water by eddy diffusion results in local concentration changes. Thus, the varying local concentration in the water passing at the sensor tip causes a decrease in reproducibility and thus increased standard deviation of the measurement. Standard deviation reaches a minimum at the sediment surface where eddy diffusion disappears. Gundersen and Jørgensen (24) observed eddy induced fluctuations even 50 µm below the sediment surface at high flow rates. The depth where fluctuations disappeared depended on flow velocity. The analysis of standard deviations may therefore slightly underestimate the exact position of the sediment in fast flowing environments. Current velocities at the sediment surface of deep lakes are quite low (in the order of 1 cm s^{-1}). Therefore we conclude that this method is probably the most reliable for detecting the sediment position and the thickness of the DBL. Furthermore, it is perfectly suited to relate concentration profiles measured in parallel simultaneously and thus at identical bottom currents. Figure 2a presents one measurement out of a time series of an O_2 profile at 35 m depth in Lake Alpnach, a mesotrophic perialpine lake in Switzerland. The curve depicts the familiar course of O_2 from the mixed zone of bottom water across a zone of linear decrease to a zone where the concentration decreases exponentially to zero. Each dot on the line represents the mean of 200 measurements that were recorded during 20 seconds with a frequency of 10 Hz at each vertical step of 30 µm. Standard deviations for each step (n=200) are given with single dots. Values varied little in the mixed zone of the bottom water. However, standard deviations increased sharply in the proximity of the DBL and decreased again at the sediment surface. The upper boundary of the DBL is determined at the crossing point of the linear extension of the zone of constant concentrations and the zone of linear decrease. The lower end of the DBL is defined at the position where the standard deviations of the measurements reach constant values. The DBL thickness here was determined as 360 µm.

The same method can also be applied with liquid membrane electrodes, though less pronounced differences in the standard deviations were observed. Figure 2b shows an example of a pH profile (Lake Zug, 40 m depth) and the corresponding standard deviations. Measurement frequency was 1 Hz during 60 s. Measurements of the first 30 s after the electrode moved vertically were omitted in order to avoid the effect of the longer electrode response times of the liquid membrane sensor compared to the O_2 microelectrodes.

Bottom water currents of eutrophic Lake Zug were found to be very small (<1 cm s^{-1} 40 cm above ground) and thus turbulence may disappear close to the sediment surface. The unexpectedly large zone of eddy induced fluctuations of 2.5 mm determined with the pH sensors in the sediments of Lake Zug may be realistic in this case. Comparative measurements with O_2 microelectrodes have to be performed, however, to determine the spatial resolution available from the miniaturized liquid membrane electrodes. In any case the method of decreasing standard deviations may be used to functionally determine the sediment surface.

Sensor Reproducibility and Sediment Heterogeneity

The reproducibility of concentration profiles measured with the LISA was tested with an experiment where all slots were equipped with pH electrodes. Out of 14 electrodes that were applied 12 kept functioning until the end of the deployment and the consecutive second calibration. The experiment was performed in Lake Zug at 40 m depth. Figure 3 shows the 12 pH profiles, slightly displaced vertically for better visibility. At the first glance the profiles in Figure 3a, however, look very similar. From pH 7.72 in the water over the sediment and a steep decrease within 2-4 mm the values approach constancy in about 40 mm depth. A closer look at the initial decrease in the top 2-4 mm in the inset of Figure 3a reveals significant differences in the slopes. Electrodes 1, 2, 4, 6, 7 show slopes of 0.045 ΔpH mm^{-1} while electrodes 5, 11, 12, 13, 14 show slopes of 0.175 ΔpH mm^{-1}. Electrodes 8 and 10 change slopes between the two extremes observed while electrodes 3 and 9 did not function. A comparison with the electrode rack in Figure 3b reveals that all profiles with the lower initial slope were situated in the hatched area. The other profiles with the steeper slopes are from the area marked gray. A difference of a factor of 4.5 in the concentration gradient of H^{+} was observed within an area of only 2 cm^{2}.

A combination of reasons could cause this pattern. As Glud et al. (25) have shown with O$_2$ microelectrodes the DBL is disturbed at a distance of up to 1 mm due to perturbations of the flow velocity field by the electrode tip. Hence, due to our compact set of sensor with diameters of 600 µm effects on the DBL may be depending on bottom water flow. The propagation of the effect from top left to bottom right (Figure 3b) also supports the idea of local heterogeneity either chemically (due to different quality of mineralizing organic material) and/or topographically (due to varying angles of approach of the electrodes over an irregular interface). The existence of local heterogeneities was shown recently by Fenchel and Glud (26) who found variations of a factor of 3 and 10 for O$_2$ uptake and release, respectively, in photosynthetic sediment over a horizontal distance of 2 mm.

While perturbations of the magnitude of the DBL are easily induced, differences in the profiles below the DBL reveal local heterogeneities in the structure of the sediment attributable to the variety of sedimenting material (leaves, needles, flakes etc.), or to effects of sediment-burrowing animals.

Seasonal patterns on the microscale

Modeling tools for the prediction of the timing and extent of hypolimnetic anoxia in lakes have been a matter of recent debate in the literature (27). A more deductive approach to model hypolimnetic oxygen profiles could benefit from

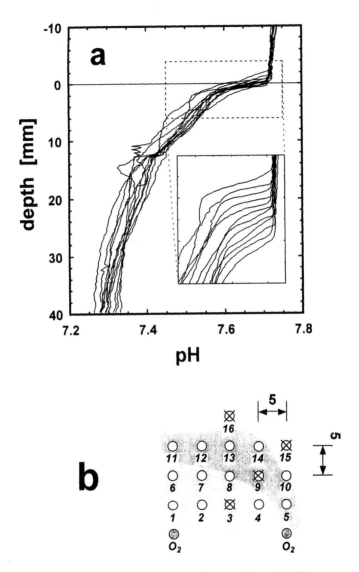

*Figure 3. **a**. Simultaneous measurement of pH profiles with 12 sensors in Lake Zug at 40 m depth (Sept 5, 2000). The inset shows the section of initial slopes spatially separated. **b**. LISA-electrode rack with the sensor slots. Sensors are 5 mm apart. The hatched area depicts higher, the shaded area lower fluxes.*

more detailed observations of respiration processes at the sediment-water interface. Seasonal data of respiration rates within the sediment could then be used together with hypolimnetic mass balances to separate oxygen consumption in the water column and in the sediment. Here we report a preliminary study of concentration profiles at the deepest site of Lake Alpnach recorded over a full annual cycle (Figure 4). This revealed the temporal changes of the biogeochemical conditions in the top sediment layer. The high resolution LISA profiles were analyzed with a simple diagenetic model developed by Epping and Helder (28). Model results of fluxes, reaction rates and penetration depths are given in Table I.

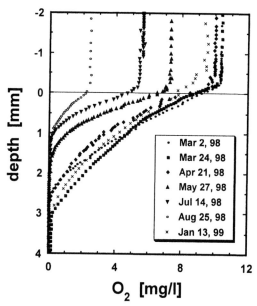

Figure 4. Seasonal profiles of O_2 in the sediment of Lake Alpnach recorded with the LISA in 35 m depth.

The annual cycle starts with bottom water close to saturation (85 %) with O_2 (February and March). During the winter season, productivity and thus sedimentation of organic carbon is low. Accordingly, O_2 penetration depths are up to 3.7 mm and correspond to O_2 fluxes of 12 to 15 mmol m^{-2} d^{-1} and low consumption rates of 3.1 to 4.6 mol m^{-3} d^{-1}. Stratification and the effect of spring algae blooms were observed at the end of May. O_2 concentration of the bottom water decreased to 7.4 mg/l. The flux had almost doubled and the consumption rate almost tripled with respect to the three preceding months. Hence, penetration depth decreased to 1.7 mm. The trend was seen to continue two months later (14.7.98) with O_2 flux and reaction rate similar to the preceding measurement.

Bottom water content and penetration depth into the sediment decreased further. At the end of August O_2 concentration in the bottom water was reduced to 2.4 mg/l. As a consequence, flux and rate were reduced to about 50 % of the July values.

Table I. Seasonal fluxes, reaction rates and penetration depths of O_2 in the sediments of Lake Alpnach

Date	Flux $mmol\ m^{-2}\ d^{1}$	Rate $mol\ m^{-3}\ d^{1}$	Penetr. depth mm
03. Feb.1998	14.7	4.6	3.2
24. Mar.1998	11.7	3.1	3.7
21. Apr.1998	13.2	4.6	2.9
27. May 1998	20.0	12.0	1.7
14. July 1998	18.6	14.3	1.3
25. Aug.1998	8.6	6.4	1.3
13. Jan.1999	11.2	3.1	3.6

In the following series of measurements (Oct. 14 and Nov. 25) O_2 concentrations above the sediment were below detection limit. However, mid-January the O_2 profiles in the sediment porewater had recovered to similar values of flux, rate and penetration depth as the year before.

Figures 5a and 5b present pH profiles of the same locations as the O_2 profiles above from selected dates. As expected, the pH of the water above the sediment decreased in the course of the year. During winter when the water was mixed to the bottom, the pH below the sediment-water interface decreased smoothly from 8.1 to 7.1 within 3-4 cm due to CO_2 released by oxic respiration. In April we observed a steep decrease of pH within the top 2 mm of the sediment. This may be an effect of a spring bloom of algae decomposing at the sediment surface at a high rate. Profiles from May and July revealed similar shaped curves, however, due to the lower pH of the bottom water the concentration gradient was smaller. In good agreement with the O_2 data these profiles showed a zone of intense oxic mineralization with a pH drop in the top mm followed by the zone of anoxic mineralization of 4 mm with constant pH, and then a smooth decrease in the zone of methanogenesis. In autumn the bottom waters of Lake Alpnach turned anoxic. As a consequence of denitrification (29) and other anaerobic processes alkalinity production in the sediment was intensified which led to a pH increase below the sediment surface. This phenomenon was observed in two consecutive years (Figure 5b). These peak profiles prevailed only for some weeks at the end of the stagnation period and quickly disappeared when the water was mixed. The continuation of the processes of decreasing pH of the bottom water increase of pH in the anaerobic part below the interface led to 'inversion' of the profiles later in the year which is represented in Figure 5b (October and November).

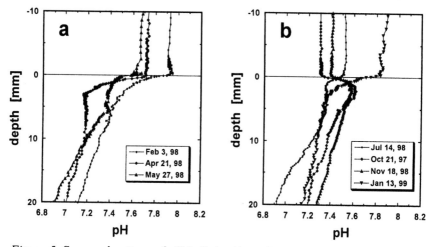

*Figure 5. Seasonal pattern of pH in Lake Alpnach at 35 m depth. **a.** pH profiles recorded from February to March. **b.** pH profiles recorded from July to January.*

In general, the lower parts of the pH profiles were shaped by the mineralization of the more refractory organic material below 4-5 mm, which led to similar pH profiles at larger depths. With a more detailed analysis of such sensor profiles, the dynamics of carbon mineralization at the bottom of lakes can now be fully resolved in space and time.

A seasonal pattern of the pH gradient occurred only in the top 6 mm. This pattern could be observed because of the spatial and temporal resolutions achieved by ion-selective electrodes.

Nitrogen turnover and carbon flux

In addition to more classical studies with O_2 and pH sensors, other available ion-selective electrodes allow analyses of specific biogeochemical processes such as denitrification (*30*) and calcite dissolution (*31*). A set of NO_3^- gradients measured in sediments from different depths from very shallow to the deepest location (64 m deep) in Lake Baldegg is given in Figure 6a. Concentrations decreased immediately below the sediment-water interface at every site. The gradients were very steep, however less than those for O_2. The course of the profiles from shallow depths indicated vertical heterogeneity of the sediment whereas in cores from lower depths the profiles run smoother. Fluxes increased with increasing depth of the lake, and penetration depths decreased accordingly (Table II). Values obtained with the Epping model (28) ranged from 0.8 to 5.5 mmol m^{-2} d^{-1}, and consumption rates from 0.1 to 4.8 mol m^{-3} d^{-1}.

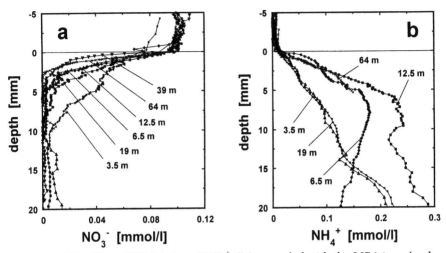

Figure 6. Profiles of NO$_3^-$ (a.) and NH$_4^+$ (b.) recorded with the LISA in a depth gradient in the sediments of Lake Baldegg (18-27 May 99).

Table II. Seasonal fluxes, reaction rates and penetration depths of NO$_3^-$ and NH$_4^+$ along a depth gradient in the sediments of Lake Baldegg

	NO$_3^-$			NH$_4^+$	
Depth *m*	*Flux* *mmol m^{-2} d^{-1}*	*Rate* *mol m^{-3} d^{-1}*	*Pen. Depth* *mm*	*Flux* *mmol m^{-2} d^{-1}*	*Rate* *mol m^{-3} d^{-1}*
3.5	0.8	0.1	14.4	-1.0	-0.1
6.5	3.1	0.6	4.8	-5.8	-2.7
12.5	3.3	0.7	4.5	-12.9	-1.3
19	2.6	0.5	5.7	-1.7	-0.1
39	5.5	4.8	5.0	-	-
64	5.0	1.7	2.9	-3.2	-0.4

The concentrations of NO$_3^-$ in oligotrophic lakes are in the lower micromolar range and thus difficult to determine accurately with ISEs. Although the detection limit of the NO$_3^-$ liquid membrane electrode is approximately 1 µM in pure solution the presence of SO$_4^{2-}$, Cl$^-$ and HCO$_3^-$ increases the detection limit to 20-40 µM. Their interference can be compensated mathematically with the Nicholsky-Eisenmann equation (extended and put on a theoretical basis by

Bakker et al. (*32*)). However, precision of the NO_3^- concentration values is hampered when the more interfering ions dominate over NO_3^-. The decreasing precision of the measurement can be seen from the increasing scatter of the NO_3^- concentrations below ~20 µM in Figure 6a. However, profiles at their upper end remain unaffected and still allow estimating NO_3^- fluxes from initial concentration gradients.

NH$_4^+$ profiles in the sediments of different depths of Lake Baldegg are shown in Figure 6b. Most profiles are parabolic in shape indicating the high mineralization rate of the new organic material within layers and the decreasing liberation of reduced N in older layers. The increase in porewater concentration can be followed over several mm and continues beyond the reach of the vertical range of our sensors (ca. 40 mm). The sink for NH_4^+ is at the sediment-water interface where nitrification with O_2 may occur. For intense nitrification within the sediment an increase of NO_3^- is expected but could not be detected in our NO_3^- profiles (Figure 6a). NH_4^+ profiles were not correlated with water depth. Concentration gradients of profiles from 3.5 m and 19 m depth were significantly smaller than from the other depths. Fluxes vary between 1-5 mmol m^{-2} d^{-1}. NH_4^+ concentrations in the deeper sediments can reach several millimoles and hence may be measured reliably. In the low NH_4^+ concentration range at the sediment-water interface and in the bottom water (<10 µM) K^+ present in concentrations of 50 - 60 µM interferes. Due to its constant vertical concentration it can be easily corrected for. In the future the combination of such ISE measurements with ^{15}N incubations and the isotope pairing technique (*30, 33*) with benthic chambers will allow us to gain detailed insight into the pathways of benthic nitrogen turnover.

The flux of inorganic carbon across the sediment-water interface is dominated by the HCO_3^- gradient. Changes of HCO_3^- concentration in the sediment occur in zones of microbial activity due to production or consumption of alkalinity. Oxidation of biomass produces mainly carbonic acid while anaerobic respiration by NO_3^-, MnO_2, $FeOOH$, and SO_4^{2-} produces alkalinity. Bicarbonate is in dynamic equilibrium with these processes. Furthermore, in hardwater lakes the dissolution and precipitation of calcite acts as an important buffer mechanism. The profile of HCO_3^- is thus an important parameter to estimate carbon flux and there are no ion-selective sensors in existence to detect the HCO_3^- ion with sufficient selectivity. Instead, we evaluated a miniaturized CO_3^{2-}-sensitive electrode with a detection limit of 1 µM and good selectivity (*7*). Combination with pH provides enough information to quantify the carbonate system.

Figure 7a depicts profiles for CO_3^{2-} and pH from Lake Alpnach. Average concentrations and standard deviations were calculated from simultaneous measurements with two sensors for each parameter. Both profiles were highly reproducible apart from some deviation of one CO_3^{2-} sensor just above the sediment-water interface that may be due to a local heterogeneity of the sediment.

HCO_3^- concentrations were calculated from adjusted profiles with a pK_2 for bicarbonate of 10.56 at 5 °C and zero ionic strength (*34*).

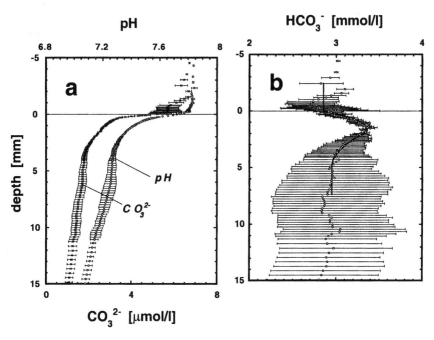

Figure 7. **a.** *Measurements of pH and CO_3^{2-} in Lake Alpnach in 35 m depth (March 3, 99). Averages and standard deviations of data from two sensors for each parameter were calculated.* **b.** *Profile for HCO_3^- as calculated from pH and CO_3^{2-}. Standard deviations were calculated combining minimum and maximum values of pH and CO_3^{2-} for each depth.*

The result is given in Figure 7b. HCO_3^- increases in the mineralizing biofilm on the top of the sediment and decreases again after 2 mm to reach approximately constant values below 4 mm depth. It is apparent from the large standard deviation of the calculated HCO_3^- profile that accuracy and precision of the measurements of pH and CO_3^{2-} must be very good and, since separate electrodes and more than one each were used for all parameters, sediment heterogeneity must be negligible. For a standard error of e.g. 10 % in the HCO_3^- concentration the sum of the errors (in %) of H^+ and CO_3^{2-} determinations must not exceed 10%. A similar conclusion was drawn by Cai and Reimers (*15*) who calculated total dissolved inorganic carbon from measurements of pH and pCO_2 and found the accuracy of the sensors limited by sensor stability, response, and calibration. De Beer et al. (*16*) in their experiments with pH and CO_2 electrodes confirmed the large uncertainties due to sensor accuracy. They showed that small uncertainties with the overlay of the two profiles to calculate HCO_3^- or TIC easily produces errors that exceeded the differences in the concentration gradients that one wants to measure. In addition, the presently available thermodynamic data used to calculate the species of the carbonate system deviate substantially and the probable errors introduced during calculations are estimated to be up to ten times larger than the errors of the selected measurements (*35*). Hence, in agreement with other researchers the measurements presented in Figure 7 suggest that also for the combination of pH and a CO_3^{2-} sensor uncertainties do not allow estimation of fluxes and production rates with confidence.

Conclusions

Liquid membrane electrodes for pH, NO_3^-, NH_4^+ and CO_3^{2-} as well as O_2 microelectrodes were applied with a submersible profiling lander (LISA) to investigate patterns of the sediment-water interface and to study seasonal variations in sediments of freshwater lakes. The versatile data collection mode of the LISA allowed the use of standard deviations of multiple measurements to determine the location of the sediment surface. Simultaneous measurements with an array of 12 pH sensors give an impression about the heterogeneity in lake sediments. The use of NO_3^- and NH_4^+ ISEs is limited to freshwater due to interferences of other ions. Combination of these sensors provides a detailed view of the benthic cycle in inland lakes and reservoirs. The combination of profiles for pH and CO_3^{2-} to calculate HCO_3^- produces large uncertainties in spite of the good reproducibility of pH and CO_3^{2-} measurements.

References

1. Urban, N.; Dinkel, C.; Wehrli, B. *Aquatic Sci.* **1997**, *59*, 1-25.
2. Sweerts, J.-P. R. A.; Bär-Gilissen, M.-J.; Cornelese, A. A.; Cappenberg, T. E. *Limnol. Oceanogr.* **1991**, *36*, 1124-1133.
3. Sweerts, J.-P. R. A.; De Beer, D. *Appl. Environ. Microbiol.* **1989**, *55*, 754-757.
4. Sweerts, J.-P. R. A.; De Beer, D.; Nielsen, L. P.; Verdouw, H.; Van den Heuvel, J. C.; Cohen, Y.; Cappenberg, T. E. *Nature* **1990**, *344*, 762-763.
5. Jensen, K.; Sloth, N. P.; Rysgaard-Petersen, N.; Rysgaard, S.; Revsbech, N. P. *Appl. Environm. Mirobiol.* **1994**, *60*, 2064-2100.
6. Lorenzen, J.; Larsen, L. H.; Kjaer, T.; Revsbech, N. P. *Appl. Environm. Microbiol.* **1998**, *64*, 3264-3269.

7. Müller, B.; Buis, K.; Stierli, R.; Wehrli, B. *Limnol. Oceanogr.* **1998**, *43*, 1728-1733.

8. Reimers, C. E.; Kalhorn, S.; Emerson, S. R.; Nealson, K. H. *Geochim. Cosmochim. Acta* **1984**, *48*, 903-910.

9. Reimers, C. E. *Deep Sea Res.* **1987**, *34*, 2019-2035.

10. Archer, D.; Emerson, S.; Smith, C. R. *Nature* **1989**, *340*, 623-626.

11. Reimers, C. E.; Fischer, K. M.; Merewether, R.; Smith, K. L. j.; Jahnke, R. A. *Nature* **1986**, *320*, 741-744.

12. Gundersen, J. K.; Jørgensen, B. B. *Kieler Meeresforsch.* **1991**, *8*, 376-380.

13. Reimers, C. E.; Glud, R. N. In *Chemical sensors in oceanography*; Varney, M., Ed.; Gordon & Breach:, 2000; Vol. 1.

14. Tengberg, A.; De Bovee, F.; Hall, P.; Berelson, W.; Chadwick, D.; Ciceri, G.; Crassous, P.; Devol, A.; Emerson, S.; Gage, J.; Glud, R.; Graziottin, F.; Gundersen, J.; Hammond, D.; Helder, W.; Hinga, K.; Holby, O.; Jahnke, R.; Khripounoff, A.; Lieberman, S.; Nuppenau, V.; Pfannkuche, O.; Reimers, C.; Rowe, G.; Sahami, A.; Sayles, F.; Schurter, M.; Smallman, D.; Wehrli, B.; De Wilde, P. *Prog. Oceanog.* **1995**, *35*, 253-294.

15. Cai, W.; Reimers, C. E. *Limnol. Oceanogr.* **1993**, *38*, 1762-1773.

16. De Beer, D.; Glud, A.; Epping, E.; Kühl, M. *Limnol. Oceanogr.* **1997**, *42*, 1590-1600.

17. Hales, B.; Emerson, S. *Geochim. Cosmochim. Acta* **1997**, *61*, 501-514.

18. Gundersen, J. K.; Jørgensen, B. B.; Larsen, E.; Jannasch, H. W. *Nature* **1992**, *360*, 454-455.

19. Kühl, M.; Steuchart, C.; Eickert, G.; Jeroschewski, P. *Aquat. Microb. Ecol.* **1998**, *15*, 201-209.

20. Müller, B.; Stierli, R. *Anal. Chim. Acta* **1999**, *401*, 257-264.

21. Müller, B.; Hauser, P. C. *Anal. Chim. Acta* **1996**, *320*, 69-75.

22. Reimers, C. E.; Ruttenberg, K. C.; Canfield, D. E.; Christiansen, M. B.; Martin, J. B. *Geochim. Cosmochim. Acta* **1996**, *60*, 4037-4057.

23. Komada, T.; Reimers, C. E.; Boehme, S. E. *Limnol. Oceanogr.* **1998**, *43*, 769-781.

24. Gundersen, J. K.; Jørgensen, B. B. *Nature* **1990**, *345*, 604-607.

25. Glud, R. N.; Gundersen, J. K.; Revsbech, N. P.; Jørgensen, B. B. *Limnol. Oceanogr.* **1994**, *39*, 462-467.

26. Fenchel, T.; Glud, R. N. *Ophelia* **2000**, *53*, 159-171.

27. Livingstone, D. M.; Imboden, D. M. *Can. J. Fish. Aquat. Sci.* **1996**, *53*, 924-932.

28. Epping, E. H. G.; Helder, W. *Continental Shelf Res.* **1997**, *17*, 1737-1764.

29. Mengis, M.; Gächter, R.; Wehrli, B. *Biogeochem.* **1997**, *38*, 281-301.

30. Mengis, M.; Gächter, R.; Wehrli, B.; Bernasconi, S. *Limnol. Oceanogr.* **1997**, *42*, 1530-1543.

31. Ramisch, F.; Dittrich, M.; Mattenberger, C.; Wehrli, B.; Wüest, A. *Geochim. Cosmochim. Acta* **1999**, *63*, 3349-3356.
32. Bakker, E.; Meruva, R. K.; Pretsch, E.; Meyerhoff, M. E. *Anal. Chem.* **1994**, *66*, 3021-3030.
33. Nielsen, L. P. *FEMS Microbiol. Ecol.* **1992**, *86*, 357-362.
34. Stumm, W.; Morgan, J. J. *Aquatic Chemistry*; 3rd ed.; Wiley Interscience, 1996.
35. Millero, F. J. *Geochim. Cosmochim. Acta* **1995**, *59*, 661-677.

Chapter 8

Exchange and Microdistribution of Solutes at the Benthic Interface: An In Situ Study in Aarhus Bight, Denmark

Ronnie N. Glud[1] and Jens K. Gundersen[2]

[1]Marine Biological Laboratory, University of Copenhagen, Strandpromenaden 5,
3000 Elsinore, Denmark
[2]Unisense A/S, Science Park Aarhus, Gustav Wieds Vej 10, 8000 Aarhus C, Denmark

In situ benthic exchange rates of O_2, DIC, NH_4^+ and NO_3^- were measured along with microprofiles of O_2, S^{2-} and pH at three coastal sites. The diffusive O_2 uptake (DOU) increased while the O_2 penetration depth decreased with increasing organic carbon content of the surficial sediment. Total O_2 uptake (TOU) showed the same pattern, but values were significantly higher than the DOU at all stations. The difference was partly due to advective porewater transport and respiration of fauna and partly due to the microtophography all leading to underestimation of the actual oxic sediment area when using a simple one-dimensional diffusive flux calculation. Sulphide could not be detected in the upper 30mm of the two oxidized sediments, while H_2S and O_2 coexisted in a narrow zone of intense H_2S oxidation just below the sediment surface at the reduced site. Nitrogen was mainly exported from the sediment as NH_4^+ and urea. The importance of urea for the total N release rate increased with the surficial organic carbon content and reached a maximum of 66% at the reduced site.

In coastal environments a significant fraction (25-50%) of the primary production reaches the seafloor (*1, 2*). Within the sediment the major fraction of the organic material is being oxidized through a complex web of degradation processes, while a minor refractory part is being permanently buried in the sediment record (*3*). The oxidation of the organic material consumes the available electron acceptors in a sequential order eliminating the energetically most favorable first (i.e. O_2, NO_3^-, MnO_x, FeO_x, SO_4^{2-}) (*4*). This leads to a laminated sequence of reduction processes within the sediment in which oxic respiration is followed by denitrification, metal reduction and sulfate reduction (*5*). Irrigating fauna may disturb this general pattern and create a more mosaic-like distribution of the various processes. The anaerobic degradation leads to the production of soluble reduced metal ions and H_2S that may interact with each other or with the various metal oxides and form reduced precipitates (e.g. FeS, FeS_2) (*6*). However, ultimately the major fraction of the reduced substances is reoxidized by an equivalent amount of O_2 (*5*). Therefore, the O_2 consumption rate of sediments has often been used as a measure for the total benthic degradation rate of organic material. During oxidation of the organic material the incorporated nitrogen is released as NH_4^+, some of this may be oxidized into NO_3^- in the oxic zone, and potentially further denitrified.

Benthic solute exchange is driven either by diffusion or advection. The diffusive exchange takes place along a linear concentration gradient within the 0.2-1.2 mm thick diffusive boundary layer (DBL) immediately above the sediment (*7, 8*). For impermeable sediments, bottom dwelling irrigating fauna mainly drives the advective component.

Most measurements of benthic mineralization and solute distribution have been performed in the laboratory on recovered sediment cores. Due to changed environment, mechanical disturbances and under representation of macrofauna such measurements may not reflect the *in situ* rates. This has lead to an ongoing development adapting submersible platforms for "state of the art" microsensors. Below we present *in situ* data on the benthic exchange of solutes most relevant for the mineralization (O_2, DIC, NH_4^+, NO_3^-, urea and H_2S) and benthic microprofiles of O_2, pH and H_2S. Data are discussed in relation to organic carbon loading and the presence of macrofauna.

Materials and Methods

Study site

Three stations located in Aarhus Bight, Denmark (*2*) were visited in early autumn (9-15 of October), station characteristics are given in (Table, Figure 1).

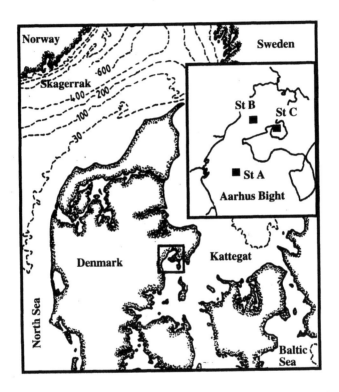

Figure 1. Study site, the enclosed panel depict an enlargement of Aarhus Bight with positions of the investigated sites. The geographic positions were A: 56.01.10N; 10.19.20E, B: 56.14.61N; 10.22.55E, C: 56.13.00N; 10.27.12E

Due to the location on the Baltic Sea-North Sea transition and calm weather, the water column was stratified in this period. Consequently, the bottom water at the deeper central sites was partly depleted in O_2 (9). Stations A and B had distinct oxidized surface layers with a thickness of 6 and 4mm, respectively. Station C was reduced to the surface and was covered by mats of the filamentous colorless sulfur bacteria *Beggiatoa* and *Thiovulum* (10,11)

The porosity and the organic carbon content of the upper 5mm increased from the oxidized to the reduced station (Table 1). Porosity was determined from the density and from the weight loss measured after 24h at 70°C, while the organic carbon content was determined as loss on ignition after 24 h at 550 °C.

Porewater profiles

Porewater microprofiles of O_2, S^{2-} and pH were measured using a benthic lander, "Profilur" (8). The central part of the lander consists of a movable cylinder equipped with six O_2, two S^{2-} and two pH microelectrodes. The positions of the sensor tips relative to each other were determined at an accuracy of 200 μm prior to deployment. This was done by lowering the sensor array in small increments towards a beaker glass with water and visual determination of when the sensor tips touched the water. The lander was lowered by wire from the ship and after it had stabilized on the seafloor (1h), the central cylinder was moved downwards in increments of 50-200 μm. Data were recorded from all sensors at each depth and were transferred via wire cable to a PC onboard the ship. A video camera mounted on the lander allowed visual inspection of the sediment surface during measurements.

The O_2 microelectrodes were of the Clark type with internal reference and a guard cathode (12). The sensing tips were 5-10 μm, stirring sensitivities <2% and the 90% response time <2s (13). The sensors had a linear response to O_2 concentrations and the obtained profiles contained two inherent calibration points: the readings in the bottom water, with a known O_2 concentration (determined on recovered water samples by Winkler titration (14)); and the constant low readings in the anoxic sediment strata.

Table I. Station characteristics	Stations		
	A	B	C
Water depth	16m	10m	11m
Temperature	11.5 °C	13.6 °C	13.6°C
Porosity (vol/vol)	0.89	0.91	0.98
Org C (dry weight)	9.9%	10.8%	14.4%
Org C (g cm^{-3})	2.37	2.53	2.63

The pH was measured by pH glass electrodes (*15,16*) with an agar coated chlorinated silver wire as reference. Tip diameters were 10-20 μm, but the length of the pH sensitive area was in the order of 200 μm, limiting the depth resolution of the profiles. Electrodes were calibrated onboard at *in situ* temperature in recovered bottom water after addition of aliquots of HCl. Independent pH measurements were obtained by a commercial pH electrode (Radiometer). The pH microsensors exhibited a log-linear response to H^+ with a slope of 48-60 mV pH^{-1} and a 90% response time of 20-30 s.

The S^{2-} sensors had outside tip diameters of 30-60 μm (*16*). Calibration was performed onboard in a dilution series of a buffered sulfide standard solution in recovered seawater made anoxic by N_2 flushing. The total sulfide concentration was subsequently determined by the methylene blue technique (*17*). The sensors are in reality sensitive to the S^{2-} ion, which stands in equilibrium with HS^- and H_2S. However, below the calibrated/recalculated data represent the total pool of sulfide (S^{2-}, HS^-, H_2S) and are referred to as H_2S (*16*). The sensors exhibited a log-linear response of 20-35 mV per decade of H_2S in the range of 0-50 μm H_2S. The H_2S profiles within the sediment were corrected for pH changes as previously described (*18*).

Total exchange rates

The benthic solute exchange was measured using a benthic chamber lander "Elinor" (*19*). During positioning of the lander, a central chamber (30 x 30 cm) was inserted in the sediment. After one hour the lid was closed and the incubation was initiated. During incubations seven water samples of 45 ml were recovered by spring-loaded syringes at predefined time intervals. The water samples were substituted by ambient water through a coil placed in the chamber lid (the dilution was corrected for). Additionally, two minielectrodes, with the same measuring characteristics as described above, continuously measured the O_2 concentration of the enclosed water (*19*). The sensors were calibrated against water samples taken from the chamber and a zero reading was recorded onboard at *in situ* temperature. During incubation the water volume was stirred by a central impeller at a speed of 8 RPM resulting in an average DBL thickness of 500-600 μm, depending on the water height and the sediment roughness (*19*). Upon completion of the incubation, a scoop closed beneath the chamber. After recovery water height inside the chamber was measured, and the sediment was sieved through a 1mm mesh screen to collect the macrofauna. The dry weight of the collected fauna was determined after 24h at 70°C, while the organic carbon content of the fauna was determined as loss on ignition after 24 h at 550 °C. The porosity and organic carbon content was determined in a subcore taken from the lander-recovered sediment. Ten ml of the recovered water samples were frozen in plastic vials for later analysis of nutrients (NO_3^-, NH_4^+, urea), while another 10 ml were transferred to gas-tight excetainers spiked with 100 µl saturated $HgCl_2$ for later analysis of dissolved inorganic carbon (DIC) on a coulometer (CM5012, UIC. Joiet, Illinois, USA). Nitrate was determined using standard techniques (20) on a flow injection analyzer (Perstop Analytical, Wilsonville, Oregon, USA) while NH_4^+ was analyzed colorimetrically (21). Urea concentrations were determined by the diacetylmonoxime method (22). Bottom water from 25 cm above the sediment was recovered *in situ* by a separate instrument consisting of a pump mounted on a frame that rested on the bottom.

Calculations

The measured O_2 microprofiles were used to determine the thickness of the diffusive boundary layer (DBL), the O_2 penetration depth and the diffusive O_2 uptake (DOU). The position of the sediment surface was estimated from a change in the O_2 gradient caused by impeded diffusion in the sediment (Figure 2) (23). From the determination of the relative position of the various sensor tips performed prior to lander deployment (see above) the position of the sediment surface in relation to the H_2S and pH microprofiles was estimated. The upper boundary of the DBL was determined as the depth where the extrapolated linear concentration gradient in the DBL reached the constant concentration in the turbulent water phase (7). The O_2 penetration depth could be determined directly as the depth where the signal of the O_2 microelectrode reached a constant low reading. The DOU was calculated by Fick's first law of diffusion: $DOU = D_0$ (dC/dZ), where D_0 is the temperature corrected diffusion coefficient in water and C is the O_2 concentration at depth Z within the DBL (24). Total exchange rates were calculated from the slope of the linear change in solute concentration versus time and the volume of the enclosed water phase. When calculating the total exchange rate of inorganic carbon and nitrogen, the C and two N atoms in urea was accounted for.

Results

The bottom water O_2 concentrations at Stations A and B were similar, but they were less than half of the concentration at Station C (Figure 2). Nevertheless, the O_2 penetration depth decreased and the DOU increased going from the open Bight (Station A) to the sheltered region of Station B and into the semi-enclosed basin of Station C (Figure 2, Table 2). The average DBL thickness was around 500 µm at Stations A and B, while it was 200 µm thicker at Station C (Figure 2, Table 2).

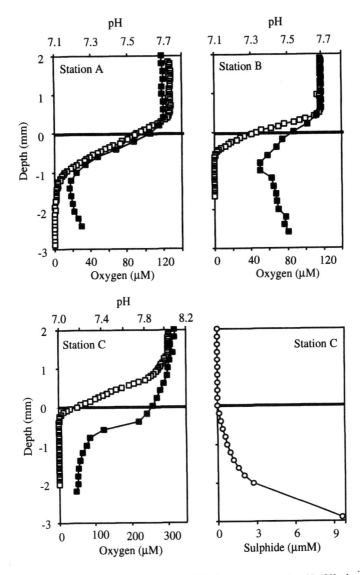

Figure 2. Representative microprofiles of O₂ (open squares), pH (filled squares) and H₂S (open circles) obtained at the three investigated stations. At station A and B, H₂S could not detected. The fat horizontal line indicates the position of the sediment surface.

Table II. Data extracted from microprofiles, measured exchange rates and fauna information

	Stations		
	A	B	C
DBL thickness (μm)	450 ± 212	490 ± 90	710 ± 168
O_2 penetration (mm)	1.35 ± 0.09	0.74 ± 0.11	0.39 ± 0.04
DOU	-15.7 ± 5.3	-20.9 ± 3.4	-46.2 ± 8.2
H_2S appearance (mm)	> 30	> 30	0.2 ± 0.2
TOU	-28.5	-39.1	-73.2
DIC flux	33.9	46.7	163.3
NH_4^+ flux	2.1	1.7	8.7
NO_3^- flux	0.0	0.7	0.0
Urea flux*	0.0	0.5	8.6
Fauna wet-weight	325	116	0.0
Fauna dry-weight	232	101	0.0
Fauna Org C	23.1	13.2	0.0

NOTE: All flux rates are in mmol m^{-2} d^{-1}
NOTE: Negative values indicate an uptake
NOTE: For O_2 microprofile data SD are included (n=4)
NOTE: Weight values are in (gm^{-2})

* For calculating urea into N fluxes both N-atoms has to be accounted for.

Free H_2S could not be detected in the upper 30mm at the two oxidized stations (data not shown). At station C, H_2S was detected from 0.2 mm below the sediment surface (Figure 2) and the concentration increased until a plateau of around 150 μM was reached at a sediment depth of 10mm (data not shown). This reflects an intense H_2S oxidation catalyzed by the well-developed bacterial mat in the surficial sediment.

At Station A and B the pH profiles showed a distinct minimum at a depth of 1.4 and 0.8mm, respectively. The minima aligned well with the O_2 penetration depth at the respective stations (Figure 2). At Station C the pH profile showed a steep decrease of almost 0.8 pH units within a horizon of <1mm immediately below the sediment surface. This coincides with the oxic anoxic boundary and the position of the microbial mat. However, instead of a distinct minimum as on the oxidized stations the pH remained at a relative low value to a sediment depth of 2.2mm (Figure 2).

The total O_2 uptake (TOU) increased moving from the open into the more protected areas of the Bight. The rates were, however, 1.6-1.9 fold higher than the DOU calculated from the microprofiles at the respective stations (Table 2). The DIC release rates also increased from station A towards station C, but whereas the RQ –ratio (CO_2/O_2 – flux) for the effluxing solutes equaled 1.19 at the two oxidized stations it increased to 2.23 at the reduced site (Table 2).

Nitrogen was mainly released to the water column as NH_4^+ and urea. Only Station B had a measurable efflux of NO_3^-. At Stations A and C, urea accounted for 29 and 66% of the total nitrogen efflux, respectively (Table 2 and Figure 3 and 4).

The infauna was dominated by bivalves (*Macoma calcarea*, *Abra alba* and *Mysella bidentata*), however, polycheates (*Nephtys* spp and, *Pectinaria* spp) and brittle stars (*Ophiura albida*) were also observed. Most of the fauna dry-weight was related to shells and organic carbon content of the fauna only accounted for 10-13% of the dry weight (Table 2). No living macrofauna was observed at Station C.

Discussion

Benthic O_2 dynamics

Both DOU and TOU increased going from the open to the sheltered regions of the Bight. However, at all sites the TOU was significantly higher than the DOU.

The DBL can impede the DOU (25). This depends on the extent by which the O_2 consumption rate within the sediment is limited by the time it takes O_2 to diffuse across the DBL (26). The impedance of the DBL towards the DOU can be estimated assuming a depth-independent specific O_2 consumption rate within the oxic zone. Thereby the DOU becomes proportional to the square root of the O_2 concentration at the sediment surface (27). The increase in DOU without the presence of a DBL is then given by the factor $(C_w/C_0)^{0.5}$, where C_w and C_0 are the O_2 concentrations in the turbulent water phase and at the sediment surface, respectively. The values for Stations A, B and C equal 1.14, 1.60, and 2.40 respectively. In other words in case that the DBL was eliminated the DOU at the three stations would temporally increase by 14, 60 and 140%, respectively. As a new steady state would establish itself the DOU would, however, gradually approach the original value. The relatively low DBL- impedance at station A and B and the fact that both DOU and TOU were obtained with a similar DBL thickness, rule out different flow scenarios during profiling and chamber incubations as an explanation for the difference between DOU and TOU at these two sites.

In oxidized impermeable sediments with low DBL impedance, the difference between TOU and DOU is a measure of the fauna mediated O_2 uptake (28, 29, 30). Calculated in this way, the O_2 consumption related to the benthic fauna accounted for 12.8 and 18.2 mmol O_2 m^{-2} d^{-1} at Stations A and B, respectively – equal to 45% of the TOU at both sites. The mass-specific metabolic rate for bivalves, the dominant fauna at the investigated sites, is in the order of 0.2 μmol O_2 g^{-1} h^{-1} (wet weight) (31). Assuming that bivalves were the only fauna at Stations A and B the respiration of macrofauna could account for 2-6% of TOU. The remaining part (39-43% of TOU) represents the fauna stimulated microbial O_2 consumption. This value may be slightly overestimated since polycheates and echinoderms have higher mass specific respiration rates (0.5-1.5 mmol O_2 g^{-1} h^{-1} (31). However, the calculation demonstrates that even though fauna mediated O_2 uptake enhanced the benthic O_2 consumption significantly, the major part was related to stimulated microbial activity in the proximity of the animals rather than the respiration of the fauna itself (13).

A seasonal study performed in the laboratory on recovered sediment cores from at Station A has previously been presented (9). Here the DOU varied between 7.5 – 15.2 mmol m^{-2} d^{-1}, (value for October = 9.8 mmol m^{-2} d^{-1}) while the TOU ranged between 5.7 – 23.5 mmol m^{-2} d$^-$ (value for October = 15.6 mmol m^{-2} d^{-1}) over the year. On the average the TOU was 45% higher than the DOU (9) – a value identical to our in situ data for Stations A and B (Table 2). However, our absolute values (Table 2) are in the high range of the laboratory obtained data, especially for the TOU. There is growing evidence, that laboratory measurements underestimate the in situ exchange rates due to poor representation of intact undisturbed macrofauna (32, 33).

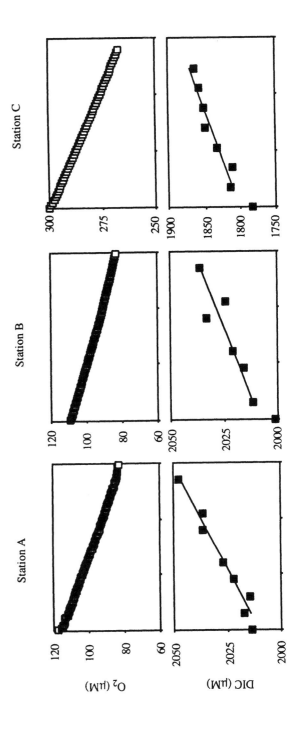

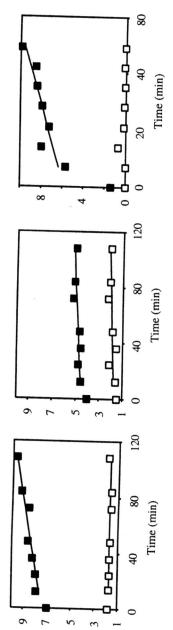

Figure 3. Solute concentration changes in the enclosed water phase of the benthic chamber at the three investigated stations. Open symbols in the lower panels represent NO_3^- concentrations. The first value in each panel was measured in an independently recovered bottom water sample. The water volume inside the chamber at station A, B and C were 9.7, 16.7 and 12.6 l, respectively.

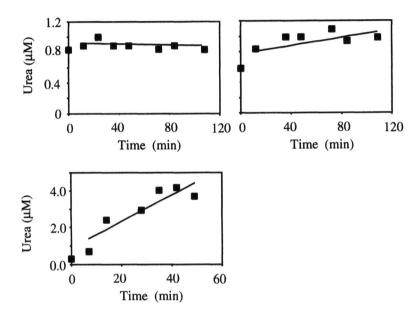

Figure 4. The change in urea concentrations in the enclosed chamber water phase at the three sites

At Station C with no macrofauna present, DOU only accounted for 63% of the TOU. However, at this site the DBL impedance was high and the DBL during profiling was thinner than during the chamber incubation (see above). Slight changes in DBL thickness covering active mats of sulfide oxidizing bacteria have major impact on the measured exchange rates (7). Another factor of importance at this site is that the one-dimensional diffusion model used to calculate the DOU assumes a plane surface. On a variable microtophographic surface, sensors would on the average approach the interface at an angle, resulting in underestimated one-dimensional fluxes. Two studies on highly variable microtophographic surfaces have shown that the one-dimensional flux approach can underestimate the true diffusive flux by a factor of 1.5-2.5 (25, 8). *Begiatoa*-mats are characterized by an extensive microtophography (34), and the one-dimensional flux-approach at station C may indeed have lead to an underestimated DOU.

The resent observation that veil architecture combined with rotation of *Thiovulum* cells can enhance the diffusive mediated solute exchange by a factor of 40 (35) could have been of importance at this site. Additionally, an advective porewater transport induced by massive occurrence of filterfeeding ciliates often observed at microaerophilic conditions could also have enhanced the exchange (36).

Benthic carbon mineralization

At steady state, benthic O_2 consumption balances the aerobic heterotrophic activity plus the oxidation of reduced solutes produced by the anaerobic degradation (i.e. assuming no burial of reduced equivalents). Consequently, the total benthic carbon oxidation is balanced by an equivalent amount of O_2 consumption (5). Assuming that the mineralized organic material consisted of carbohydrates, the RQ-ratio for the effluxing solutes should be 1. For coastal settings RQ-values between 1-2 are typical (37, 38, 39). The values of 1.2 at the two oxidized stations indicate incomplete reoxidation of reduced solutes from the anaerobic degradation. Dinitrogen could only have been of minor importance since denitrification in Danish coastal sediment only accounts for a few percent of the benthic carbon mineralization (2). Precipitating sulfide is more likely to be responsible for the mismatch. Measurements have shown that 15% of the annually produced sulfide is permanently buried below the bioturbated horizon at Station A (6). It can, however, not be excluded that dissolution of carbonate shells have added to the DIC release rates at all sites. For the central parts of Aarhus Bight it has been estimated that dissolution of shells leads to an average DIC release rate of 3 mmol m^{-2} d^{-1} (2).

The benthic mineralization in Aarhus Bight does show seasonal (and inter annual) variations (2), and our data cannot be extrapolated beyond the measuring period. However, for comparison the yearly average primary production and sedimentation rate in the central part of Aarhus have been measured to be 60, and 27 mmol m^{-2} d^{-1}, respectively. Our measurements suggest that a significant fraction of this material is mineralized in the sediment.

At Station C, oxygen was mainly consumed by the chemolithotropic sulfur bacteria, gaining energy from the oxidation of H$_2$S. However, H$_2$S is often only oxidized to S^0, which is stored internally as sulfur-granules (34). The incomplete reoxidation of H$_2$S caused a high RQ –ratio for effluxing solutes at this station.

The pH profiles at all stations reflected a decrease around the oxic/anoxic interface. Aerobic respiration as well as the reoxidation of reduced components (e.g. H$_2$S, Mn^{2+}, Fe^{2+}) lowers the pH (40). The distinct minimum strongly suggests that a major part of the O$_2$ consumption was related to reoxidation. The gradual increase in pH in the anoxic sediment strata could be related to reduction of various metal oxides.

Benthic nitrogen mineralization

Nitrate concentrations in the bottom water were low (<2µM), consequently denitrification based on nitrate from the overlying water was insignificant. Coupled denitrification (denitrification based on nitrification) was measured *in situ* at Station A shortly after the present study and equaled 0.3 mmol m^{-2} d^{-1} (41) equivalent to 15% of the ammonia flux (Table 2). Due to the small O$_2$ penetration depth at Stations B and C, coupled denitrification at these sites was probably even lower than at Station A. Ignoring the contribution of dinitrogen the C/N ratio of the effluxing solutes equaled 16.1, 12.9 and 6.6 at Stations A, B and C, respectively. That indicates degradation of more labile organic material at the reduced sites. Along with decreasing C/N ratio of the effluxing solutes the organic content of the surficial sediment increased and the relative importance of urea as a nitrogen carrier increased from 0% at Station A to 29% and 66% at Stations B and C, respectively.

Urea is an intermediate product from the degradation of organic material and an important precursor for NH$_4^+$ (42). In benthic systems the turn over constant is in the order of minutes to hours (43). The importance of urea as nitrogen carrier across the benthic interface is only marginally investigated. A seasonal study has suggested a positive correlation between urea concentration (and urea turn-over) and the concentration or the lability of organic carbon (44). The present data supports this idea and demonstrate that large effluxes of urea do occur *in situ* also in sediments without macrofauna present.

Acknowledgements

Sten P. Nielsen, Fritz Hansen, Anni Glud, Lars B Pedersen and Einer Larsen are thanked for their skillful technical assistance. Pete Sampou is thanked for help during the DIC determinations. "Skipper Hans" is thanked for his patience and skillful maneuvering of RV "Genetica II". This work was supported by the National Agency of Environmental Protection through the Marine Research Program 90. A preliminary version of the data has previously been included in a Danish language report for the Marine Research Program 90 (ISBN 87-7810-300-2).

References

1. Wollast, R. In: *Ocean margin processes in global change*; Mantoura, R. F. C.; Martin, J.M.; Wollast, R., Eds.; John Wiley & Sons. **1991**, pp 365-381.

2. Jørgensen, B. B. In *Eutrophication in coastal marine ecosystems*; Jørgensen, B. B.; Richardson, K., Eds.; American Geophysical Union: Washington D. C. **1996**, pp. 137-155.

3. Berner, R.A. *Early diagenesis: A theoretical approach*. Princeton University Press. **1980**, pp 1-241.

4. Froelich, P. N.; Klinkhammer, G. P.; Bender, M. L.; Luedtke, N. A.; Heath, G. R.; Cullen, D.; Dauphin, P.; Hammond, D.; Hartman, B.; Maynard, V. *Geochim Cosmochim. Acta.* **1979**, *34*, 1075-1090.

5. Canfield, D.E.; Jørgensen, B.B.; Fossing, H.; Glud, R.N.; Gundersen, J.K.; Thamdrup, B.; Hansen, J.W.; Nielsen, L.P.; Hall, P.O.J.. *Mar. Geol.* **1993**, *113*, 27-40.

6. Thamdrup, B.; Fossing, H.; Jørgensen, B. B. *Geochim. Cosmochim. Acta.* **1994**, *58*, 5115-5129.

7. Jørgensen, B. .B.; Revsbech, N,. P. *Limnol. Oceanogr.* **1985**, *30*, 111-122.

8. Gundersen, J.K.; Jørgensen, B. B. *Nature* **1990**, *345*, 604-607.

9. Rasmussen, H.; Jørgensen, B. B. *Mar. Ecol. Prog. Series.* **1992**, *81*, 289-303.

10. Larkin, J. M.; Strohl, W. R. *Ann. Rev. Microbiol.* **1983**, *37*, 341-367.

11. Wirsen, C. O.; Jannasch, H. W. *J. Bacteriol.* **1978**, *136*, 765-774.

12. Revsbech, N. P. *Limnol Oceanogr.* **1989**, *34*, 474-478.

13. Glud, R. N.; Gundersen, J. K.; Ramsing, N. B. In *In situ analytical techniques for water and sediment.* Buffle, J.; Horvai, G., Eds.; **2000**, pp 19-73.

14. Strickland, J. D.; Parson, T. R. *A practical handbook of seawater analysis*, 2nd. ed. Bulletin of the Fish Research Board of Canada. **1972**, *167*, pp 1-310.

15. Revsbech, N. P.; Jørgensen, B.B. In *Advances in Microbial Ecology*, Marshall K. C. Ed.; Plenum, New York. **1986**, 9, pp 293-352.

16. Kühl, M.; Revsbech, N. P. In *The Benthic Boundary Layer*. Boudreau, B. P.; Jørgensen, B. B., Eds.; Oxford University Press. **2000,** pp 180-211.

17. Cline, J. D. *Limnol. Oceanog.* **1969,** *14*, 454-458.

18. Kühl M.; Lørgensen, B. B. *Appl. Environ. Microbiol.* **1992,** *58*, 1164-1174.

19. Glud, R. N.; Gundersen, J. K.; Revsbech, N. P.; Jørgensen, B. B.; Huettel, M. *Deep Sea Res.* **1995,** *42*, 1029-1042.

20. Grasshoff, K.; Erhardt, M.; Kremling, K. *Methods of Seawater Analysis,* 2.nd ed. Verlag Chemie, Weinheim, Germany, 1983, pp 1-414.

21. Bower, E.; Hansen, H. A. *Can. J. Fish. Aquat. Sci.* **1980,** *37*, 794-798.

22. Price, N. M.,; Harrison, P.J. *Mar. Biol.* **1987,** *94*, 307-317.

23. Glud, R. N.; Jensen, K.; Revsbech, N. P. *Geochim. Cosmochim. Acta.* **1995,** *59*, 231-237.

24. Crank, J. *The Mathematics of Diffusion*. Clarendon Press, Oxford, **1997**, pp 1-414.

25. Jørgensen, B. B.; Des Marais, D. J. *Limnol. Oceanogr.* **1990,** *35*, 1343-1355.

26. Jørgensen, B. B. In *the Benthic Boundary Layer*. Boudreau, B. P.; Jørgensen, B. B. Eds. Oxford University Press. **2000,** pp 348-367.

27. Bouldin, D. R. *J. Ecol.,* **1968,** *56*, 77-87.

28. Archer, D.; Devol, A. *Limnol. Oceanogr.* **1992,** *37*, 614-629.

29. Glud, R. N.; Gundersen J. K.; Jørgensen, B. B., Revsbech, N.P.; Schulz, H.D. *Deep Sea Res.* **1994,** *41*, 1767-1788.

30. Glud, R. N.; Gundersen, J. K.,; Holby, O. *Mar. Ecol. Prog. Series.* **1999,** *186*, 9-18.

31. Piepenburg, D.; Blackburn, T. H.; von Dorrien, C. F.; Gutt, J.; Hall, P. O. J.; Hulth, S.; Kendall, M. A.; Opalinski, K. W.; Rachor, E.; Schmid, M. K. *Mar Ecol Prog Series.* **1995,** *118*, 199-213.

32. Devol, H.A.; Christensen, J.P. *J. Mar. Res.* **1993,** *51*, 345-372

33. Glud, R. N.; Holby, O.; Hofmann, F.; Canfield, D. E. Benthic mineralisation in Arctic sediments (Svalbard). *Mar. Ecol. Prog. Series.* **1998,** *173*, 237-251.

34. Møller, M. M.; Nielsen, L. P.; Jørgensen, B. B. *Appl. Environ. Microbiol.* **1985,** *50*, 373-382.

35. Fenchel, T.; Glud, R. N. *Nature* **1998,** *394*, 367-369.

36. Glud, R. N.; Fenchel , T. *Mar. Ecol. Prog. Series.* **1999**, *186*, 87-93.
37. Hargrave, B. T.; Phillips G. A. *Estuar. Coast. Shelf Sci.* **1981**, *12*, 725-737.
38. Boucher, G.; Clavier, J.; Garrigue, C. *Mar. Ecol. Prog. Series.* **1994**, *107*, 185-193.
39. Therkildsen M. S.; Lomstein, B. A. *FEMS Microbiol. Ecol.* **1993**, *12*, 131-142.
40. Boudreau, P. B. *Geochim. Cosmochim. Acta.* **1991**, *55*, 145-159.
41. Nielsen, L. P.; Glud, R. N. *Mar. Ecol. Prog. Series.* **1996**, *137*, 181-186.
42. Lomstein, B. A.; Blackburn, T. H.; Henriksen, K. *Mar. Ecol. Prog. Series.* **1989**, *57*, 237-247.
43. Lund, B. A.; Blackburn, T. H. *J. Microbiol. Meth.* **1989**, *9*, 287-308.
44. Therkildsen, M. S.; Lomstein, B. A. *Mar. Ecol. Prog. Series.* **1994**, *109*, 77-82.
45. Reimers, C.; Glud R. N.   In *Chemical sensors in Oceanography*, Gordon and Breach Science Publishers M Varney, Ed.; **2000**, pp 249-282.

Chapter 9

Benthic Carbon Mineralization in Sediments of Gotland Basin, Baltic Sea, Measured In Situ with Benthic Landers

Frank Wenzhöfer[1,2], Oliver Greeff[1], and Wolfgang Riess[1]

[1]Max Planck Institute for Marine Microbiology, Celsiusstraße 1,
D–28359 Bremen, Germany
[2]Current address: Marine Biological Laboratory, University of Copenhagen,
Strandpromenaden 5, 3000 Helsingør, Denmark

Benthic carbon mineralization was studied in situ at 4 stations of the Gotland Deep (Baltic Sea). Three separate benthic lander systems measured the diffusive oxygen uptake (DOU), the total oxygen uptake (TOU), and the sulfate reduction rate (SRR). At the three stations with oxygenated bottom water DOU varied between 6.4 and 9.3 mmol m^{-2} d^{-1}, while TOU varied between 7.3 and 10 mmol m^{-2} d^{-1}. SRR's were moderately high for coastal settings (0.7 - 9.0 mmol m^{-2} d^{-1}). Laboratory determined sulfate reduction rates were significant higher than in situ rates at two out of the four stations. In general, sulfate reduction was the major benthic carbon mineralization pathway of the benthic community. Carbon mineralization rates estimated from TOU and integrated SRR's agreed within a factor of 2.

Introduction

About 90% of the world's annual organic carbon input to marine sediments is deposited in areas with oxygen penetration depths ≤ 2 cm (*1, 2*). Microbial mineralization under oxic or anoxic conditions recycles a significant fraction of carbon and nutrients of the sedimenting organic matter, while the remainder is permanently buried. This buried fraction is positively correlated to the sediment accumulation rate and consequently it is high in deposition areas (e.g. *3, 4*). The quantitatively most important electron acceptors for benthic mineralization of organic carbon have proven to be oxygen and sulfate (*5, 6*), with Fe-oxides and Mn-oxides being of importance in some sediments (e.g. *7*). The contribution of oxygen to overall degradation in coastal marine sediments (with oxygenated bottom water and high deposition rates) is on the order of 5 - 20% (*8*), while sulfate reduction usually accounts for the major fraction. Total oxygen uptake of sediment includes two fractions: 1) direct heterotrophic respiration, and 2) oxygen consumed by reoxidation of reduced products from anaerobic degradation pathways. Consequently, the benthic oxygen consumption is often used as a measure of the total mineralization rate (*9*). In euxinic and semi-euxinic environments, however, integrated sulfate reduction rates (SRR) alone are considered a good measure of the total carbon degradation (*3, 10*).

A fundamental understanding of biogeochemical processes in sediments requires a quantitative assessment of process rates. *In situ* high-resolution measurements of pore water concentration profiles together with *in situ* benthic flux chamber incubations have provided an effective way to quantify the benthic flux across the sediment-water interface (*11, 12, 13, 14, 15*). This information can be used to estimate the rates of metabolic and chemical reactions in sediments and provide important insights into balance of processes that control sediment biogeochemistry. Another type of benthic lander for *in situ* radiotracer measurements of sulfate reduction to a sediment depth of 60 cm was recently developed (*16*). The success of *in situ* microprofiling, chamber incubation and radiotracer injection is generally attributed to the fact that sediment disturbance is minimized. While laboratory measurements require recovering and disturbance of the sediment, benthic lander measurements leave the sediments in place and thereby the natural gradients of metabolites and substrates intact (e.g. *11, 17*).

In this study we present *in situ* data on benthic mineralization processes with three autonomous benthic landers capable of measuring different key parameters of carbon mineralization (diffusive and total oxygen uptake, and sulfate reduction). The results are used to investigate the importance of oxygen and sulfate as degradative pathways. All 3 landers were deployed at 4 stations along a transect from the oxic slope of Gotland Basin into its anoxic center.

Materials and Methods

Study area and sediment description

The Baltic Sea is the largest brackish water body in the world, covering an area of approx. 372,000 km^2. A permanent halocline at approx. 60 - 80 m depth prevents vertical mixing between the almost homohaline surface waters and deeper water masses of higher salinity; additionally, a seasonal thermocline develops during the summer months in the surface water body, producing an intense stratification (18). During stagnation periods the deep water can be subdivided into two bodies: a subhalocline layer with a low and constant oxygen concentration, and the bottom water below approx. 170 m that is frequently anoxic (19, 20). Gotland basin is a flat basin with gentle slopes and a maximum depth of 249 m that displays almost permanently anoxic conditions in its deepest parts. Annual sediment accumulation rates of 1.0 - 1.3 mm yr^{-1} have been estimated in the central Gotland Basin, but on the slopes these values are lower and resuspension occasionally occurs, which cause lateral transport of the suspended material (21). The Holocene muds in the Central Baltic have very high organic matter concentration (10 - 15% dry weight; 21). The surface sediments of the investigated stations were soft, black, and organic-rich with high porosities.

Station locations and sampling methods

Four stations along a transect from the oxic into the anoxic part of Eastern Gotland Basin, Baltic Sea, were visited in August 1996 (Figure 1). CTD profiling revealed a stratified water column with thermocline and pycno-/oxycline at each station (data not shown). Basic station information is summarized in Table I. Sediment was sampled at each station with a multiple corer for laboratory incubation measurements (22). Porosity (vol/vol) was determined at a resolution of 1 cm from the water content of a core subsampled from a multicorer core (24 hour drying to constant weight at 70°C). All three benthic landers were deployed at each station in order to measure oxygen-microprofiles (PROFILUR; 14, 23), total oxygen uptake (ELINOR; 24) and sulfate reduction rates (LUISE; 16) in situ.

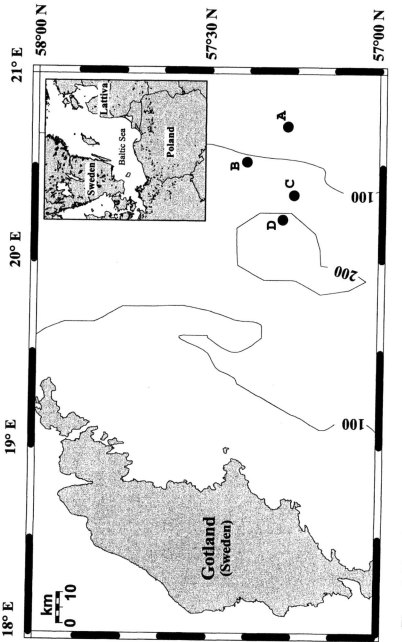

Figure 1. Map of the study area and position of the stations along the transect into Gotland Basin (Baltic Sea).

Table I. Station characteristics and macroinfauna. Oxygen data are from water samples taken by landers, other hydrographical data are from CTD casts and refer to bottom water.

	Station			
	A	*B*	*C*	*D*
Position	57°16 N 20°43 E	57°18 N 20°32 E	57°15 N 20°21 E	57°17 N 20°17 E
Depth [m]	75	115	155	210
Bottom water: O_2 [µM]	112	78	40	0
% air saturation	29.4	18.1	10.6	0
Temperature [°C]	4.5	4.5	4.3	4.0
Salinity [‰]	8.5	11	11	12
pH	6.98	7.06	7.04	6.96
Macoma balthica [individuals m^{-2}]	30 - 40	0	0	0
Monoporeia affinis [individuals m^{-2}]	200 - 250	0	0	0

Measurements of diffusive oxygen uptake (DOU) rates by the benthic profiling lander PROFILUR

The profiler is a pre-programmed free falling benthic lander system (*14, 23, 25*) designed to measure *in situ* microprofiles. During this cruise the profiling lander was equipped with six oxygen microelectrodes and the depth resolution was set at 100 μm. The oxygen microelectrodes were of Clark type with a guard cathode, internal reference, and an outer tip diameter of 10 - 30 μm (*26*). Response time was $t_{90} < 2$ sec, and stirring sensitivity was 1 - 2% (*27*). An *in situ* two-point calibration was performed using the constant sensor signal in the overlying water and in the anoxic part of the sediment. The lander was equipped with a Niskin bottle to collect bottom water, which was used for electrode calibration by Winkler titration. All electrodes were checked for drift by comparing *in situ* readings in bottom water before and after the profiling. Typically 4 parallel microprofiles were obtained at each oxic station (2 electrodes did not function or broke during the deployment). The position of the sediment-water interface was determined directly from the measured microprofiles. The transition between the homogeneous oxygen concentration in the turbulent bottom water and the linear gradient in the diffusive boundary layer (DBL) was used as the upper DBL boundary (*28*). The lower DBL boundary (sediment-water interface) was determined from the change in slope of the oxygen concentration gradient across the sediment-water interface due to the impeded diffusion in the sediment relative to that in the water (*29*). Due to the fluffy sediment surface with high porosities (between 0.93 and 0.99 in the upper first centimeter) at our investigated stations, it was difficult to measure the DBL accurately. Consequently we defined the oxygen penetration depth (OPD) as an oxygenated zone, starting from the first change in oxygen concentration measured by the microelectrodes. This includes the DBL as well as the oxygenated sediment horizon. To calculate the diffusive oxygen uptake (DOU) using Fick's first law of diffusion we therefore choose two different approaches: 1. Assuming a DBL above the sediment interface, DOU was calculated from the first sensor reading showing a decrease in signal (Figure 2a): $DOU = (-) D_0$ dC/dz, where D_0 = seawater diffusion coefficient and dC/dz = the linear oxygen gradient in the DBL. 2. Due to difficulties in accurately defining the DBL we also calculated the oxygen uptake using Fick's first law of diffusion as well as the empirical relations for the sediment diffusion coefficient (Fig 2b): $DOU = (-)$ $\Phi D_S dC/dz$ and $D_S = D_0^{m-1}$, where Φ is the porosity, D_S the sediment diffusion coefficient and $m = 3$ (*30, 31*). The seawater diffusion coefficient (D_0) for oxygen given in (*32*) was corrected for salinity and *in situ* temperature by the Stokes-Einstein equation. Oxygen consumption rates in discrete sediment layers were modeled from the oxygen profiles assuming constant porosity in the oxic zone and zero-order kinetics for the oxygen consumption reaction (*33*).

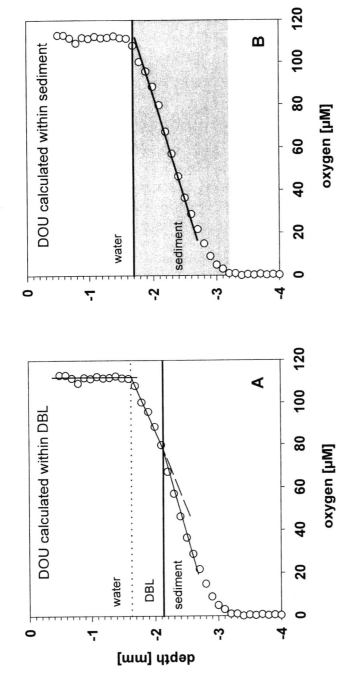

Figure 2. *Example of the treatment of an oxygen microprofile to calculate the diffusive oxygen uptake (DOU). Graph A shows the approach using the linear gradient in the diffusive boundary layer (DBL). Graph B shows the gradient used when calculating DOU directly below the sediment surface; the gray box indicates the oxygenated zone. For more details see text.*

Measurements of total oxygen uptake (TOU) rates by the benthic chamber lander ELINOR

The chamber lander is a free falling lander system equipped with a chamber that incubates 30 x 30 cm of sediment. After landing at the seafloor the chamber penetrates 20 to 30 cm into the sediment and after a resting period of 1 – 2 hours a lid closes the chamber (*11*). Two Clark type oxygen mini-electrodes are mounted in the lid to monitor the decrease of oxygen in the enclosed bottom water during the incubation. The electrodes were calibrated from an onboard zero reading at *in situ* temperature and the constant reading in the bottom water. During the incubation the water was gently mixed by a central stirrer (8 rpm), which creates an average DBL of approx. 500 μm (*24*), depending on the roughness of the sediment surface. At the end of the incubation, a shovel closes, isolating the incubated sediment from the sea floor. After recovery of the lander, the sediment was sieved and analyzed for macroinfauna.

Measurements of sulfate reduction rates (SRR) by the benthic lander LUISE

The autonomous benthic lander LUISE was used to determine sulfate reduction rates (SRR) *in situ*, by incubating six sediment cores (max. 60 cm length and 36 mm I.D.) with radiolabeled sulfate ($^{35}SO_4^{2-}$). After reaching the sea floor all six cores were inserted into the sediment. At a pre-programmed time, the enclosed sediment of three cores was vertically injected with radiolabeled sulfate (0.5 ml diluted in bottom water, which equals 150 kBq $^{35}SO_4^{2-}$ per cm sediment). The incubation times at the investigated stations was chosen between 13 and 18 hours. Shortly before the return of the lander the remaining three cores were also injected. The turnover of these three cores are subtracted from the turnover of the remaining cores which have been incubated *in situ* for a given incubation time to overcome the artefacts of transient warming and decompression during ascent and recovery of the lander, which may add to the sulfate reduction rates (for a detailed description of the lander operation see *16*). Upon recovery, the cores from LUISE were immediately sliced in 1 cm intervals and fixed in 20 ml 20% zinc acetate to stop metabolic activity (slicing procedure was approx. 20 min per core at room temperature).

In order to evaluate potential differences between *in situ* and laboratory studies, the injection and incubation pattern of the lander was replicated in the laboratory with different cores at *in situ* temperature and with the same incubation time. However, these cores were recovered by a multiple corer and were injected sideways through silicone-stoppered ports.

In the laboratory the samples collected *in situ* and onboard were distilled in an acid Cr(II) solution to volatilise and trap the reduced sulfur species *(34)*. The hydrogen sulfide evolved from the total reduced inorganic sulfur was precipitated in 5% ZnAc. After distillation, the ZnS suspension was mixed with scintillation fluid. All samples were counted in a liquid scintillation counter. Sulfate reduction rates were calculated according to *(35)*. Sulfate was measured in pore water samples fixed in ZnAc by ion chromatography (WATERS).

Results

Oxygen dynamics

The two different methods for calculating the diffusive oxygen uptake (DOU) did not differ significantly (Table II). The reason for this are the high porosity values in the first 1 cm, ranging from 0.93 to 0.99. In the following discussion we therefore use only the rates calculated from the DBL approach. At the oxic stations (A, B, C) diffusive oxygen uptake (DOU) varied between 6.4 and 9.3 mmol m^{-2} d^{-1} (Table II), and the thickness of the oxygenated zone was 1.6, 1.4 and 0.9 mm at Station A, B, C, respectively (Figure 3). At the anoxic Station D no oxygen could be measured with the microelectrodes (Figure 3, Table II). Total oxygen uptake (TOU) as measured by *in situ* uptake rates in the benthic chamber decreased with increasing water depths (Figure 4; Table II). The total rate was 20 % higher than DOU at Station A, the only station with benthic macroinfauna. While at the other two oxygenated stations (B and C) TOU and DOU were similar (Table II). There was no measurable oxygen in the benthic chamber at Station D.

Sulfate reduction rates

The measured sulfate reduction rates (SRR) of Stations A to D are presented in Figure 5. The *in situ* incubations from Stations B - D showed the maximum activity at the sediment surface and a rapid decline with depth. In most cores the SRR was negligible below a depth of approx. 15 cm. This pattern was not due to sulfate limitation in the deeper sediment strata, as concentration was > 2 mM (data not shown). The depth integrated SRR activity (0 - 15 cm) was highest in cores at Station A (75 m water depth) - where there was a distinct secondary subsurface peak at approx. 8 cm sediment depth. The lowest SRR were measured

Table II. Calculated oxygen fluxes from *in situ* microprofiles (diffusive oxygen uptake, DOU) and benthic chamber lander incubations (total oxygen uptake, TOU) and depth-integrated sulfate reduction rates (SRR; 0 - 15 cm) as determined by *in situ* and laboratory whole core incubation. +/- indicates fluxes out of and into the sediment, respectively. Standard deviation given in parentheses (n=4 for DOU and n=3 for SRR).

	Station			
	A	B	C	D
in situ DOU_{DBL} [1] [mmol m^{-2} d^{-1}]	-8.2 (± 1.80)	-9.3 (± 3.68)	-6.4 (± 0.45)	0
in situ DOU_{Sed} [2] [mmol m^{-2} d^{-1}]	-7.8 (± 1.69)	-7.5 (± 2.96)	-6.2 (± 0.44)	0
in situ TOU [mmol m^{-2} d^{-1}]	-10.0	-8.9	-7.3	0
in situ SRR [mmol m^{-2} d^{-1}]	8.99 (± 2.36)	7.52 (± 1.46)	0.80 (± 0.39)	0.69 (± 0.15)
laboratory SRR [mmol m^{-2} d^{-1}]	4.66 (± 0.82)	10.55 (± 1.93)	1.01 (± 0.18)	4.21 (± 0.60)

[1] DOU = diffusive oxygen uptake calculated from the oxygen gradient in the diffusive boundary layer (DBL) using the diffusivity coeficient in water (D_0).

[2] DOU = diffusive oxygen uptake calculated directly below the sediment-water interface using the diffusivity coeficient in the sedimet (D_S).

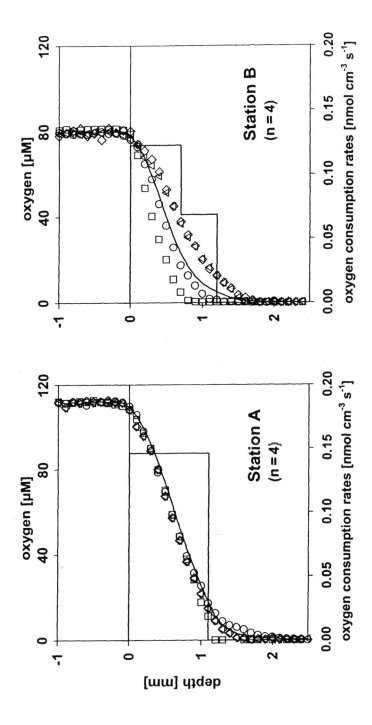

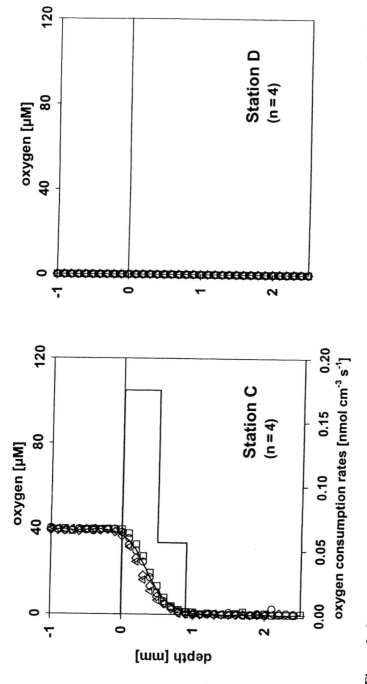

Figure 3. Oxygen microprofiles obtained by the profiling lander. Symbols represent the single profiles, while the line shows the average profile. Oxygen consumption rates were calculated from the average profile. Values in parentheses indicate the number of profiles.

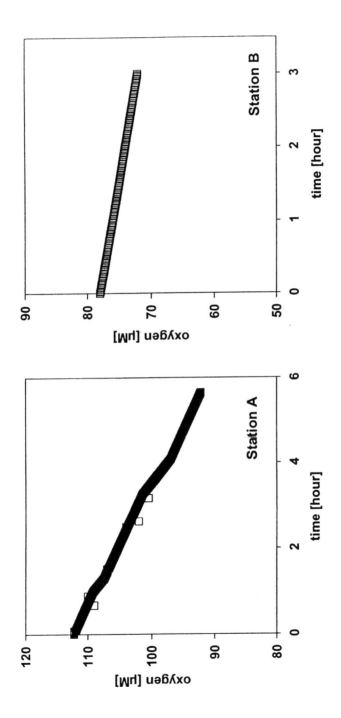

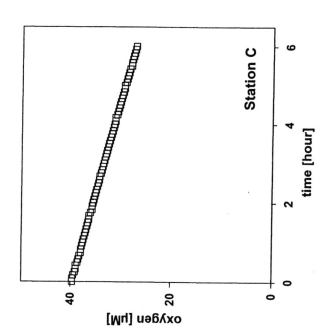

Figure 4. Oxygen concentration changes in the enclosed bottom water of the benthic chamber at the three oxygenated stations.

176

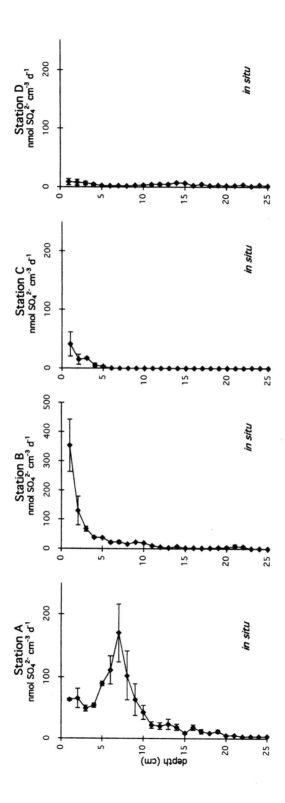

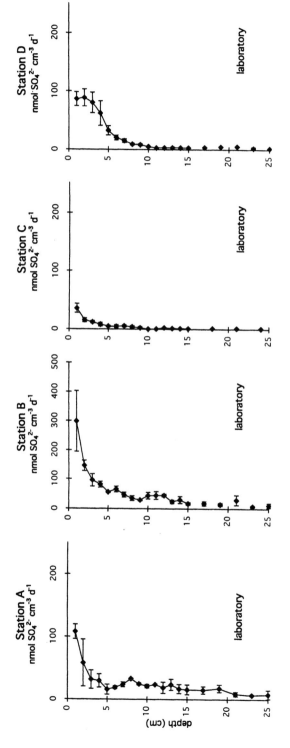

Figure 5. Sulfate reduction rates in sediments from Gotland Basin. Upper panel shows results from in situ incubations, lower panel those from laboratory incubations of all stations. Each data point represents triplicate measurements. The error bar indicates the standard error of the mean value (= mean value/(SQR(n)). For information on incubation time and calculation method see text.

at Station D (210 m; Table II). At Station B and C the SRR profiles obtained in the laboratory and *in situ* were comparable, and exhibited no significant differences (Figure 5; Table II). However, at Station A and D, *in situ* and laboratory results differed. The distinct subsurface peak between 4 and 10 cm observed in all *in situ* cores at Station A could not be found in the laboratory profiles. As a result the depth integrated *in situ* rate from Station A was higher by a factor of 2 as compared to the laboratory rate (Table II). Station D, however, exhibited the opposite picture in all laboratory replicates; rates at a given horizon derived from laboratory incubations were higher by a factor of up to 9 than the corresponding *in situ* incubations, resulting in 6 times higher depth integrated rates (Table II).

Discussion

Oxygen dynamics

At the oxygenated stations oxygen is rapidly consumed within the upper sediment horizon, resulting in low oxygen penetration depths (Figure 3). Oxygen is utilized either through mineralization of organic matter and/or reoxidation of reduced species (e.g. NH_4^+, Fe^{2+}, Mn^{2+}, H_2S) diffusing upwards from the anoxic zone of the sediment. The profile gradients result in fluxes of 6 - 9 mmol m^{-2} d^{-1} of O_2 into the sediment and are in the range of *in situ* values reported from coastal sediments of Svalbard (4 – 10 mmol m^{-2} d^{-1}; *12*) and the Washington shelf and slope (3 – 10 mmol m^{-2} d^{-1}; *36*). Generally, oxygen microprofiles indicated high uptake rates at the sediment surface, where the most reactive organic carbon is deposited (Figure 3). However, at the fauna-rich Station A, the specific oxygen activity was constant suggesting a more even distribution of the oxygen consumption processes (Figure 3).

Comparison of the *in situ* DOU and TOU of the oxygenated stations show a higher TOU value for Station A (+20%) caused by fauna related irrigation and respiration. Similar rates were obtained at the deeper Station B and C without any fauna present (Table I and II). Benthic macroinfauna increases TOU because it consumes oxygen in the benthic chamber (respiration) and because it brings oxygen in the sediment (irrigation). The difference between the DOU and TOU measurements is the benthic fauna related oxygen uptake, which in our case is 1.8 mmol m^{-2} d^{-1} (Station A). This includes the metabolic respiration of the fauna as well as the sediment uptake through irrigation.

For studies of the sediment oxygen dynamic and to calculate uptake rates it is necessary to measure the diffusive as well as the total uptake rates. Since oxygen microsensors resolve an one-dimensional picture, benthic fauna and sediment topography can alter the oxygen penetration and uptake rate (*12, 37*). Sediment surface topography can increase the oxygen flux across the sediment-water interface up to 50 % relative to the one-dimensional diffusion flux calculated from the vertical oxygen gradient (*23, 38*). On the other hand chamber incubations consider the enclosed sediment area as a black box. Instead microsensors measure the vertical oxygen concentration in the sediment and define the oxygenated sediment horizon. From the obtained vertical oxygen microprofiles it is further possible to model the consumption rate and to define distinct zones with different microbial activities.

Sulfate reduction

In this study laboratory and *in situ* SRR differed significantly at two stations (Table II). At both stations the temperature difference between the bottom water and the surface water was around 15°C. Using Q_{10}-values (temperature coefficient of sulfate reduction) for sulfate reducing bacteria of approx. 3 (*39, 40, 41, 42*), the exposure time of the multicorer samples (laboratory measurement) to increased temperature (approx. 1 h) was not sufficient to stimulate fast growth and/or enhanced metabolic activity. It was calculated that only a maximum increase of 11 - 15 % could have occurred. Comparison between *in situ* and laboratory determined benthic O_2 uptake rates for deep-sea environments have demonstrated significant overestimation of mineralization rates in laboratory-incubated cores (*11, 14, 17*). This is typically ascribed to transient heating during the retrieval of samples. However, at shallow water depths such effect seems to be of minor importance due to the fact that steady state conditions are quickly reestablished (*12*). Station A showed a subsurface peak in the *in situ* profiles that was absent in the laboratory profiles, causing a higher *in situ* depth integrated rate at this site. Station D represent the opposite end in the range of SRR, where the integrated *in situ* sulfate reduction rate was only one third of the laboratory value. Because of the fluffy sediment surface at this setting we cannot exclude that a bow wave effect disturbed the upper centimeter of the cores. This, however, would apply to both the *in situ* and laboratory measurements. However, sediment heterogeneity and sampling artefacts may result in a more complex picture, in which the true differences between *in situ* and laboratory measurements are hard to resolve on the basis of 4 stations. Recent investigations on *in situ* and laboratory obtained sulfate reduction rates confirm our observation. Measurements in water depths between 50 and 2050 m showed that rates from *in situ* incubations were 60 – 80 % lower than laboratory incubations (*43*). Nevertheless, depth integrated SRR determined in this study (Table II) are similar to rates found in a study on sediments from an anoxic basin in Costa Rica ranging between 2.4 and 6.0 mmol m^{-2} d^{-1} (44) and in shelf sediments from the Skagerrak (1.5 – 4.0 mmol m^{-2} d^{-1}; 8). Whereas anoxic settings from the upwelling region off Chile exhibited far higher rates. SRR from similar water depths were 6 - 30 times higher than our rates (45).

Carbon mineralization rates

Applying the total oxygen uptake rate (TOU) as an integrated measure for the total mineralization rate, the relative role of sulfate reduction for the overall mineralization of organic carbon can be estimated (Table III). At steady state conditions TOU balances the aerobic respiration as well as the oxidation of reduced inorganic solutes produced by anaerobic mineralization processes (12, 46). To account for the stoichiometry of sulfate reduction (2 moles of CO_2 are produced per mole of sulfate reduced; 10, 47) the molar rates of sulfate reduction have to be multiplied by 2. At Station A and B, SRR accounts for almost twice the amount of carbon degraded as inferred from TOU (Table III). The discrepancy in SRR and TOU most likely is the consequence of accumulating reduced sulfur complexes within the sediment (S^0, FeS and FeS$_2$). Seasonal studies on anaerobic mineralization processes in coastal marine sediments have shown that reduced sulfur compounds accumulate within the sediment when mineralization processes are enhanced (48, 49). On a yearly budget most of the reduced complexes are reoxidized, leaving only 15 % of the precipitated sulfide for permanent burial (49). In other words the e$^-$-equivalents from the anaerobic mineralization processes are not "paid back" but accumulate during this period of enhanced sedimentation and mineralization. As a consequence SRR at our investigated Stations A and B is a better measure for total carbon mineralization at this seasonal period. However, on an annual time scale with a cycle of precipitation and reoxidation of the reduced solutes the TOU rates might be a good measure for the carbon mineralization. Another factor accounting for the discrepancy between SRR and TOU could be spatial heterogeneity of the sediment, since all three *in situ* rates were determined with different landers separated several meters from each other.

Taking these processes of mineralization, reoxidation and precipitation into account, sulfate reduction seems to be the major mineralization process at Station A and B. This is in good agreement with data from the Kattegat, were (8) found that reoxidation processes of reduced metabolic products to be more important in the consumption of oxygen than aerobic respiration. In contrast to these stations, SRR account only for 22 % of the total mineralization at Station C. At Station D, however, all of the carbon most probably was degraded by SRR.

Table III. Comparison of carbon oxidation rates as calculated from benthic oxygen uptake (TOU) and integrated *in situ* sulfate reduction rates (SRR; 0 - 15 cm). A ratio of O_2 : TCO_2 of 1 : 1 has been assumed (e.g., *46*), while SRR were converted to carbon equivalents by multiplication by 2 (*47*).

Station	Carbon oxidation as measured by [mmol m^{-2} d^{-1}]	
	TOU	*SRR*
A	10.0	17.98
B	8.9	15.40
C	7.3	1.60
D	-[1]	1.38

[1] no oxygen in bottom water

The annual primary production in the surface waters above the thermocline at 10 m depth of Central Gotland Basin is estimated at approx. 15 mol C m^{-2} yr^{-1} (50, 51). Assuming our *in situ* TOU measurements represent the yearly average annual benthic carbon mineralization rate, then between 1 to 4 mol C m^{-2} yr^{-1} is mineralized in the sediment. This accounts for around 18 - 25 % (Stations A - C) and 7 % (Station D) of the primary production, leaving 75 to 93 % for water column respiration and permanent burial. Using the SRR rates at Station A and B the annual benthic carbon mineralization rates is equivalent to 7 and 6 mol C m^{-2} yr^{-1}, respectively, which accounts for 37 – 44 % of the primary production. Since our measurements were performed during the summer, when the sediment can be assumed to be enriched in organic carbon, these values most likely represent an overestimation. A phytoplankton bloom was indeed observed during the cruise period. *In situ* measurements have demonstrated that benthic oxygen uptake rates respond to the seasonal input of organic carbon, even when the response is significantly reduced (52, 53).

Conclusion

The presented results from Gotland Basin give rise to the following conclusions:

- Of a primary production of approx. 15 mol C m^{-2} yr^{-1}, between 7 – 25 % were degraded by benthic activity using TOU as an integrated measure for the total annual carbon mineralization.
- Oxygen microprofiles by themselves only account for the diffusive uptake ignoring the fauna mediated uptake and therefore underestimate the carbon mineralization rates. But they visualize the oxygen distribution within the sediment and the obtained micoprofiles can be used to calculate consumption zones. In combination with total oxygen uptake rates (TOU) measured with benthic chambers the total carbon mineralization can be determined.
- Our results indicate that SRR measurements provide a better measure of carbon mineralization rates for Station A and B than TOU at the investigated measering period. Sulfate reduction was generally the most important process in the benthic mineralization of organic matter; the observed results do not reflect steady-state conditions of sulfide production and reoxidation.
- As a final conclusion this study shows how important the use of different *in situ* methods is to describe the benthic mineralization processes accurately, since each single method themselves do not predict carbon mineralization rates accurately at certain sites. Only all three techniques together resolve the complex carbon mineralization system in the sediment.

Acknowledgements

Andreas Weber and Ola Holby are thanked for their help during the lander deployments and sulfate reduction measurements. We wish to thank Ronnie N. Glud, Bo Thamdrup and Susan Boehme for helpful discussions and valuable comments during the preparation of the manuscript, which was further improved by three anonymous reviewers. We are also indebted to Anja Eggers, Gabi Eickert and Vera Hübner for assembling the electrodes used. The assistance of the crew of R.V. „Professor Albrecht Penck" is greatly appreciated. This work was supported by the Max Planck Society.

References

1. Berner, R.A. *Am. J. Sci.* **1982**, *282*, 451-473.
2. Hedges, J.I.; Keil, R.G. *Mar. Chem.* **1995**, *49*, 81-115.
3. Canfield, D.E. *Deep-Sea Res.* **1989**, *36*, 121-138.
4. Henrichs, S.M. *Mar. Chem.* **1992**, *39*, 119-149.
5. Bender, M.L.; Heggie, D.T. *Geochim. Cosmochim.* **1984**, *48*, 977-986.
6. Henrichs, S.M.; Reeburgh, W.S. *Geomicrobiol. J.* **1987**, *5*, 191-237.
7. Canfield, D.E.; Thamdrup, B.; Hansen, J. W. *Geochim. Cosmochim.* **1993**, *57*, 3867-3883.
8. Canfield, D.E.; Jørgensen, B.B.; Fossing, H.; Glud, R.; Gundersen, J.; Ramsing, N.B.; Thamdrup, B.; Hansen, J.W.; Nielsen, L.P.; Hall, P.O.J. *Mar. Geol.* **1993**, *113*, 27-40.
9. Smith, K.L. jr.; Hinga, K.R. In *The sea*; Rowe, G.T., Ed.;. John Wiley & Sons: New York, 1983; Vol.8, pp 331-370.
10. Westrich, J.T. Ph.D. thesis, Yale University, New Haven, CT, 1983
11. Glud, R.N.; Gundersen, J.K.; Jørgensen, B.B.; Revsbech, N.P.; Schulz, H.D. *Deep-Sea Res.* **1994**, *41*, 1767-1788.
12. Glud, R.N.; Holby, O.; Hoffmann, F.; Canfield, D.E. *Mar. Ecol. Prog. Ser.* **1998**, *173*, 237-251.
13. Wenzhöfer, F.; Holby, O.; Glud, R.N.; Nielsen, H.K.; Gundersen, J.K. *Mar. Chem.* **2000**, *69*, 43-54.
14. Wenzhöfer, F.; Holby, O.; Kohls, O. *Deep-Sea Res.* **2001**, *48*, 1741-1755.
15. Wenzhöfer, F.; Adler, M.; Kohls, O.; Hensen, C.; Strotmann, B.; Boehme, S.; Schulz, H.D. *Geochim. Cosmochim.* **in press**.
16. Greeff, O.; Glud, R.N.; Gundersen, J.K.; Holby, O.; Jørgensen, B.B. *Cont. Shelf Res.* **1998**, *18*, 1581-1594.
17. Glud, R.N.; Gundersen, J.K.; Holby, O. *Mar. Ecol. Prog. Ser.* **1999**, *186*, 9-18.

18. Kullenberg, G. In *The Baltic Sea*; Voipio, A., Ed.; Elsevier: Amsterdam, 1981; pp 135-182.
19. Grasshoff, K.; Voipio, A. In *The Baltic Sea*; Voipio, A., Ed.; Elsevier: Amsterdam, 1981; pp 183-218.
20. Stigebrandt, A.; Wulff, F. *J. Mar. Res.* **1987**, *45*, 729-760.
21. Winterhalter, B.; Flodén, T.; Ignatius, H.; Axberg, S.; Niemistö, L. In *The Baltic Sea*; Voipio, A., Ed.; Elsevier: Amsterdam, 1981; pp 1-122.
22. Barnett, P.R.O.; Watson, J.; Connelly, D. *Oceanol. Acta* **1984**, *7*, 399-408.
23. Gundersen, J.K.; Jørgensen, B.B. *Nature* **1990**, *345*, 604-607.
24. Glud, R.N.; Gundersen, J.K.; Revsbech, N.P.; Jørgensen, B.B.; Hüttel, M. *Deep-Sea Res.* **1995**, *42*, 1029-1042.
25. Reimers, C.E. *Deep-Sea Res.* **1987**, *34*, 2019-2035.
26. Revsbech, N.P. *Limnol. Oceanogr.* **1989**, *34*, 474-478.
27. Glud, R.N.; Gundersen, J.K.; Ramsing, N.B. In *In situ analytical techniques for water and sediment*; Buffle, J.; Horvai, G., Eds.; Wiley: Chichester, 2000; pp 19-74.
28. Jørgensen, B.B.; Revsbech, N.P. *Limnol. Oceanogr.* **1985**, *30*, 111-122.
29. Sweerts, J.-P.R.A.; St. Louis, V.; Cappenberg, T.E. *Freshw. Biol.*, **1989**, *21*, 401-409.
30. Berner, R.A. *Early diagenesis: a theoretical approach.* Princeton University Press, 1980.
31. Ullman, W.J.; Aller, R.C. *Limnol. Ocenaogr.* **1982**, *27*, 552-556.
32. Li, Y.-H.; Gregory, S. *Geochim. Cosmochim.* **1974**, *38*, 703-714.
33. Nielsen, L.P.; Christensen, P.B.; Revsbech, N.P.; Sorensen, J. *Microbial Ecology* **1990**, *19*, 63-72.
34. Fossing, H.; Jørgensen, B.B. *Biogeochem.* **1989**, *8*, 205-222.
35. Jørgensen, B.B. *Geomicrobiol. J.* **1978**, *1*, 11-27.
36. Archer, D.; Devol, A. *Limnol. Ocenaogr.* **1992**, *37 (3)*, 614-629.
37. Glud, R.N.; Ramsing, N.B.; Gundersen, J.K.; Klimant, I. *Mar. Ecol. Prog. Ser.* **1996**, *140*, 217-226.
38. Jørgensen, B.B. ; Des Marais, D.J. *Limnol. Oceanogr.* **1990**, *35*, 1343-1355.
39. Jørgensen, B.B. *Limnol. Oceanogr.* **1977**, *22*, 814-832.
40. Skyring, G.W.; Chambers, L.A.; Bauld, J. *Australian J. Mar. Freshw. Res.* **1983**, 34, 359-374.
41. Westrich, J.T.; Berner, R.A. *Geomicrobiol. J.* **1988**, *6*, 99-117.
42. Isaksen, M.F.; Jørgensen, B.B. *Appl. Environ. Microbiol.* **1996**, *62*, 408-414.
43. Weber, A. ; Riess, W. ; Wenzhöfer, F. ; Jørgensen, B.B. *Deep-Sea Res.* **in press**.
44. Thamdrup, B.; Canfield, D.; Ferdelman, T.; Glud, R.N.; Gundersen, J.K. *Rev. Biol. Trop.* **1996**, *44*, 19-33.
45. Thamdrup, B.; Canfield, D.E. *Limnol. Oceanogr.* **1996**, *41*, 1629-1650.

46. Canfield, D.E. In *Interactions of C, N, P and S. Biogeochemical cycles and global change*; Wollast, R.; Chou, L.; Mackenzie, F., Eds.; Springer Verlag: Berlin, 1993; pp. 333-363.
47. Froelich, P.N.; Klinkhammer, G.P.; Bender, M.L.; Luedtke, N.A.; Heath, G.R.; Cullen, D.; Dauphin, P.; Hammond, D.; Hartmann, B.; Maynard, V. *Geochim. Cosmochim.* **1979**, *43*, 1075-1090.
48. Thamdrup, B.; Fossing, H.; Jørgensen, B.B. *Geochim. Cosmochim.* **1994**, *58*, 5115-5129.
49. Jørgensen, B.B. In *Eutrophication in Coastal Marine Ecosystems*; Jørgensen, B.B.; Richardson, K., Eds.; Coastal and Estuarine Studies 52; American Geophysical Union: Washington, DC, 1996; pp 115-136.
50. Lassig, J.; Leppäner, J.-M.; Niemi, A.; Tamelander, G. *Finnish Mar. Res.* **1978**, *244*, 101-115.
51. Schulz, S.; Kaiser, W.; Breuel, G. *Biol. Oceanogr. Com.* **1992**, *L:19*, 1-5.
52. Smith, K.L.jr; Kaufmann, R.S.; Baldwin, R.J. *Limnol. Oceanogr.* **1994**, *39*, 1101-1118.
53. Smith, K.L.jr; Glatts, R.C.; Baldwin, R.J., Beaulieu, S.E.; Ulman, A.H.; Horn, R.C.; Reimers, C.E. *Limnol. Oceanogr.* **1997**, *42*, 1601-1612.

Sediment Porewaters
and Microbial Mats

Chapter 10

Porewater Redox Species, pH and pCO_2 in Aquatic Sediments: Electrochemical Sensor Studies in Lake Champlain and Sapelo Island

Wei-Jun Cai[1], Pingsan Zhao[1], Stephen M. Theberge[2], Amy Witter[2], Yongchen Wang[1], and George Luther, III[2]

[1]Department of Marine Sciences, University of Georgia, Athens, GA 30602
[2]College of Marine Studies, University of Delaware, Lewes, DE 19958

This paper reports millimeter depth resolution microelectrode-based porewater profiles of O_2, Mn^{2+}, Fe^{2+}, pH and pCO_2 in sediments from Lake Champlain in the northeastern US and from a creek bank in Sapelo Island in the southeastern US. Such fine scale profiles of multiple redox species measured together with pH and pCO_2 have not been reported previously for lake or salt marsh creek bank sediments. This paper discusses the relationship between redox reactions and the porewater pH values based on micro-profiles and diagenetic mechanisms from both fresh and salt water systems. The microelectrode data clearly show that the very sharp pH minimum is a result of Mn^{2+}, Fe^{2+} and NH_4^+ oxidation at the zone near the O_2 penetration depth as well as CO_2 release during organic matter decomposition. In the freshwater sediment, an overlapping of O_2 and Mn^{2+} profiles is observed indicating a close coupling between O_2 usage and Mn^{2+} oxidation. This is not the case in the marine system except when biological disturbance is serious. A laboratory experiment supports an earlier hypothesis that an alternative Mn^{2+} oxidation mechanism such as via NO_3^- reduction may be important in marine systems unless biological irrigation and /or resuspension bring Mn^{2+} in direct contact with O_2.

Introduction

On the surface of the earth, solar energy is captured in organic matter and free oxygen is released as a result of photosynthesis. This thermodynamically unfavorable process (increase in Gibbs free energy) provides energy for all the spontaneous, but slow redox reactions between the ultimate reductant, organic carbon, and the ultimate oxidant, free oxygen (1). The decomposition process of dead organic matter is the driving force for most elemental cycling in aquatic sediments. Organic matter (OM) decomposition reactions and other physical and biological changes that occur at surface sediments after deposition are called early diagenesis of sediments (2).

Redfield (3) has determined the average elemental ratio of C, N, P in living organic matter (marine phytoplanktons) as $(CH_2O)_{106}(NH_3)_{16}(H_3PO_4)$. During organic matter decomposition, oxygen is consumed and other elements are released according to the ratio, $O_2/C/N/P = 138/106/16/1$ (4, 5). However, sediment organic matter often has a higher C to N or C to P ratio due to the preferential loss of N and P during early diagenesis. A recent summary shows that lake sediments may have a wide range of C/N ration with an average about 12/1 (6). In addition, a small amount of S (with P/S ratio of 1/0.035, (7)) and trace amounts of metals are also released during organic matter diagenesis, but they are rarely accounted for in the OM stoichiometric formula. Sulfur and metals are brought into rivers, lakes and oceans mainly from the weathering of the land surface (other major sources to the ocean include hydrothermal and atmospheric inputs)(8).

While O_2 is the ultimate oxidant during OM decomposition, a number of intermediate oxidants (bound oxygen), including NO_3^-, MnO_2, FeOOH, and SO_4^{2-} are found to be important players of early diagenesis. These oxidants may be used to oxidize OM in a sequence according to the maximum yield of Gibbs free energy and the availability of the oxidants. (Oxidants are also termed electron acceptors and OM electron donors because of electron transfer from OM to the oxidants.) Table I summarizes these primary redox reactions. The general sequence and the stoichiometry of these reactions have been well developed for marine sediments (9, 10). Recently, this reaction series has been incorporated into comprehensive diagenetic models for marine sediments (*11, 12*) and for lake sediments (*13*). As given in Table I, aerobic respiration and methanogenesis will result in a porewater pH decrease due to the large release of CO_2 and MnO_2 and FeOOH reductions will result in a porewater pH increase due to the large increase in total alkalinity (TA). Denitrification may decrease pH slightly while sulfate reduction does not affect porewater pH greatly (i.e., it will buffer porewater pH value around 7.0 (*11, 14*)). The TA gain/consumption per mole of organic carbon decomposition for these reactions are –0.17, +0.93, +4.13, +8.13, +1.13 and +0.13 respectively.

Table I. Organic Matter Oxidation Reactions (Primary Redox Reactions) in Order of Decreasing Yield of Gibbs Free Energy

$OM + 138O_2 \Rightarrow 106CO_2 + 16HNO_3 + H_3PO_4$

$OM + 0.8 \times 106NO_3^- + 0.8 \times 106\ H^+ \Rightarrow 106CO_2 + 0.4 \times 106N_2 + 16NH_3 + H_3PO_4$

$OM + 2 \times 106MnO_2 + 4 \times 106\ H^+ \Rightarrow 106CO_2 + 2 \times 106Mn^{2+} + 16NH_3 + H_3PO_4$

$OM + 4 \times 106FeOOH + 8 \times 106\ H^+ \Rightarrow 106CO_2 + 4 \times 106Fe^{2+} + 16NH_3 + H_3PO_4$

$OM + 106/2SO_4^{2-} + 106\ H^+ \Rightarrow 106CO_2 + 106/2H_2S + 16NH_3 + H_3PO_4$

$OM \Rightarrow 106/2CO_2 + 106/2CH_4 + 16NH_3 + H_3PO_4$

NOTE: OM = $(CH_2O)_A(NH_3)_B(H_3PO_4)_C$ or $(CH_2O)_{106}(NH_3)_{16}(H_3PO_4)_1$ with a Redfield ratio. FeOOH is used to represent reactive Fe-oxides. H_2O is omitted.

In freshwater sediments, the redox sequences with depth are not necessarily the same. It appears that in Lake Sempach, central Switzerland, FeOOH, SO_4^{2-} and MnO_2 reductions occur simultaneously (*13*). In marine sediments, O_2 and SO_4^{2-} are the dominant oxidants whereas in freshwater systems O_2 respiration and CH_4 generation are dominant pathways for OM decomposition.

During this sequential OM decomposition, a number of reduced chemicals can be generated. These reduced species can then be oxidized when encountered with O_2 or other oxidants (e.g. MnO_2). Such reactions are called secondary redox reactions (*12*) and are summarized in Table II. It is seen from Table II most of the secondary redox reactions generate protons and thus should reduce porewater pH. An interesting feature of the secondary redox reactions is that the reduced species (electron donors) produced at depth in anoxic environments usually diffuse up to the more oxidized environments in surface sediments. These oxidized species can then serve as oxidants again for anaerobic respiration of OM; thus forming redox cycles. In addition to these redox reactions, precipitation and dissolution reactions also affect porewater pH (Table III).

Table II. Oxidation of Reduced Diagenetic By-Products (Secondary Redox Reactions)

$NH_4^+ + 2O_2 \Rightarrow NO_3^- + H_2O + 2H^+$

$Mn^{2+} + 0.5O_2 + H_2O \Rightarrow MnO_2 + 2H^+$

$Fe^{2+} + 0.25O_2 + 1.5H_2O \Rightarrow FeOOH + 2H^+$

$H_2S + 2O_2 \Rightarrow SO_4^{2-} + 2H^+$

$CH_4 + 2O_2 \Rightarrow CO_2 + 2H_2O$

$Fe^{2+} + 0.5MnO_2 + H_2O \Rightarrow 0.5\ Mn^{2+} + FeOOH + H^+$

$H_2S + MnO_2 + 2H^+ \Rightarrow Mn^{2+} + S^0 + 2H_2O$

$CH_4 + SO_4^{2-} + 2H^+ \Rightarrow H_2S + CO_2 + 2H_2O$

Table III. Precipitation and Dissolution Reactions

$Fe^{2+} + HS^- \Leftrightarrow FeS + H^+$

$Mn^{2+} + CO_3^{2-} \Leftrightarrow MnCO_3$

$Fe^{2+} + CO_3^{2-} \Leftrightarrow FeCO_3$

$CaCO_3 \Leftrightarrow Ca^{2+} + CO_3^{2-}$

The top few centimeters of sediments are often the zone of very rapid microbial mineralization of labile OM. O_2 is depleted quickly in the top few cm of sediments as a result of primary and secondary redox reactions (13, 15-18). However, oxidation of reduced species such as Mn^{2+}, Fe^{2+}, NH_4^+ and CH_4 by O_2 are a major reason for extremely sharp O_2 profiles that may exist in lake and coastal marine sediments (19).

An important result of the secondary redox reactions is the formation of a sharp pH minimum zone around the O_2 penetration depth as opposed to a broad pH increase in the zone below it. This very sharp pH minimum was largely unknown before several microelectrodes were applied. Revsbech and Jørgensen (20) demonstrated such sharp pH changes at the O_2/H_2S boundary in a microbial mat. Recently such a sharp pH minimum around the redox boundary was measured in a few coastal marine sediments (21-24).

In order to correlate the pH minimum to redox reactions, sensors are required for detecting redox species with 0.5 to 1 millimeter depth resolution or better. For example, the depth of Mn^{2+} appearance was often used to indicate the O_2 penetration depth in marine systems (25). However, it was found that the depth of Mn^{2+} appearance was 1.5 cm deeper than the O_2 penetration depth in the California Borderland Basin (15, 26). Due to the relatively coarse spatial resolution of porewater Mn^{2+} profiles, the issue whether the depth of Mn^{2+} appearance can indicate O_2 penetration depth was not resolved (17). In Lake Sempach, a porewater study using *in situ* peepers with only a depth resolution of

5-15 mm did not reflect the rapid processes occurring at the sediment surface (*18*). For example, the O_2 profile could not be discerned, and the depths of Mn^{2+} and Fe^{2+} appearance appeared to start at the sediment-water interface while other evidence indicated O_2 penetration to a depth of more than 10 mm (*13*).

Our understanding of biogeochemical processes near the sediment-water interface has been limited seriously by the ability to accurately measure such sharp gradients of porewater constituents with fine depth resolution without significantly disturbing the sediment environments during core recovery and processing. Such disturbances include temperature and pressure artifacts (*27*), gas (i.e., O_2 and CO_2) loss to or exchange with the atmosphere or N_2 gas in a glove bag, and other effects (*28*). The situation has improved greatly since the introduction of microelectrodes, which allow us to measure the concentration of solutes at a millimeter (or sub-mm) scale and *in situ* (O_2:(*29, 30*); O_2/pH/H_2S: (*31, 32*); O_2/pH/pCO_2:(*21*); O_2/Mn/Fe/HS: (*33*); new pCO_2: (*34, 35*)). For example, a voltammetry sensor for O_2/Mn^{2+}/Fe^{2+}/S(II) measurements became available only recently (*33, 36*). It soon became a powerful tool for redox chemistry studies in aquatic sediments, particularly when combined with pH microelectrodes (*23*). With this sensor, Luther et al. (*36*) showed clearly that O_2 and Mn^{2+} do not overlap in most marine sediments.

This paper will demonstrate the correlation between porewater pH profiles and redox reactions with examples from Lake Champlain sediments and sediments from a salt marsh creek bank at Sapelo Island. We will also compare the similarities and differences of diagenetic processes in freshwater and salt marsh systems.

Experiment section

Preparation of Microelectrodes

Gold amalgam glass electrodes were constructed as described in (*33*) with modifications to insure a waterproof seal. Briefly, the end of a 15 cm section of 4 mm-diameter glass tubing is heated in a small flame and the tip pulled to a diameter of less than 0.4 mm for a length of about 3-5 cm. A non-conductive epoxy is used as a fill. In this work, we used West System 105 epoxy resin and 206 hardener to form a high-purity, optical-grade, nonconductive fill. The epoxy is injected into the glass, which contained the gold wire that was previously soldered to the conductor wire of the BNC cable but which was not sealed at the tip. The epoxy has a moderate setting time (~1 hr) and drains slowly through the open tip. On setting, the epoxy seals the tip and the top end can be refilled with epoxy. Then the top end is coated with Scotchkote (3M) electrical coating and Scotchfil (3M) electrical insulation putty. After final setting of the epoxy, the tip is sanded and polished. Once cooled the excess glass is sanded away with 400 grit sandpaper on a polisher to expose the gold wire. Once constructed each electrode surface is polished and plated with Hg by reducing $Hg(II)$ from a 0.1 N Hg / 0.05 N HNO_3 solution, for 4 minutes at a potential of -0.1 V, while purging with N_2. The mercury/gold (Au/Hg) amalgam interface is conditioned using a 90-second -9 V polarization procedure in a 1 N NaOH solution. The electrode is then run in linear sweep mode from -0.05 to -1.8 V versus a Saturated Calomel Electrode (SCE) or Ag/AgCl electrode several times in oxygenated seawater to obtain a reproducible O_2 signal.

The preparation and properties of the neutral carriers-based PVC liquid membrane pH microelectrodes and the pCO_2 microelectrode using this pH microelectrode as an internal sensor are detailed in references (*34, 37, 38*). The tip diameter of the pH microelectrode is about 15 μm.

The pCO_2 microelectrode consists of an outer pipette with a gas-permeable silicone membrane in the very tip and an inner pH microelectrode positioned at the center of the outer pipette and at a distance of ~ 5-30 μm from the silicone membrane. The internal filling solution is 2 mM $NaHCO_3$ and 0.5 M NaCl. The tip diameter of the pCO_2 microelectrode is 100-250 μm. Three CO_2 gas standards (0.1%, 0.5% and 2.5% balanced with N_2) are used for calibration. Due to the high impedance of the microelectrodes and the electrical noises in an average lab, laboratory measurements of micro-profiles are normally conducted in a Faraday cage. A typical setting of pH and pCO_2 measurements is illustrated in Figure1.

The pH measurement has a standard deviation of 0.01-0.02 units while the pCO_2 microelectrode has a standard deviation of 3-5%. Evaluation of the electrode performance in marine sediments can be found in (*24*). Fine scale porewater DIC profiles calculated from pH and pCO_2 microelectrode *in situ* measurements compared well with the coarse scale DIC profile measured in extracted porewater (*24*). Comparison of a pH microelectrode with a glass mini-electrode showed that pH profiles measured by both electrodes were in reasonable agreement. However, the down core pH signal changes measured by the microelectrode were sharper than that by the mini-electrode. This was interpreted as a mixing of the sediment particles by the larger glass mini-electrode (tip diameter =2-3 mm) (*24*).

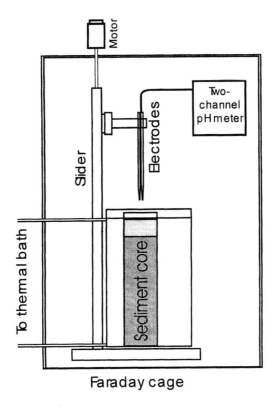

Figure 1. Illustration of microelectrode measurements in a Faraday cage. The motor and the slider can be controlled manually or by a program outside the Faraday cage. The two-channel pH meter is connected to a computer.

Site Description and Sample Collections

Lake Champlain is located in the Northwest of New Hampshire, USA. A study of the role of intermediate oxidants ($Mn(IV)/Fe(III)/NO_3^-/SO_4^{2-}$) on rates and pathways of carbon cycling was conducted at sites 19 (location: main lake, $44°28.26'$ N and $73°17.95$; water depth 100 m) and 21 (location: Burlington Bay, $44°28.49'$ N and $73°13.90$; water depth 15 m) in the middle part of the Lake in summer 1997. Bottom water temperature in site 19 and site 21 were 12.6 and 24.0 °C respectively. This paper concentrates on the results collected with microelectrodes. Two other papers (*39, 40*) will summarize the entire project including numerical modeling, rate and flux measurements, and laboratory manipulation of sediment cores with several electron acceptors.

In both sites, undisturbed sediments were collected by a non-contaminating box core system (AC-6, Fabau Inc., Massena, NY) operating from a boat. The 6"x6" corer box was subcored with a 4" inner diameter PVC corer. The cores were store in coolers filled with water from the site and were brought back to the lab at the Ecosystem Center, Marine Biological Laboratory at Woods Hole, MA. Cores were kept at *in situ* temperature. Porewaters were also collected using a whole core squeezer modified from (*41*). Porewaters were expressed directly into syringes and had no contact with air.

Salt marsh creek bank sediment cores were collected from a tidal creek near the University of Georgia Marine Institute on Sapelo Island, GA, southeastern US (see (*42*) for a description of the site and the area). This sample was visibly rich in Fe-oxides with yellowish coloration around roots and crab burrows. Another distinguishing feature of the creek bank sediment was severe bio-disturbance by fiddler crab activities and many crab burrow openings. The biogeochemistry of these sediments has been studied by several workers (see refs in (*42*)). Our microelectrode study was conducted when P. Van Cappellen and T. DiChristina conducted a biogeochemical and microbial study program in the same marsh-creek system (*43, 44*). Intact cores were collected and pH, pCO_2, O_2, Mn^{2+}, Fe^{2+} and S(-II) profiles were measured with microelectrodes immediately in the field within a few hours of sample collection.

Results

Lake Champlain

Several major redox species, pH and pCO_2 micro-profiles were measured in the Lake Champlain sediments during August 1997 (Figure 2 and Figure 3). O_2 was depleted to undetectable level (~1 μM) at around 6-mm depth. Mn^{2+} appeared right around the O_2 penetration depth. There was a slight spatial overlap between O_2 and Mn^{2+} profiles. Fe^{2+} increased slowly below the O_2 minimum depth and a spatial separation of a few mm between the appearance of Fe^{2+} and the O_2 penetration depth exists at both sites. The major difference between the two sites was the relative shape of the Mn^{2+} and Fe^{2+} profiles. For site 21 (15 m water depth and 24 °C), porewater Mn^{2+} concentrations were higher than the Fe^{2+} concentrations while for site 19 (100 m water depth and

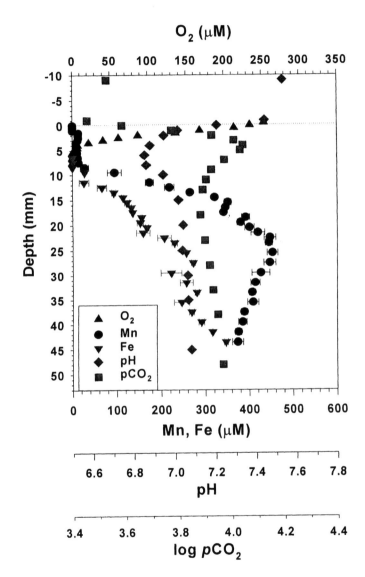

Figure 2. Profiles of redox species measured with a Au-Hg microelectrode (site 21, Lake Champlain).

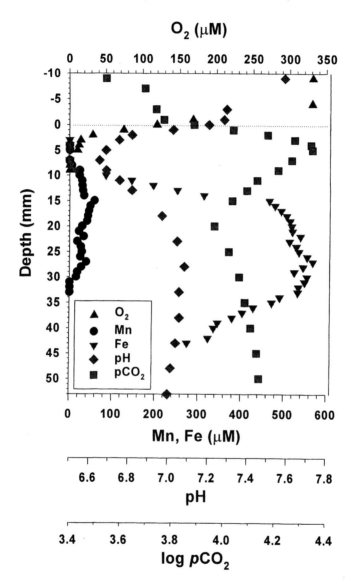

Figure 3. Profiles of redox species measured with a Au-Hg microelectrode (site 19, Lake Champlain).

12.6 °C) porewater Mn^{2+} profiles leveled off quickly at a concentration below 40 μM. However, porewater Fe^{2+} concentrations at site 19 were much higher than the Mn^{2+} concentrations and were also much higher than Fe^{2+} concentrations at site 21. The Fe^{2+} profile sharply decreased below 3.5 cm depth at site 19. S(-II) was not detected (<0.5 μM) in these sediments.

At the O_2 penetration depth, a pH minimum was seen at both sites. At site 21, a 0.65 unit decrease in pH occurred at a depth of only 8 mm. Corresponding to this sharp pH decrease, pCO_2 increased sharply over this depth zone. The increase in $logpCO_2$ is also about 0.60 logarithmic unit. At the deep site (site 19), the change in pH and $logpCO_2$ are even larger (0.9-1.1 logarithmic unit). Below the pH-minimum depth, pH increased gradually while pCO_2 decreased. DIC profiles increased from 900-1000 μM in the overlying water to 2100-2400 μM at 12 cm depth (39). The sharpest DIC increase occurred at the surface 2 cm (from 1000 to 1900 μM). Below 4 cm, the DIC increase became minor and gradual. OM content and HCl extractable Mn and Fe solids are abundant in Lake Champlain sediments (45).

Salt marsh creek bank

In the salt marsh creek bank sediment of Sapelo Island, O_2 was depleted to zero in less than 1 mm (Figure 4). Mn^{2+} increased sharply to 180 μM at less than 10 mm and then quickly decreased to less then 50 μM below 20 mm. Fe^{2+} was present below Mn^{2+}, and quickly surpassed Mn^{2+}, reaching a peak value as high as 300-400 μM. Total S(-II) was measurable below 20 mm but was only significant at depths greater than 80 mm.

In the creek bank sediment, a broad pH minimum and pCO_2 maximum occurred at depths between 10-30 mm depth (Figure 5 and Figure 6). The change is approximately over 1 unit for both pH and $logpCO_2$. The pH minimum value is around 6.4-7.0. As will be argued later, this broad pH minimum may be related to strong biological irrigation in this sediment.

Below 5 cm in the creek bank sediment , the porewater has TA and DIC values as high as 6000 μM (Figure 7). The lower values near the surface are partly a result of contamination by overlying water during the whole-core squeezing operation and partly a result of natural bio-irrigation. This was because the sediments were channelized by fiddler crab burrows and overlying water introduction was evidenced by a significant shrinking of the overlying water column during squeezing. Also significant heterogeneity was observed in TA and DIC for four cores collected in an area less than 30 m^2 of the creek bank.

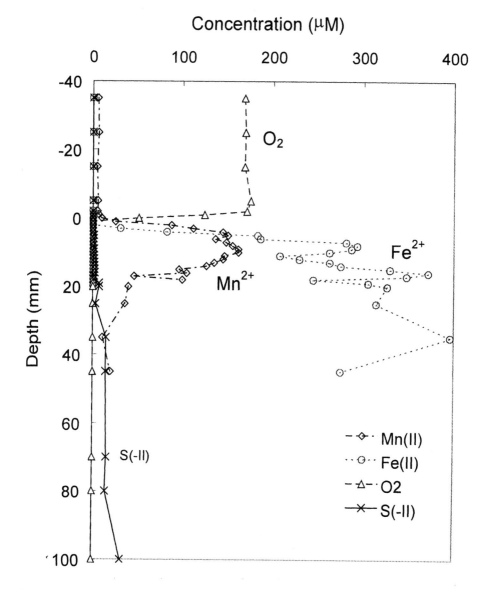

Figure 4. Redox species measured with a Au-Hg microelectrode on a sediment core taken from salt marsh creek bank of Sapelo Island in May 20, 1997. Measurements were conducted within a few hours of sample collection.

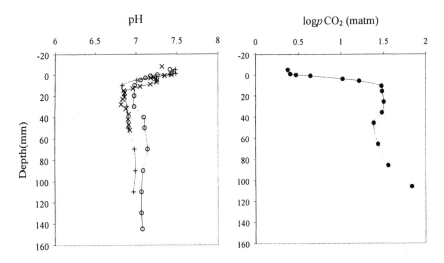

Figure 5. pH and pCO₂ profiles measured in May 1997 on sediment cores from a creek bank at Sapelo Island. Symbols with x and + were measured with pH microelectrodes. Symbols with open circles were measured with a glass pH mini-electrode.

Discussion

Diagenetic Reactions In Fresh and Salt Water Sediments

This is the first time such a fine scale and complete set of redox species, pH and pCO_2 profiles were measured together in either lake or salt marsh creek bank sediments, although redox species have been measured *in situ* together with pH at a coastal marine site (*23*). The good quality of the data is not only shown by the analytical standards set in earlier works (*24, 33*), but also because the data make sense geochemically. The distributions of redox species measured by microelectrodes are consistent with the sequence of diagenetic reactions (Table I). These results clearly demonstrate the power of microelectrode applications in the study of early diagenesis.

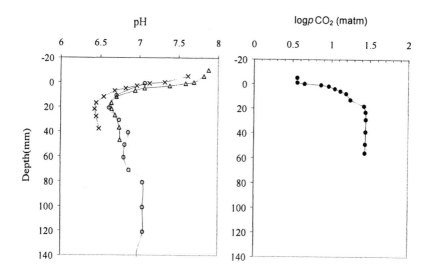

Figure 6. pH and pCO₂ profiles measured in August 1997 on sediment cores from a creek bank at Sapelo Island. Symbols with x were measured with pH microelectrodes. Open circles and squares were measured with a glass pH mini-electrode.

We have constructed a multi-component transport and reaction model similar to those of (*11, 12*) to interpret microelectrode measurements in Lake Champlain (*39*). While this section is mostly qualitative and is presented according to well known diagenetic reactions (Table I and Table II), some of the model conclusions are used here to facilitate the discussion. According to the diagenetic sequence, oxygen and NO_3^- should be depleted first (Table I) in both fresh and salt marsh sediments. When O_2 and NO_3^- are depleted, MnO_2 becomes the major oxidant. In Lake Champlain sediments, Mn^{2+} quickly increased to 300-400 μM at a depth of a few cm and then decreased as a result of $MnCO_3$ precipitation and the depletion of reactive MnO_2 (site 21). Simultaneously, but with a slight inhibition relative to MnO_2, FeOOH and SO_4^{2-} started to react with OM. SO_4^{2-} was depleted to nearly zero in less than 5 cm in Lake Champlain (*45*) and played an important role in OM oxidation (*39, 40*).

While FeOOH played a more important role than MnO_2 in OM oxidation, Fe^{2+} increased slowly in Lake Champlain sediment (site 21). As pointed out by Van Cappellen and Wang (*12*), MnO_2 oxidation of Fe^{2+} becomes a major source in generating Mn^{2+}. The initial gap between Mn^{2+} and Fe^{2+} profiles is caused by this reaction. FeS precipitation is less significant. $FeCO_3$ precipitation is also minor in Lake Champlain sediments. One distinct difference between lake and marine sediments is that FeS precipitation has a relatively small role in Fe cycling in lakes. This is because the S(-II) generating source ($[SO_4^{2-}]$) is limited in lake water. Instead Fe^{2+} in porewater is largely controlled by the reaction with MnO_2.

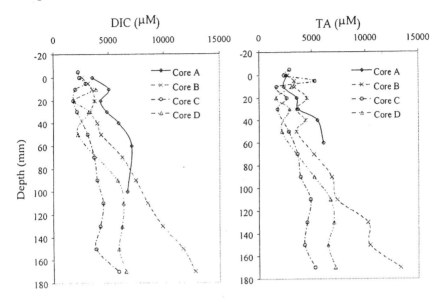

Figure 7. Porewater DIC and TA values in Sapelo Island salt marsh creek bank sediment.

The Lake Champlain results may be compared with the porewater data from Lake Sempach. In the deep site (50 m water depth), both Mn^{2+} and Fe^{2+} increased to 100-200 µM within a few cm below the sediment-water interface, although the relatively coarse scale depth resolution excluded a conclusion for the depths of Mn^{2+} and Fe^{2+} appearance in the porewater. However, Mn^{2+} and Fe^{2+} concentrations were relatively low in the littoral site. It has been recognized that pelagic lake sediments are enriched with OM and solid Mn and Fe oxidesdue to the focusing of resuspended fine sediments from nearshore to pelagic areas (*18, 46*).

In the salt marsh creek bank sediment, porewater Mn^{2+} concentration increased to around 180 µM at only 10 mm depth. The rapid decrease in Mn^{2+} concentration below that is most probably a result of $MnCO_3$ (or $CaCO_3+MnCO_3$) precipitation and the depletion of reactive MnO_2. The creek bank porewater has higher TA and DIC values and, therefore, a higher rate of $MnCO_3$ precipitation may be expected. Porewater Fe^{2+} concentration surpassed Mn^{2+} at a depth of only a few mm and continued to increase to 300-400 µM. Fe^{2+} leveled off below 5 cm as a result of both FeS and $FeCO_3$ precipitation. Due to the larger source of S(-II) in marine environments, FeS and pyrite will play a more significant role in the salt marsh than in the freshwater system. In addition, due to the higher TA, $FeCO_3$ precipitation in the creek bank sediment may also be more significant as compared to minor $FeCO_3$ precipitation in the Lake Champlain sediment. While FeS precipitation and pyrite formation might significantly reduce S(-II) concentration in the salt marsh porewaters (47), S(-II) was measurable below 2-cm depth and increased significantly below 8 cm, indicating the importance of SO_4^{2-} reduction as an electron acceptor for OM decomposition in salt marsh sediments.

Microelectrode measurements in the creek bank are consistent with microbial studies in the same salt marsh system. Lowe et al. (43) reported that culturable Fe-reducing bacterial (FeRB) density in the Sapelo salt marsh sediment was high and reached a maximum in the 3-4 cm depth. Culturable FeRB density decreased greatly below a depth of 6 cm and was not detectable below 10 cm, corresponding to a decrease in porewater Fe^{2+} concentration below 6 cm. In contrast, culturable sulfate-reducing bacterial (SRB) densities increased nearly 20-fold below a depth of 8 cm, corresponding to an increase of porewater S(-II) concentration. The microbial and geochemical depth profiles collected by Lowe et al. (43) indicate that sulfate reduction is the major OM degradation pathway at depths of 6-12 cm.

Another major difference in sediment biogeochemistry between salt marsh system and the Lake Champlain system is the timing of SO_4^{2-} reduction. In Lake Champlain, SO_4^{2-} reduction must occur simultaneously with FeOOH and MnO_2 reduction since porewater SO_4^{2-} concentration was depleted to nearly zero at 5 cm. However, in the creek bank sediment, SO_4^{2-} reduction leading to dissolved sulfide became significant only after reactive MnO_2 and perhaps even reactive FeOOH were largely consumed, as indicated by porewater Mn^{2+}, Fe^{2+}, and SO_4^{2-} profiles and by the densities of FeRB and SRB.

The Coupling of O_2 Consumption and Mn^{2+} Oxidation

One very clear conclusion from our data is that in Lake Champlain sediments there is no spatial separation between the O_2 penetration depth and the depth of Mn^{2+} appearance. Instead there is a partial overlap between O_2 and Mn^{2+} profiles. No such overlap is seen with the O_2 and Fe^{2+} profiles. This partial overlap suggests a strong coupling of Mn^{2+} oxidation and O_2 consumption, providing a mechanism for Mn^{2+}-MnO_2 recycling. Existing data from other lake sediments do not have the spatial resolution to illustrate this point (18). This result may be unique for lake sediments. Luther et al. (36) clearly showed the existence of a spatial separation of several mm between the O_2 penetration depth and the depth of Mn^{2+} appearance in marine sediments. They suggested an alternative redox pathway of Mn^{2+} oxidation by NO_3^- to interpret this spatial separation. A significant overlap of Mn^{2+} and O_2 profiles was assumed for marine sediments in the models presented by Van Cappellen and Wang (*12*). However, their assumption cannot be validated with field data due to the coarse scale depth resolution of the data.

The overlap of O_2 and Mn^{2+} profiles were determined by microelectrodes in the Sapelo Island and in a Delaware salt marsh sediment (*33*). We view the partial overlap of O_2 and Mn^{2+} profiles in salt marsh sediments to be the consequence of severe bio-disturbance and a highly compressed redox zonation rather than a fundamental difference between salt marsh and other coastal marine systems, where Luther et al. (*23, 36*) clearly showed no overlap between O_2 and Mn^{2+}. In the Sapelo Island creek bank, the O_2 penetration depth is only 0-1 mm. This indicates the existence of a highly compressed redox zonation when compared to the Lake Champlain system (6-8 mm) or those coastal marine sediments studied by Luther et al. (*23, 36*) or by Cai and Reimers (*21*) (6 mm in San Diego Bay). In salt marsh sediments, the alternative redox pathway suggested in (*36*) may still be operating, but simply has no time to completely oxidize Mn^{2+} before it diffuses molecularly and mixed biologically into the oxygenated zone. Molecular diffusion of 2 mm for Mn^{2+} is in the time scale of a few hours (i.e., roughly, Time $\approx$ Distance^2/(Diffusion Coefficient) = $0.2^2/(10^{-5})$ = 4×10^3 seconds).

Biological disturbance is very significant in this salt marsh creek site. The creek banks of the Sapelo Island salt marsh are densely populated by benthic macrofauna, particularly fiddler crabs (*48, 49*). The average density of burrow openings at the surface of creek bank sediments is 1040 m^{-2} (*48*). The bio-irrigation at this site has recently been studied by Van Cappellen's group (*44*). A fiddler crab was actively disturbing the sediment-porewater system when our profiles were being measured, and we believe that bio-disturbance (here irrigation and resuspension) is the main cause of Mn^{2+} export into the surface of sediment and the overlying water. Biological mixing was also important in the Delaware salt marsh sediment.

To investigate the existence of this mechanism, we conducted a laboratory experiment in which creek bank sediment was sealed in a plastic bag and kept anoxic in a refrigerator for a few weeks to kill all benthic animals. Then dead animals and plant debris were removed, and the homogenized sediment was incubated with overlying water from the site. Redox profiles were measured with the same type of Au-Hg microelectrodes. In this artificial sediment core, O_2 penetration depth was 1-2 mm while the depth of Mn^{2+} appearance at or slightly below 4 mm (Figure 8), with a clear separation of 2 to 2.5 mm between them. This spatial separation of O_2 penetration depth and depth of Mn^{2+} appearance suggests that the alternative oxidation mechanism of Mn^{2+} oxidation by another oxidant such as NO_3^- (*36*) for coastal marine systems may also work in salt marsh creek bank sediment.

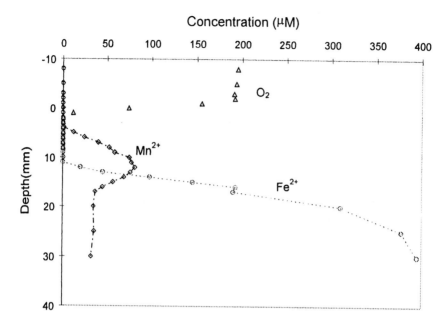

Figure 8. Redox species measured in an incubated salt marsh sediment core with no biological disturbance.

It is also interesting to note that Fe^{2+} appeared around 11 mm and quickly surpassed Mn^{2+} by several folds in concentration in the incubated creek bank sediment. The Fe^{2+} profile is similar in both incubated and field sediments except that the depth of Fe^{2+} appearance is much shallower in the field sediment, which is most probably also due to strong benthic animal activities in the field. Smooth profiles collected in the laboratory experiment also support the view that the fairly large "signal noises" seen in the field profiles are a result of biological disturbances of the sediment-porewater system.

A unique feature of the secondary redox reactions is that the reduced species diffuse from anoxic environments at depth to the more oxidized environments in surface sediments. The recycling of elements, such as Mn, Fe and S, between the oxidized and reduced states and between the solid and dissolved states has important geochemical consequences (*50, 51*). The redox front developed near the O_2 penetration depth can also influence the fate of other trace metals that may undergo redox reactions, form precipitates, induce complexation with reduced species, or strongly adsorb to oxides (*26, 52*). The importance of such processes and reactions has been observed by many, but is far from being well understood (*50*). Our microelectrode-based measurements and modeling exercises (*39*) demonstrate clearly the close coupling between cycles of various redox species.

The Coupling Between Redox Reactions and the pH Balances

As a result of CO_2 release and proton generation from the oxidation of OM, Mn^{2+}, NH_4^+ and other reduced species in the aerobic surface sediment, the pH profile decreased quickly below the sediment-water interface. Near the O_2 penetration depth, a pH minimum was reached. A total of 0.6-1.1 pH unit changes occured in only a few mm depth in Lake Champlain and in the salt marsh creek bank sediment. Below the pH minimum, pH increased gradually as a result of anaerobic oxidation of OM by MnO_2, FeOOH, and the oxidation of S(-II) by MnO_2. The precipitation of FeS, $MnCO_3$ and $FeCO_3$, and the oxidation of Fe^{2+} by MnO_2, however, decreased the pH. Finally, the CO_2 released during methanogenesis in lake sediments also lowered the pH at depth. This may represent a major difference in control mechanism of porewater pH between marine and lake systems.

A three-layer porewater pH structure is often seen in organic matter-rich coastal marine sediments: a pH minimum near the O_2 penetration depth, a pH maximum below the minimum, and another pH decrease at the bottom of the profile that is not always recorded (21-23, 53). Depending on the strength of various reactions and the sizes of the pH electrodes used, the sharpness of the measured pH variations may be different. The first layer (pH decrease) is in the oxic zone with a large amount of CO_2 generation from OM degradation and a large amount of proton generation from the oxidation of reduced species which diffused upward from depth. Model exercises also show that the oxidation of Mn^{2+}, Fe^{2+} and NH_4^+ is the main cause of large pH decreases in freshwater and coastal marine sediments while the release of CO_2 plays a smaller role in decreasing porewater pH (39). The second layer (pH increase) represents a proton consumption zone during anaerobic decomposition of OM. The relative importance of MnO_2 or FeOOH may be different in different systems. Porewater pH profiles reflect a delicate proton balance among various reactions of proton consumption (MnO_2 and FeOOH reduction, and S(-II) oxidation by MnO_2) and proton generation (Fe oxidation by MnO_2 and FeS and carbonate precipitation). In the third layer, proton generation due to carbonate precipitation and CO_2 formation by methanogenesis decreases again the porewater pH.

In Lake Sempach, the pH measured in the peeper-based porewater decreased continuously from 7.9 in the overlying water to 7.5 at about 8 cm. However, no minimum or maximum was measured due to the coarse scale depth resolution of the sampling method. The multiple-box model did not generate a pH minimum either because Mn^{2+}, Fe^{2+}, NH_4^+ and CH_4 oxidation in the oxic zone was not included in their model (13).

Therefore, in addition to the recycling of redox species, the other important geochemical consequence of the diffusion of reduced species from deep anoxic environments to the more oxidized surface sediments is the spatial separation of proton transfer reactions. This spatial separation of proton transfer reactions is often not discussed in the literature.

Summary

We report the first set of fine scale and multiple redox species profiles simultaneously with pH and $p\mathrm{CO_2}$ profiles that are measured together in any lake or salt marsh sediment. A partial overlap of $\mathrm{O_2}$ and $\mathrm{Mn^{2+}}$ profiles were measured, indicating a strong coupling of Mn recycling to the $\mathrm{O_2}$ consumption. No such overlap was seen between $\mathrm{O_2}$ and $\mathrm{Fe^{2+}}$ profiles in these sediments. The slight overlap of $\mathrm{O_2}$ and $\mathrm{Mn^{2+}}$ in the salt marsh creek bank sediment was a result of severe bio-disturbance. In an artificially incubated creek bank sediment devoided of bioturbations, a spatial separation of $\mathrm{O_2}$ and $\mathrm{Mn^{2+}}$ profiles was observed. This result supports the earlier hypothesis that an alternative oxidation mechanism of $\mathrm{Mn^{2+}}$ by $\mathrm{NO_3^-}$ may exist in marine systems that have no significant biological irrigation and/or resuspension.

Microelectrode-based profiles of $\mathrm{O_2}$, $\mathrm{Mn^{2+}}$, $\mathrm{Fe^{2+}}$, pH and $p\mathrm{CO_2}$ are geochemically consistent. The microelectrode data and model exercises (*39*) showed that the very sharp pH decrease was largely a result of $\mathrm{Mn^{2+}}$, $\mathrm{Fe^{2+}}$ and $\mathrm{NH_4^+}$ oxidation near the $\mathrm{O_2}$ penetration depth and partly a result of $\mathrm{CO_2}$ release during OM oxidation. Porewater pH microelectrode profiles provide a valuable diagnostic check on the diagenetic reactions.

Acknowledgements

Logistical supports from A. Giblin and J. Cornwell are greatly appreciated. Andrew Bono (Luther's lab) and Feng Lin (Cai's lab) helped with Au-Hg microelectrode works. Financial support is provided through NSF DEB-9612291 for W.-J. Cai and through NSF DEB-9612293 and NOAA Sea Grant # NA16RG0162-03 for G. Luther.

References

1   Stumm, W.; Morgan, J. J. *Aquatic Chemistry*, 2nd ed.; John Wiley and Sons: New York, 1981.

2   Berner, R. A. *Early Diagenesis: A Theoretical Approach*; Princeton University Press: Princeton, N.J., 1980.

3   Redfield, A. C. *Am. Sci.* **1958**, *46*, 206-226.

4 Richards, F. A. In *Chemical Oceanography*; Riley, J. P., Skirrow, G., Eds.; Academic Press: New York, 1965, pp 611-645.

5 Richards, F. A.; Cline, J.; Broenkow, W. W.; Atkinson, L. P. *Limnol. Oceanogr. (suppl.)* **1965**, *10*, R185-R201.

6 Guildford, S. J.; Hecky, R. E. *Limnol. Oceanogr.* **2000**, *41*, 1213-1223.

7 Mucci, A.; Sundby, B.; Gehlen, M.; Arakaki, T.; Zhong, S.; Silverberg, N. *Deep-Sea Res. II* **2000**, *47*, 733-760.

8 Chester, R. *Marine Geochemistry*; Chapman & Hall: London, 1990.

9 Froelich, P. N.; Klinkhammer, G. P.; Bender, M. L. *Geochim. Cosmochim. Acta* **1979**, *43*, 1075-1090.

10 Emerson, S.; Jahnke, R.; Bender, M.; Froelich, P. *Earth Planet. Sci. Lett.* **1980**, *49*, 57-80.

11 Boudreau, B. P. *Comp. Geosci.* **1996**, *22*, 479-496.

12 Van Cappellen, P.; Wang, Y. *Amer. J. Sci.* **1996**, *296*, 197-243.

13 Furrer, G.; Wehrli, B. *Geochim. Cosmochim. Acta.* **1996**, *60*, 2333-2346.

14 Ben-Yaakov, S. *Limnol. Oceanogr.* **1973**, *18*, 86-93.

15 Reimers, C. E.; Jahnke, R. A.; McCorkle, D. C. *Global Biogeochem. Cycles* **1992**, *6*, 199-224.

16 Archer, D.; Devol, A. *Limnol. Oceanogr.* **1992**, *37*, 614-629.

17 Cai, W.-J.; Sayles, F. L. *Mar. Chem.* **1996**, *52*, 123-131.

18 Urban, N. R.; Dinkel, C.; Wehrli, B. *Aquat. Sci.* **1997**, *59*, 1-25.

19 Canfield, D. E., B. B. Jorgensen, H. Fossing, R. Glud, J. Gundersen, N. B. Ramsing, B. Thamdrup, J. W. Hansen, L. P. Nielsen, and P. O. J. Hall *Mar. Geol.* **1993**, *113*, 27-40.

20 Jorgensen, B. B.; Revsbech., N. P. *Appl. Environ. Microbiol.* **1983**, *45*, 1261-1270.

21 Cai, W.-J.; Reimers, C. E. *Limnol. Oceanogr.* **1993**, *38*, 1776-1787.

22 Komada, T.; Reimers, C. E.; Boehme, S. E. *Limnol. Oceanogr.* **1998**, *43*, 769-781.

23 Luther, G. W.; Reimers, C.; Nuzzio, D. B.; D., L. *Environ. Sci. Technol.* **1999**, *33*, 4352-4356.

24 Cai, W.-J.; Zhao, P.; Wang, Y. *Mar. Chem.* **2000**, *70*, 133-148.

25 Sayles, F. L.; Livingston, H. D. In *Oceanic Processes in Marine Pollution*; I.W. Duedall, Kester, D. R., Park, P. K., Eds.; Robert E. Krieger Publishing Co.: Molabor, FL,, 1987, pp 175-195.

26 Shaw, T.; Gieskes, J. M.; Jahnke, R. A. *Geochim. Cosmochim. Acta* **1990**, *54*, 1233-1246.

27 Murray, J. W.; Emerson, S.; Jahnke, R. *Geochim. Cosmochim. Acta* **1980**, *44*, 963-972.

28 Schulz, H.; Zabel, M. *Marine Geochemistry*; Springer: Berlin, 2000.

29 Revsbech, N. P.; Sorensen, J.; Blackburn, T. H.; Lomholt, J. P. *Limnol. Oceanogr.* **1980**, *25*, 403-411.

30 Reimers, C. E. *Deep-Sea Res.* **1987**, *34*, 2019-2035.
31 Revsbech, N. P.; Jorgensen, B. B.; Blackburn, T. H. *Limnol. Oceanogr.* **1983**, *28*, 1062-1074.
32 Archer, D.; Emerson, S.; Reimers, C. E. *Geochim. Cosmochim. Acta* **1989**, *53*, 2831-2845.
33 Brendel, P. J.; Luther, G. W. *Environ. Sci. Technol.* **1995**, 751-761.
34 Zhao, P.; Cai, W.-J. *Anal. Chem.* **1997**, *69*, 5052-5058.
35 de Beer, D.; Glud, A.; Epping, E.; Kühl, M. *Liminol. Oceanogr.* **1997**, *42*, 1590-1600.
36 Luther, G. W.; Sundby, B.; Lewis, B. L.; Brendel, P. J.; Silverberg, N. *Geochim. Cosmochim. Acta.* **1997**, *61*, 4043-4052.
37 Zhao, P.; Cai, W.-J. *Anal. Chim Acta* **1999**, *395*, 285-291.
38 Cai, W. J.; Reimers, C. E. In *In-situ Monitoring of Aquatic System: Chemical Analysis and Speciation*; Buffle, J., Horvai, G., Eds.; John Wiley & Sons Ltd.: London, 2000.
39 Cai, W.-J.; Luther, G. W.; Cornwell, J.; Giblin., A. *in preparation* **2001**.
40 Giblin, A.; Cornwell, J.; Cai, W.-J.; Luther, G. W. *in preparation* **2001**.
41 Jahnke, R. A. *Limnol. Oceanogr.* **1988**, *33*, 483-487.
42 Cai, W.-J.; Pomeroy, L. P.; Moran, M. A.; Wang, Y. *Limnol Oceanogr.* **1999**, *44*, 639-649.
43 Lowe, K. L.; DiChristina, T. J.; Roychoudhury, A. N.; Cappellen, P. V. *Geomicrobiol. J.* **2000**, *17*, 163-178.
44 Meile, C.; Koretsky, C. M.; Van Cappellen, P. *Limnol. Oceanogr.* **2001**, *46*, 164-177.
45 Cornwell, J. *personal communication* **2001**.
46 Davison, W.; Heaney, S. I.; Talling, J. F.; Rigg, R. *Schweiz. Z. Hydrol.* **1980**, *42*, 196-224.
47 Howarth, R. W.; Giblin., A. *Limnol. Oceanogr.* **1983**, *28*, 70-82.
48 Basan, P. B.; Frey, R. W. In *Trace Fossils*; Crimes, T. P., Harper, J. C., Eds.; Steel House: London, 1977, pp 41-70.
49 Teal, J. M. *Ecolo.* **1958**, *39*, 185-193.
50 Aller, R. *J. Mar. Res.* **1994**, *52*, 259-295.
51 Howarth, R. W. *Biogeochem.* **1984.**, *1*, 5-27.
52 Brown, E. T., L.L. Callonne and C.R. German *Geochim. Cosmochim. Acta* **2000**, *64*, 3515-3525.
53 Zhao, P. Ph.D. Dissertation, University of Georgia, Athens, 2000.

Chapter 11

Probing Zinc Speciation in Contaminated Sediments by Square Wave Voltammetry at a Hg/Ir Microelectrode

Melissa A. Nolan and Jean-François Gaillard*

Department of Civil Engineering, Northwestern University, 2145 Sheridan Road, Evanston, IL 60208-3109

Square wave voltammetry at a mercury plated iridium (Hg-Ir) microelectrode was used to probe directly the chemical speciation of zinc in contaminated sediments. The electrochemically labile fraction of zinc was quantified in sediments from Lake DePue (IL), a lake heavily impacted by former zinc smelting activities. Eight samples were analyzed across a metal contamination gradient where total zinc concentrations ranged from 6,000 to 190,000 ppm and the corresponding peak currents from 0.1 to 62 nA. Under the analytical conditions chosen, the Zn^{2+}/Zn system behaved reversibly, as demonstrated by the forward and reverse current signals in square wave. The presence of the sediment had little effect on the electrochemical response, and Hg recovery experiments demonstrated that the mercury film remained on the electrode and that it was not fouled during analysis. An external calibration was performed using a background electrolyte that mimicked *in situ* conditions and in presence of a 10 mM TES buffer. The concentrations of labile zinc varied according to the locations and ranged from 0.8 to 9 ppm that represented between 50 to 90 % of the total dissolved zinc present in the porewater.

Introduction

The contamination of freshwater aquatic systems by metals is an important environmental problem because, unlike organics, these compounds cannot be degraded. However, depending on their chemical form, metals can be more or less harmful *(1-3)*. Sediments represent the ultimate repository of these elements and understanding how metals are transformed in these systems is fundamental for the delineation of remediation actions. In this case, the chemical speciation of metals is modified as a result of diagenetic processes that are mediated primarily by microorganisms. As a consequence, there is a growing need to monitor the speciation of metals *in situ* by means of appropriate analytical techniques such as voltammetry. This technique is powerful and versatile since it can analyze for trace level concentrations, can distinguish between weakly complexed or free metal ions (labile) and strongly bound metals (inert) - the latter fraction is thought to be of lesser environmental concern since it is most likely unavailable to organisms -, can be performed directly in the field, and when used in conjunction with microelectrodes allows investigation of chemical gradients.

Amongst the various voltammetric methods, square wave voltammetry (SWV) *(4-7)* is one of the most sensitive (DL 10^{-8} M) and rapid techniques (scans take only a few seconds). Other advantages of SWV are that it discriminates against background currents and the interference caused by O_2 reduction is reduced. The combination of SWV with microelectrodes is ideal for the analysis of complex environmental samples because microelectrodes are characterized by enhanced mass transport, leading to high current densities, and negligible influence of convective processes. In addition, since the charging current is reduced the signal to noise ratio is improved, allowing for better sensitivity. More important for freshwater systems low overall currents are obtained therefore allowing analysis to be performed in rather resistive solutions without adding a supporting electrolyte, i.e., modifying the sample's matrix. This limits sources of error and contamination, but more importantly reduces the possibility of altering the chemical speciation.

In electroanalysis, the most commonly used electrode is mercury. Routinely, the mercury is deposited on a substrate such as *e.g.*, carbon, gold, iridium, platinum or silver. However not all of these substrates provide the best operating conditions for a mercury electrode, since mercury can migrate into them or they can dissolve into the mercury resulting in the formation of inter-metallic compounds that perturb the signal. Carbon is often used since it is inert, but it has been shown to form unstable mercury films *(8)*. On the other hand, Ir is characterized by a low solubility (below 10^{-6} wt. %) and allows forming stable mercury hemisphere *(9-12)*. It is therefore one of the most suitable substrate.

Mercury plated iridium (Hg-Ir) microelectrodes have previously been used to analyze for metals in environmental samples. Tercier *et al.* used this microelectrode in combination with square wave anodic stripping voltammetry

(SWASV) to determine the concentrations of Cd and Pb in river water *(12)*, and Wang *et al.* measured U and Cr by adsorptive stripping methods. *(13)* More recently, Hg-Ir microelectrode arrays have been utilized to analyze a wide variety of environmental samples *(14-18)*. To prevent fouling and also allow the determination of the diffusible metal fraction only, agarose covered Hg-Ir microelectrodes were designed either in an array or single format *(19-21)*.

Voltammetric analysis on sediment porewater samples has focused on major redox species such as oxygen, iron, manganese and sulfide. Luther *et al.* has concentrated on porewater profiles of marine sediment cores. Initial experiments *(22)* were performed on cores that were analyzed aboard ship or in the laboratory using an Au/Hg microelectrode (diameter of 100 μm). Recently this technology was performed in situ in Raritan Bay, NJ. *(23)* Recently, Tercier-Weber *et al.* *(24)* have developed a sediment-water interface voltammetric in situ profiling (SIVIP) system that consisted of 64 lines of 3 iridium microelectrodes with diameters of 5 μm covered with an agarose gel. The SIVIP system was used in combination with SWASV at a liquid-liquid interface for lead determination *(25)*.

In this research the analytical performance of a single Hg-Ir microelectrode was assessed for studying zinc speciation in contaminated sediments. The specific goals of this research were to (1) determine if the presence of the sediment was detrimental to the electrochemical response, (2) quantify of the electrochemical response, and (3) compare the labile and total dissolved porewater zinc concentration in the sediment samples.

Experimental

Apparatus, Reagents, and Protocols.

Chronoamperometry, square wave voltammetry (SWV) and linear scan voltammetry were performed with a μAutolab Type II potentiostat/galvanostat (Eco Chemie BV). All voltammetric experiments were performed in a two-electrode system consisting of the Ir microelectrode (diameter = 7 μm) (Idronaut) and a double junction Ag/AgCl reference electrode (Orion) that also functioned as the counter electrode. All potentials are reported relative to the Ag/AgCl reference electrode. Flame atomic absorption spectroscopy (FAAS) was performed using a GBC 932. Cl^- and SO_4^{2-} concentrations were measured by capillary electrophoresis using a Waters Capillary Ion Analyzer. All analyses were performed in triplicate. No data manipulation was performed on any of the voltammograms shown and no preamplification was used.

All solutions were prepared using 18 MΩ de-ionized water from a Milli-Q RG system (Millipore Corporation). Zinc standard solutions were prepared from 99.999% zinc chloride (Aldrich), the mercury plating solution was prepared from 99.9995% $HgNO_3$ (Alfa Aesar), and trace metal grade nitric acid (Fisher). All other solutions were prepared with ACS grade reagents.

Mercury was deposited from an 8×10^{-3} M Hg^{2+} in 0.1 M HNO_3 solution at a potential of –0.4 V until 16.1 μC was reached. At the end of an experiment the mercury was removed by linear scan voltammetry, scanning from -0.3 to 0.3 V at a scan rate of 10 mV/s in 1 M KSCN stripping solution. Prior to each experiment, the Ir microelectrode was polished with 0.25 μm diamond paste (Buehler) and then sonicated for 5 minutes.

For the analysis of the samples and the calibration standards the following SWV parameters were used: initial potential –1.7 V, final potential –0.1 V, frequency 240 Hz, pulse height 25 mV, and step increment 2 mV.

Site Characterization and Sample Collection.

Sediment samples were collected from Lake DePue (DePue, IL) located about 170 km west of Chicago. More details about this lake and the extent of zinc contamination can be found in Webb *et al.* (26). In June 1999, the spatial distribution of total zinc concentrations in surficial sediments was established based on about thirty sediment samples collected over a wide area of Lake DePue (Figure 1A). Three different zinc regions were then defined: $[Zn]_T <$ 15,000 ppm, 15,000 < $[Zn]_T$ < 30,000 ppm and $[Zn]_T$ > 60,000 ppm.

For electrochemical analysis, sediment samples (Figure 1B, D1-D7) were collected near the three highest contaminated regions (S15, S28 and S33). A reference sediment sample was taken from the far end of the lake (D8 - low in zinc contamination). Samples were collected with an Ekman dredge and stored in plastic containers, on ice, and under N_2 atmosphere to maintain anoxic conditions. At all times extreme care was taken to insure that the sediments were not exposed to oxygen.

The samples were then brought back to the laboratory where they were handled in an anaerobic chamber. Electrochemical analyses were performed in 12 mL LDPE beakers closed with a tight lid that contained the two-electrode system. The bulk sediments were poured into the beaker and kept under a N_2 atmosphere for the analyses that were performed outside the anaerobic chamber. To determine the total dissolved zinc in the associated pore water, bulk sediment were placed, in the anaerobic chamber, in 50 ml tubes that were sealed and centrifuged at 4000 rpm for 5 minutes. The pore water was then collected by carefully withdrawing the supernatant that was then filtered through 0.45 μm, and preserved by acidification for further analysis by FAAS.

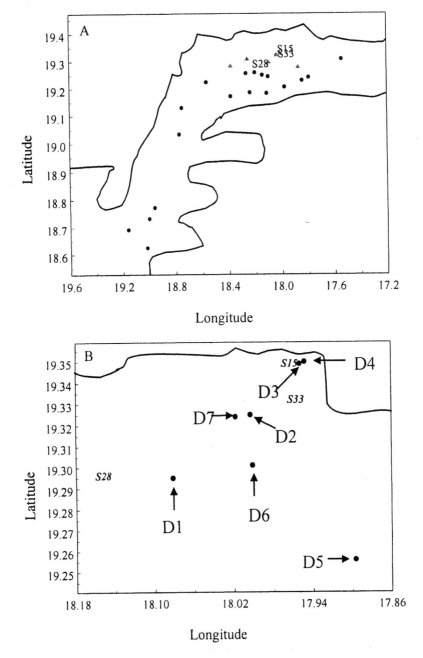

Figure 1 – (A) Zinc profile of the lake. (•) < 15,000 ppm, (▲) 15,000 to 30,000 ppm and (S15, S28, S33) > 60,000 ppm. (B) Location of electrochemical samples (D1-D7).

Results and Discussion

Zinc Response in the Presence of Sediment

Initial experiments were performed to determine if an electrochemical response could be obtained when placing the Hg-Ir microelectrode and reference electrode in direct contact with the sediment sample. Concerns were that the sediment might interfere with the mercury electrode possibly by knocking the mercury off the electrode surface, or that the adsorption of interfering species onto the mercury surface would affect the signal. Initially, the use of an Hg-Ir microelectrode in direct contact with sediment was tested by performing three consecutive scans in eight different samples (Table I lists the peak potential and current for the three consecutive measurements). The peak potential did vary slightly between sites (-1.016 V to -1.031 V), however it was very consistent within a site. Variations between sites are most likely due to changes in solution composition such as pH or ionic concentration. The peak potential ranged from 0.08 to 62.0 nA, and the relative standard deviation of each series of three measurements was below 10%, with the exception of sample D2 whose signal was the lowest. No zinc peak was observed in the reference sediment (D8).

Table I. Zinc Peak Potential and Current for Three Consecutive Runs

Sampling Site	Peak Potential (V)	Peak Current (nA)
D1	-1.030	0.601 ± 0.04
D2	-1.016	0.08 ± 0.02
D3	-1.028	2.02 ± 0.02
D4	-1.016	62.0 ± 0.7
D5	-1.031	0.20 ± 0.01
D6	-1.018	11.6 ± 0.2
D7	-1.031	0.47 ± 0.01
D8	No peak	

Voltammograms (Figure 2) demonstrated that peak positions were consistent between three consecutive runs, and that the peak current was reproducible. In addition, the peaks are Gaussian, as it is expected in classical SWV responses. Two additional peaks are observed for some of the samples. One at –1.6 V, which was observed in the voltammograms for all the sites, and that could be attributed to the Fe^{2+}/Fe system, and another at –0.6V that was only observed in D4 and that could be attributed to Cd. Figures 2 (A & B) present raw data without any filtering, showing that a low signal some noise was present.

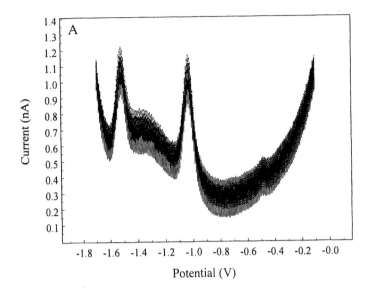

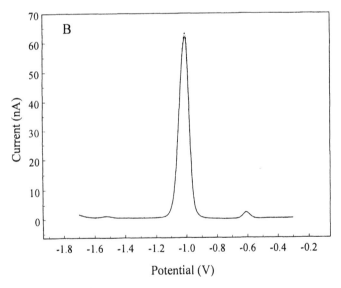

Figure 2. Three consecutive voltammograms for (A) D1 and (B) D. The voltammograms were acquired by performing anodic scans, without any electro-deposition step at the initial potential of – 1.7 V. Note the peaks at –1.6 V in (A) and – 0.6 V in (B) that could be attributed to Fe(II) and Cd respectively Given the level of Zn concentrations in the pore waters, the concentrations of Fe(II) and Cd are likely to be large.

As mentioned earlier, SWV provides the opportunity to examine the reversibility of the electrochemical system by observing at the forward and reverse currents. To obtain the best sensitivity the currents must be opposite of each other. In the case of a glassy carbon microelectrode Dewald *et al. (8)* showed that the zinc response was sensitive to instabilities in the mercury film. Consecutive measurements were affecting the reversibility of the system, leading to additional peaks and potential shifts. These variations were attributed to the morphology of the mercury film at the carbon electrode. In the case of the Hg-Ir microelectrode in direct contact with sediments, no comparable effects were observed indicating that a stable mercury film was obtained with the iridium substrate. The voltammograms (Figure 3) gave classical forward and reverse currents indicating that under the experimental conditions selected the reactions at the electrode surface were reversible.

When one compares the zinc peak current presented in Table I to the location where the sediment samples were obtained, the intensity of the peak current showed the following trend across the sites: D4 > D6 > D3 > D1 > D7 > D5 > D2 > D8. The highest peak current was observed for the sediment sample (D4), which was located the closest to the source of contamination (Figure 1B). The next two highest currents were observed at D3 and D6 located a little away from D4. As for the background reference sample (D8), located in a relatively low zinc contamination zone, no zinc peak was obtained. Low peak currents were obtained for four other sediment samples but were not quantified, since they were close to the method's detection limit.

Effect of Sediment on the Mercury Electrode

Since the initial experiments demonstrated that an electrochemical response was obtained with the electrode in direct contact with the sediment more extensive studies were performed to determine the stability and reproducibility of the mercury film. The two major concerns were that the mercury would come off the electrode or that the mercury surface would become fouled while in contact with the sediment. The influence of the sediment on the mercury film was examined by performing stability studies and by monitoring the deposition and removal of the mercury on the iridium electrode.

The influence of the sediment on the electrochemical response was investigated by performing analyses in presence and absence of sediment. After three consecutive scans with the electrode system directly in contact with the sediment the porewater was extracted by centrifugation and three additional scans were performed in the porewater (Figure 4). In the presence of sediment the average peak current was 47.5 ± 0.9 nA (RSD 1.9%) while in the porewater an average peak current of 46.2 ± 0.6 nA (RSD 1.3%) was obtained. The peak potential, its shape, and the reversibility of the system were similar in the two cases. This shows that the presence of sediment leads to negligible changes in the voltammetric signal. In addition, the diffusion of the zinc species towards

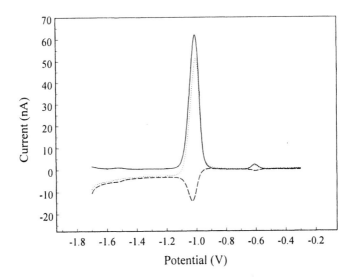

Figure 3. Net (solid line), forward (dotted line) and reverse (broken line) currents for D4.

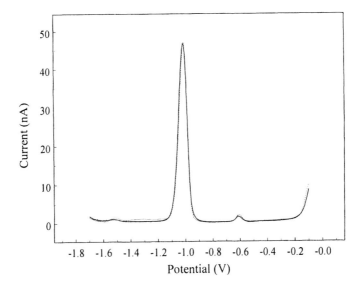

Figure 4. Electrochemical response in the presence (solid line) and absence (dotted line) of sediment.

the electrode was not affected by the presence of the sediment. Therefore, quantification of the zinc response in sediment samples is feasible.

The stability of the electrochemical response was investigated by performing twenty consecutive scans in two different sediment samples. A high and a low concentration sample were used to determine if interferences would influence one greater than the other. For the twenty consecutive scans at high and low concentrations, the current responses were respectively (D4) 61 ± 1 nA with a RSD of 1.6% and (D3) 1.9 ± 0.1 nA with a RSD of 5.3%. Overall, an excellent stability was observed over twenty scans indicating that the mercury surface was not being fouled over time.

The previous experiments provided indirect evidence that the mercury film was not altered when in contact with the sediment. This can be further verified by monitoring closely the deposition and recovery of Hg before and after measurements. After analysis, either in presence or absence of sediment, the mercury was removed by linear scan voltammetry in potassium thiocyanate, which strongly complexes mercury. Voltammograms for the removal of mercury in these two cases are shown in Figure 5. The resulting curves are essentially similar, the peak in the absence of sediment being slightly higher.

The amount of mercury on the electrode was quantified by calculating the percent recovery from the reduction and oxidation charge of the mercury. Since reduction charge was kept constant at 16.1 μC, variations in the percent recovery occurred due to variations in the oxidation charge, which depended on the mercury that was removed from the electrode after analysis. For eight analyses in the porewater only, the average oxidation charge was 14.2 ± 0.1 μC and the percent recovery was 88.3 ± 0.5%, whereas with sediment present the average oxidation charge was 13.9 ± 0.1 μC and the percent recovery was 86.3 ± 0.7%. The slight decrease in the oxidation charge when the electrode was in contact with the sediment could be due in part to the vigorous rinsing required to remove the sediment from the electrode before starting the Hg removal step. However, the difference in the percent recovery (2%) is minimal and demonstrates that the mercury remains on the electrode surface.

Quantification of the Electrochemical Response

Quantification of the electrochemical response is possible because the peak current is directly proportional to the analyte concentration. However, the electrochemical signal is also affected by the diffusion of the electrochemical species. Since it was previously shown that the electrochemical responses were identical in the presence and absence of sediment, processes at the electrode should remain similar in these two cases, and therefore the electrode response can be quantified by means of an external calibration.

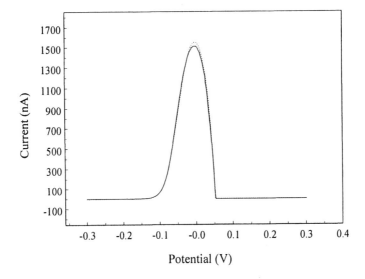

Figure 5. LSVs for the removal of mercury when analysis was performed in the absence (dotted line) and presence of sediment (solid line).

The use of calibration curves poses some difficulties because generally the electrochemical response depends upon the chemical composition of the background electrolyte. Part of the problem is overcome by the use of a microelectrode since the signal shows little dependence on variations in the ionic strength of the solution. However, the electrochemical signal is still dependent on other conditions such as pH and the chloride concentration. The porewater samples were analyzed for pH, chloride, and sulfate and the results are reported in Table II. External calibration curves were then obtained in solutions that mimicked the sediment pore water's chemical composition at each site, the pH being adjusted using TES partially neutralized HNO_3. From the data presented in Table II it is interesting to note that D4, which gave the highest zinc peak current and is the closest from the source of contamination, has the lowest pH and also has the highest sulfate concentration.

The raw calibration results, *i.e.,* normalized slope and regression coefficient, are given in Table III. Different concentration ranges were used for each site depending on the peak current that was obtained in the sediment sample. The calibration standards were chosen to bracket the peak current obtained in the sediment. Each calibration curve was obtained with five standard solutions,

Table II. Composition of the Sediment Sample

Sampling Site	pH	[Cl⁻] (mM)	[SO₄²⁻] (mM)
D1	7.63 ± 0.06	1.62 ± 0.06	0.744 ± 0.004
D2	7.57 ± 0.09	1.18 ± 0.03	
D3	7.53 ± 0.10	1.28 ± 0.01	1.03 ± 0.02
D4	7.11 ± 0.02	1.38 ± 0.01	5.4 ± 0.2
D5	7.49 ± 0.08	1.81 ± 0.02	
D6	7.43 ± 0.06	1.43 ± 0.02	
D7	7.53 ± 0.07	1.65 ± 0.01	1.68 ± 0.02
D8	7.44 ± 0.10	1.36 ± 0.02	

NOTE: The detection limit of sulfate was 0.0012 mM (S/N = 3).

which were run in triplicate. The calibration curves were then normalized for the amount of mercury that was removed from the electrode surface after the calibration was performed. This was effected because variations in the percent recovery occur with the deposition of each mercury film.

Table III. Calibration data for quantification of the Sediment Samples

Sampling Site	Zinc Concentration Range (ppm)	Normalized Slope (nA/ppm)	R^2
D3	0.5 to 2.5	2.8 ± 0.1	0.998
D4	5 to 25	8.3 ± 0.3	0.993
D6	1 to 5	12.6 ± 0.4	0.996

NOTE: The electrode was regenerated for every calibration.

To verify that the solutions used for the each calibration curve were actually mimicking the composition of the sediment samples their electrochemical signals were further compared (Figure 6). The shape and the width at half-peak height of the zinc peak were very similar and the peak potentials were identical. Hence, we can conclude that the background electrolyte solutions matched.

Comparing the Labile, Dissolved and Total Zinc Concentrations

Once the method established and tested, the primary goal of this research became to compare the concentration of labile zinc, as determined by electrochemical analysis, to the total dissolved porewater zinc concentration and establish whether there was any correlation with the total zinc concentration in the sediments. The results are presented in (Table IV).

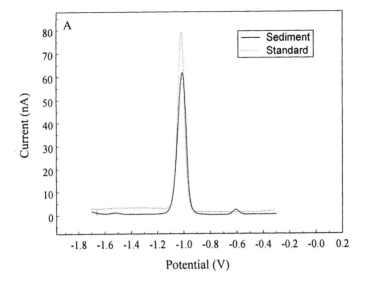

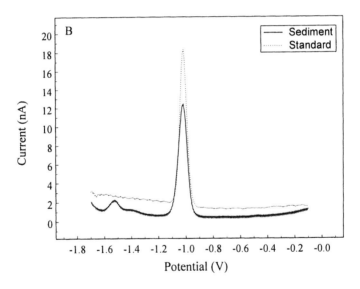

Figure 6. Electrochemical response for (A) D4 and (B) D6 solutions in the sediment sample (solid line) and in the calibration solution (dotted line). Calibration solution contains (A) 9.66 mM TES, pH 7.11, [Cl⁻] 1.41 mM, [SO₄²⁻] 5.39 mM and 9.14 ppm Zn²⁺ and (B) 10.1 mM TES, pH 7.44, and [Cl⁻] 1.47 mM and 1.98 mM Zn²⁺.

Table IV. The labile, dissolved and total zinc concentration for the eight sampling sites.

Sampling Site	Labile (ppm)[a]	Dissolved (ppm)[b]	Total (%)[c]
D1	DL	0.042 ± 0.004	2.02 ± 0.05 (20,200 ppm)
D2	DL	0.065 ± 0.003	4.12 ± 0.09
D3	0.84 ± 0.02	0.928 ± 0.002	9.40 ± 0.70
D4	9.0 ± 0.6	17.1 ± 0.3	19.0 ± 1.2
D5	DL	DL	1.30 ± 0.02
D6	1.59 ± 0.03	1.785 ± 0.007	2.66 ± 0.08
D7	DL	DL	2.90 ± 0.23
D8	No Peak	DL	0.62 ± 0.02

a: D1, D2, D5 and D7 were not quantifiable but a zinc peak was obtained.

b: D5, D7 and D8 were below the detection limit of AAS (~30 ppb).

c: one value in ppm, mg/kg dry sediment, is provided for comparison.

DL: Values are below or equal to detection limits

NOTE: All analyses were performed in triplicate.

The labile zinc concentrations are always lower than the total dissolved concentrations, indicating that a fraction of the zinc present in the porewater was strongly complexed. The concentrations of zinc followed the same trend for both the labile and dissolved concentrations: D4 > D6 > D3. However, the percents of labile zinc varied depending on the site; they were 53%, 89%, and 91% for D4, D6, and D3, respectively. At the site characterized by the highest total zinc concentration in either the bulk sediments or the porewater, the fraction of labile zinc is the lowest. The nature of Zn complexation at this site needs to be investigated in greater details to determine the processes that may control metal speciation. In any case, the above analyses clearly demonstrate the necessity for the development of direct analytical techniques and methods to establish the speciation of metals in contaminated sediments, if one is interested in their fate and impact on the overall aquatic system.

Conclusion

The three major objectives of this research were obtained, and it was demonstrated that the combination of SWV and the Hg-Ir microelectrode could be used reliably for assessing labile zinc concentrations in sediments. In the presence of sediment the electrochemical signal was reproducible and stable for twenty consecutive runs, the mercury film remained on the electrode and was not fouled, and, more importantly, there was no significant difference between the electrochemical response in presence or absence of sediment. Hence, the electrochemical signal can be quantified by means of an external calibration, in solutions that need to match porewater composition. The comparison of the labile, total dissolved, and total zinc concentrations indicated that significant zinc complexation was occurring in the highly impacted site. The present limitation of this method resides in its rather high detection limit (*ca.* 0.3 ppm for Zn), but this could be improved significantly by using a pre-amplifier.

Our future analytical objectives for the utilization of microelectrodes *in situ* are focused on three main areas: (1) improving detection limits, (2) expanding analytical possibilities to other metals such as Cd, Cu, and Pb, and (3) performing metal depth profiles. A preliminary experiment, performed on a core sampled near S33 (Figure 7) demonstrated that depth profiling was indeed possible with the Hg-Ir microelectrode. Calibration of the electrochemical signal becomes an issue in this case since pH, and the background electrolyte composition change as a function of depth in the core.

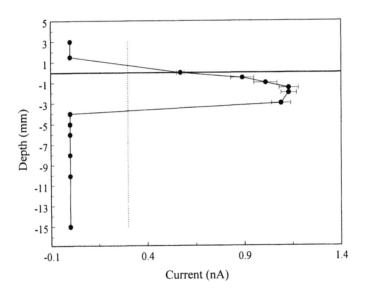

Figure 7. A preliminary depth profile for Zn in the sediments of Lake DePue. The vertical dashed line shows the current detection limit.

Acknowledgements

This work was supported in part by grants from the National Science Foundation (MCB-9807697) and the DOE-NABIR program. The authors would like to thank Amy Dahl, Heidi Gough and Ted Peltier for their assistance in obtaining the field samples, Dr. Deanna Hurum for helpful discussions, and two anonymous reviewers and the editors for their comments on the manuscript.

References

1. Florence, T. M. *Analyst* **1986**, *111*, 489-505.
2. Florence, M.; Batley, G. *Chem. Aus.* **1988**, 363-366.
3. Lund, W. *Fresenius J. Anal. Chem.* **1990**, *337*, 557-564.
4. Kounaves, S.; O'Dea, J.; Chandreselhar, P.; Osteryoung, J. *Anal. Chem.* **1986**, *58*, 3199-3202.
5. Osteryoung, J.; Osteryoung, R. *Anal. Chem.* **1985**, *57*, 101A-105A.
6. Osteryoung, J.; O'Dea, J. J. In *Electroanalytical Chemistry*; Bard, A. J., Ed.; Marcel Dekker, Inc.: New York, 1986; Vol. 14, pp 209-308.
7. Wikiel, K.; Osteryoung, J. *Anal. Chem.* **1989**, *61*, 2986-2092.
8. Petrovic, S. C.; Dewald, H. D. *Anal. Chim. Acta* **1997**, *357*, 33-39.
9. Kounaves, S.; Buffle, J. *J. Electroanal. Chem.* **1988**, *239*, 113-123.
10. Kounaves, S. *Platinum Metals Rev.* **1990**, *34*, 131-134.
11. Kounaves, S.; Deng, W. *J. Electroanal. Chem.* **1991**, *301*, 77-85.
12. Tercier, M.-L.; Parthasarathy, N.; Buffle, J. *Electroanalysis* **1995**, *7*, 55-63.
13. Wang, J.; Wang, J.; Tian, B.; Jiang, M. *Anal. Chem.* **1997**, *69*, 1657-1661.
14. Kounaves, S. P.; Deng, W.; Hallock, P. R.; Kovacs, G. T. A.; Storment, C. W. *Anal. Chem.* **1994**, *66*, 418-423.
15. Kovacs, G. T. A.; Storment, C. W.; Kounaves, S. P. *Sens. Actuators B* **1995**, *23*, 41-47.
16. Reay, R. J.; Flannery, A. F.; Storment, C. W.; Kounaves, S. P.; Kovacs, G. T. A. *Sens. Actuators B* **1996**, *34*, 450-455.
17. Herdan, J.; Feeney, R.; Kounaves, S. P.; Flannery, A. F.; Storment, C. W.; Kovacs, G. T. A.; Darling, R. B. *Environ. Sci. Technol.* **1998**, *32*, 131-136.
18. Belmont, C.; Tercier, M.-L.; Buffle, J.; Fiaccabrino, G. C.; Koudelka-Hep, M. *Anal. Chim. Acta* **1996**, *329*, 203-214.
19. Tercier, M.-L.; Buffle, J. *Anal. Chem.* **1996**, *68*, 3670-3678.
20. Tercier, M.-L.; Buffle, J.; Graziottin, F. *Electroanalysis* **1998**, *10*, 355-363.
21. Belmont-Hebert, C.; Tercier, M. L.; Buffle, J.; Fiaccabrino, G. C.; de Rooij, N. F.; Koudelka-Hep, M. *Anal. Chem.* **1998**, *70*, 2949-2956.
22. Brendal, P. J.; Luther III, G. W. *Environ. Sci. Technol.* **1995**, *29*, 751-761.
23. Luther, G. W. I.; Reimers, C. E.; Nuzzio, D. B.; Lovalvo, D. *Environ. Sci. Technol.* **1999**, *33*, 4352-4356.
24. Tercier-Weber, M.-L.; Buffle, J.; Confalonieri, F.; Riccardi, G.; Sina, A.; Graziottin, F.; Fiaccabrino, G. C.; Koudelka-Hep, M. *Meas. Sci. Technol.* **1999**, *10*, 1202-1213.

25. Tercier, M. L.; Buffle, J.; Koudelka-Hep, M.; Graziottin, F. In Electrochemical Methods for the Environmental Analyses of Trace Element Biogeochemistry, Taillefert, M.; Rozan, T., Eds. American Chemical Society Symposium Series; American Chemical Society: Washington, D. C., 2001, this volume, Chapter 2.

26. Webb, S. M.; Leppard, G. G.; Gaillard, J. F. *Environ. Sci. Technol.* **2000,** *34,* 1926-1933.

Chapter 12

Microsensor Studies of Oxygen, Carbon, and Nitrogen Cycles in Lake Sediments and Microbial Mats

Dirk de Beer

Max-Planck-Institute for Marine Microbiology, Celsiusstrasse 1,
D–28359 Bremen, Germany

Microsensors for O_2, pH, CO_2, NH_4^+ and NO_3^- are described. Measurements with these sensors in sediments and microbial mats were performed to investigate the interrelation of the oxygen-, nitrogen- and carbon cycles. The effect of light was studied in lake sediments, obtained from 1 and 2.5 m depth. In the dark O_2 penetrated ca 1 mm and in the oxic zone NO_3^- was formed. However, nitrate transport rates from the waterphase into the sediment was higher than the nitrification rates. Nitrification was restricted to the zone below the photosynthetic layer. In the presence of light, photosynthesis induced a threefold increase in O_2 penetration depth and a proportional increase in nitrification. In the light, NO_3^- was consumed by both denitrification and assimilation in the photic zone. NO_3^- assimilation was particularly high in the sediments from shallow parts of the lake that were low in NH_4^+ content. In the deeper parts of the lake NH_4^+ was used for assimilation. The NH_4^+ profiles did not reflect the nitrification rates, because NH_4^+ production and reversible binding to the sediment buffered NH_4^+. In microbial mats, CO_2 is depleted in the photosynthetic zone during illumination. The absence of nitrification in the top layer of sediments is explained by competition (between phototrophs and nitrifiers) for NH_4^+ and CO_2.

Introduction.

In aquatic systems microbial conversions mainly take place in the sediments, microbial mats and biofilms covering the sediments and solid surfaces. Conversions in the overlaying waterphase are often of less quantitative importance. Due to mass transfer limitations inside these structures, gradients of substrates and products are established, the concentrations of substrates being lower and the concentration of products higher than in the overlying water. The slope of the concentration gradients depends on the conversion rates and mass transfer rates. Significant changes can occur within 10 microns in highly active biofilms while in less active deep-sea sediments significant changes occur typically in millimeters to centimeters. In such systems porewater analysis has limitations. The extraction of porewater may influence the concentration profiles and in case of biofilms its spatial resolution is insufficient. The best technique available nowadays is the use of microsensors, needle shaped devices with a tip size of 1-20 μm which can measure the concentration of a specific compound. Highly localized measurements are possible, since the spatial resolution is approximately equal to the tip-size of the sensor. There are indications from theoretical and experimental studies that microsensors can influence the concentration profiles. The evidence is conflicting: while the theory predicts underestimation especially by sensors larger than 10 μm[1], experiments with O_2 microsensors in a biofilm showed an overestimation of local concentrations by sensors larger than 16 μm[2]. Microsensors may change the local concentrations by the consumption of substrate (only amperometric sensors), by compression of the local matrix [3], by changing the diffusion field (blocking diffusion by the sensor body) [1], or by compressing the boundary layer [4]. Although microelectrodes have a small influence on structures and processes, it is the best choice for direct concentration measurements inside mats and sediments.

Microsensors were introduced in microbial ecology by Bungay et al (1969) [5] who measured O_2 profiles in biofilms. The technique was improved by N.P. Revsbech who constructed reliable O_2 microsensors for profiling sediments and biofilms.[6, 7] More microsensors relevant for microbial ecology were developed and used, such as for N_2O [8], pH [9], NH_4^+ [10], NO_3^- [11-13], S^{2-} [14], H_2S [15], NO_2^- [16], CH_4 [17], Ca^{2+} [18] and CO_2 [19].

At the moment there are 4 principally different microsensor types available: amperometric, potentiometric, optical [20] and voltammetric [21-23]. While the first 3 categories are compound specific, with a voltammetric sensor a wide array of compounds can be measured. Here one amperometric (O_2) and four potentiometric (pH, CO_2, NO_3^- and NH_4^+) microsensors will be described.

The O_2 microsensor is based on current measurements induced by the electrochemical reduction of O_2 in the tip, with a rate proportional to its concentration. O_2 diffuses into a sensor through the silicon membrane in the 2-10 µm tip, and is reduced at a cathode close to the tip. The O_2 microelectrode is the oldest microsensor used in microbial ecology, and in its present form (with internal reference) the most reliable. This microsensor is described in detail elsewhere (7). Other useful sensors based on the amperometric principle are for N_2O, H_2S and HClO. O_2 microsensors have been used by many research groups to study photosynthesis and respiration in sediments (24-31), microbial mats (6, 14, 32, 33), biofilms (34-37), activated sludge flocs (38), marine snow (39-41), corals (42, 43) and foraminifera (27, 44).

The selectivity of amperometric sensors is based on the applied potential difference, permeability of membranes in the tip or specificity of redox mediators. For example, the potential of -0.8 V of an O_2 sensor is insufficient for reduction of N_2O which needs -1.2 V. The HClO sensor operates at 0.2 V, too low for O_2 reduction, while HClO is not sensed by an O_2 microsensor because it does not permeate through its silicon membrane. N_2O sensors are sensitive for O_2 and must be used in absence of O_2 or in a combined sensor in which O_2 cannot reach the N_2O cathode. The selectivity of the H_2S microsensor is based on the specific chemistry of ferricyanide as the redox mediator (15).

Also a cathode coated with an O_2 permeable resin has been used for O_2 microprofile measurements. The catalytic site of this sensor is in close contact with the sample, which should be low in calcium and magnesium ions, and have a pH above 6 (45). If these conditions are not fulfilled (sea water, acidic springs) the calibration will change gradually during the measurements. Despite these problems valuable measurements have been done in marine sediments with sturdy designs of this type (46).

Liquid membrane microsensors. The liquid ion-exchange (LIX) membrane microsensor technique was developed by cell physiologists for intracellular measurements (mostly H^+, CO_3^{2-}, Mg^{2+}, Ca^{2+}, Li^+, Na^+ and K^+) (47, 48). These sensors can be very small, with a tip diameter of less than 1 µm, i.e. the size of a bacterial cell. Relevant liquid membrane microsensors for use in microbial ecology are NH_4^+, NO_3^-, NO_2^-, H^+, Ca^{2+}, and CO_3^-.

The potentiometric microsensors can be divided into those with classical ion-exchanging membranes and carrier based membranes. The ion-exchanging membranes contain hydrophobic ions (ion-exchangers) and the analyte counter-ion. Carrier based membranes contain complexes between specifically binding hydrophobic agents (also called ionophores) and the analyte ion. The NH_4^+ and pH microelectrodes used in this study are examples of the carrier LIX electrode.

Classical ion-exchanging membranes are based on lipophilic salts (e.g. quaternary ammonium salt) that act as exchanger site. In these membranes complexation between the anions and cationic sites in the membrane phase is negligible, therefore, all anions are freely dissolved in the membrane. They all exhibit the same selectivity sequence with a preference for lipophilic over hydrophilic ions, following the Hofmeister series:

$$ClO_4^->SCN^->I^->NO_3^-> NO_2^->Cl^->HCO_3^->SO_4^{2-}>HPO_4^{2-}$$

The selectivity is determined by the distribution coefficient of the anions between sample solution and membrane. Although the membrane composition has an effect on the selectivity coefficients, the selectivity sequence remains unchanged. As consequence, the NO_3^- sensor is the best sensor (based on this principle) for environmental application. Also carbonate sensors have been described (49) for use in animal physiology. Due to their sensitivity to NO_3^- their application can be problematic in aquatic environments. Innovations for anion sensors based on ion-exchange membranes (deviations from the Hofmeister series) can only be expected from development of anion selective carriers.

The low selectivity of liquid membrane sensors for NH_4^+ towards Na^+ and K^+ and of those for NO_3^- and NO_2^- towards Cl^- prohibits measurements in seawater. Fortunately, NO_3^- can be measured in seawater with the recently developed biosensor (50). However, for studies in freshwater, liquid membrane sensors may be preferred because of the easy preparation.

The pH microsensor is the most selective liquid membrane sensor available and can be used in seawater and even hypersaline environments (51, 52). A good alternative is the full glass pH sensor (27, 47, 53, 54), however, the spatial resolution is not too good, due to the sensing area of ca 50 micron. The CO_2 microsensor is a special electrode, with a liquid membrane pH sensor as transducer (19). As it is a gas sensor, where the electrochemical parts are shielded from the environment by a gas permeable membrane, the sensor can be used in marine environments. Interference can occur by other gases that induce a pH change in the electrolyte. H_2S does interfere, consequently, the CO_2 profiles must be corrected using measured H_2S microprofiles. Ammonia and volatile fatty acids (acetate, propionate) do not interfere at near neutral pH (6-8).

New voltammetric microsensors. Recently new developments were obtained with microsensors based on voltammetry. The advantage of these sensors is that many chemical species can be analyzed with one single sensor. For detailed description of these sensors the reader is referred to the relevant chapters in this volume (55, 56)

In this study we report on O_2, NO_3^- and NH_4^+ laboratory measurements in sediment samples from two depths from an oligotrophic lake. In addition, O_2, pH and CO_2 profiles in a microbial mat are reported.

The measurements were aimed to compare O_2 uptake rates of different sites, and to understand the coupling of the benthic nitrogen cycle to mineralization and photosynthesis.

Methods

Study sites and sampling

Sediment samples were obtained in the spring from the oligotrophic Lake Veluwemeer (The Netherlands). The lake is shallow (up to ca 1 m) with 2.5 m deep channels for ship traffic. Cores with a diameter of 10 cm were taken by scuba diving from the bottom of a channel (Station 17) and from 1 m depth (Station 14). The sediments consisted of organic rich fine silt. The cores were used for measurements within 4 days.

The microbial mats were obtained from Solar Lake (Sinai) from a depth of 60 cm. After sampling they were transported to the laboratory in Bremen, and maintained at a salinity of 7.5% and a day-night light regime. Measurements were completed within 2 months.

Microsensor measurements and profile analysis

O_2, pH, CO_2, NO_3^- and NH_4^+ sensors were prepared and used as described earlier (*7, 10, 11, 16, 19*). The sensors were mounted on a motor driven micromanipulator. Positioning was done using a dissection scope, and the surface was used as the reference depth.

The cores were incubated at 18°C and stirred using an impeller (*57*), so that the flow velocity was ca 6 cm/s at 5 cm above the sediment. Air saturation was maintained by gentle aeration. The overlaying water was replaced daily with stored (2°C) lake water to prevent significant changes in NO_3^- and NH_4^+ concentrations. The microsensor measurements were done in homogeneous parts of the sediments, i.e. burrows were avoided.

A piece (5x5 cm) of microbial mat was embedded in agar in a flow cell and hypersaline water was recycled over the mat with a velocity of ca 5 cm/s (*32*). The measurements were done within one day.

Illumination was done by a Schott halogen light source with fiber optic connection. Dark profiles were recorded after at least 1 hour darkness.

The steady state profiles were assumed to be a function of combined diffusion and conversion processes. The local microbial activities were calculated from the concentration profiles with Fick's first law by use of a spreadsheet (*58, 59*). The diffusion coefficients used were for oxygen 1.8×10^{-9} $(m^2 s^{-1})$ for nitrate 1.3×10^{-9} $(m^2 s^{-1})$ and for ammonium 1.35×10^{-9} $(m^2 s^{-1})$ (*24, 57*). The concentration profiles were smoothed by averaging over 3 adjacent points, before calculating the local conversions. Conversions can be either consumption processes (rate values are positive numbers) or production processes (rate values are negative numbers) of a compound. The interfacial flux was calculated from the interfacial substrate gradient. Areal rates were calculated by summation of the local conversion rates after division by the thickness of the considered layer. Areal rates could thus be calculated for different depth intervals.

Results

Lake sediments

The profiles of each compound showed excellent reproducibility. In Figure 1 and 2 representative profiles measured in sediments of both stations are shown, incubated in dark and after 2 hours in the presence of light. The concentration profiles are given in the left panels, the corresponding local microbial activities are given in the right panels. The areal conversion rates, calculated from the profiles, reflect the importance of individual processes for the total budget and their distributions. It should be noted that the NH_4^+ and NO_3^- balances are not completely closed, due to the relatively low signals. The areal budgets are given in Tables I and II.

The profiles of O_2, NO_3^- and NH_4^+ were strongly influenced by light, in both the shallow and the deeper station (Figure 1 and 2). In the dark O_2 penetrated ca 1 mm, NO_3^- penetrated ca 2 mm in both stations. In both stations a clear NO_3^- production zone could be observed at ca 1 mm depth. NO_3^- was consumed at the sediment surface and in the anoxic zone. Whereas the O_2 and NO_3^- dark profiles were comparable, the NH_4^+ profiles were different in both stations.

Dark. In the shallow station 14 NH_4^+ diffused upwards from the deep sediment into the waterphase. A little NH_4^+ production was detected at the surface (areal rate 0.1 x 10^8 mol/m^{2}s). A high NH_4^+ production was calculated at the sediment surface of station 17 (areal rate 2 x 10^8 mol/m^{2}s). In this station NH_4^+ showed a peak at sediment-water interphase, from which it diffused both downwards and upwards. With the local microbial activity analysis no clear NH_4^+ consumption zones could be detected. The calculated areal rates confirm that the NO_3^- conversions seem much higher than the NH_4^+ conversions (Table I and II). In the shallow station, ca 1.2 x 10^8 mol/m^{2}s NO_3^- was formed, in the same zone only 0.08 x 10^8 mol/m^{2}s NH_4^+ was consumed. Here nitrification contributed significantly to the denitrification. In the deeper station, nitrification was lower (0.3 x 10^8 mol/m^{2}s) than the NH_4^+ consumption (1 x 10^8 mol/m^{2}s), however, the NH_4^+ profile clearly showed that the NH_4^+ concentrations were not in steady state, but determined by recent organic depositions. In both stations, less than 10% of the O_2 was consumed by nitrification.

Table I. Areal conversion rates (x 10^8 mol/m^{2}s) at station 14.

depth (mm)	O_2 dark	O_2 light	NO_3^- dark	NO_3^- light	NH_4^+ dark	NH_4^+ light
influx	**10**	**-30**	**1.3**	**3.2**	**-0.2**	**0**
0-0.75	10	-60	0.26	4.4	-0.1	-0.008
0.75-1.5		30	-1.2	-2.2	0.08	
1.5-5	---	0	2	1.5	-0.3	0.5
5-10	---	----	---	---	---	-0.4

TableII. Areal conversion rates (x 10^8 mol/m^{2}s) at station 17.

depth (mm)	O_2 dark	O_2 light	NO_3^- dark	NO_3^- light	NH_4^+ dark	NH_4^+ light
influx	**14**	**-21**	**2.1**	**0.14**	**1.1**	**0.03**
0-0.75	15	-48	1.3	0.25	-2	-0.05
0.75-2		25	-0.3	-1	1.5	
2-5	---		0.7	1.5		0.6
5-10	---	----	---	---	---	-0.5

NOTES: Negative numbers indicate production, positive consumption, --- relevant compound is absent.

Influx is the flux measured in mass boundary layer above sediments.

Areal rates are calculated from the profiles in Figure 1 and 2.

Vertical line indicates that areal conversion rates (on the right) are averaged over more depth intervals. The criteria for the choice of intervals were based on the main zones of activity, i.e. where the conversion rates were either positive or negative.

Light. Upon illumination, new steady state O_2 and NO_3^- profiles were obtained within 30 minutes, but NH_4^+ profiles were still changing after 2 hours of illumination. Photosynthesis occurred in both stations, at comparable maximum volumetric (1-1.5 x 10^{-4} mol/m^{3}s) and areal rates (48-60 x 10^{-8} mol/m^{2}s), and depths (0-1 mm). Approximately half of the produced O_2 was exported to the waterphase, the rest was used for sediment respiration. The effect on the nitrogen compounds was different depending on the site.

-    In the shallow site 14 a strong NO_3^- concentration decrease occurred at the surface. The surfacial consumption rate was close to 1 x 10^{-4} mol/m^{3}s (4.4 x 10^{-8} mol/m^{2}s areal rate in the top 0.75 mm). At 1mm depth a peak was observed due to nitrification, followed by a gradual decrease to zero.

234

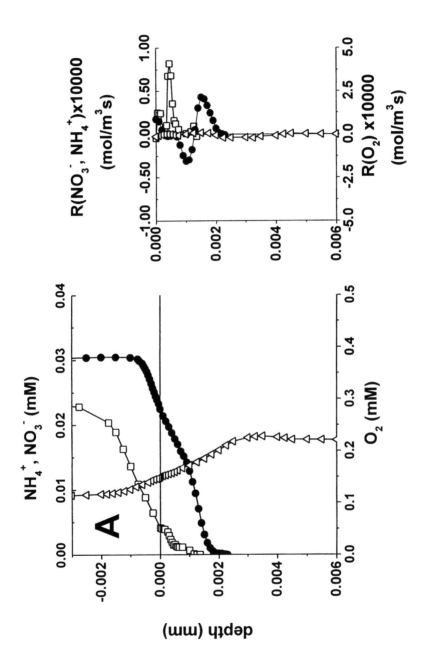

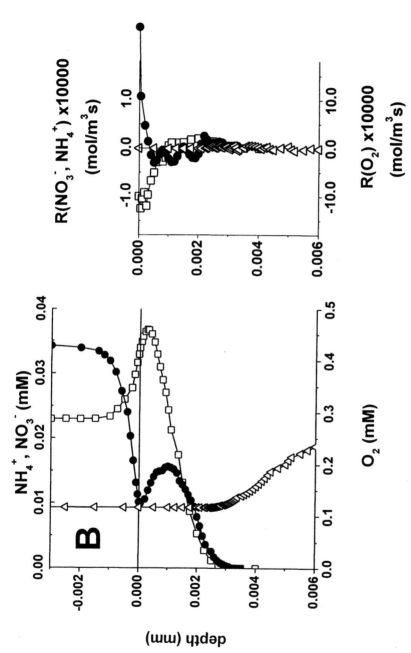

Figure 1. Microprofiles of O_2 (□), NO_3^- (●) and NH_4^+ (△) in sediments from Station 14 (shallow site). (A) dark incubation, (B) after 2 hours illumination. The concentration profiles are in the left panels. The right panels represent the volumetric conversion rates, calculated from the profiles.

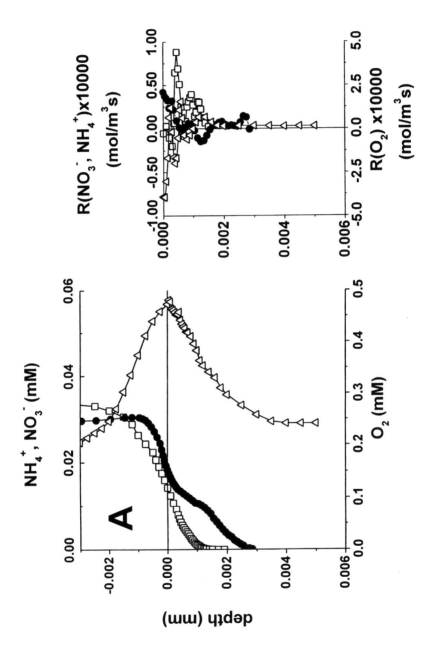

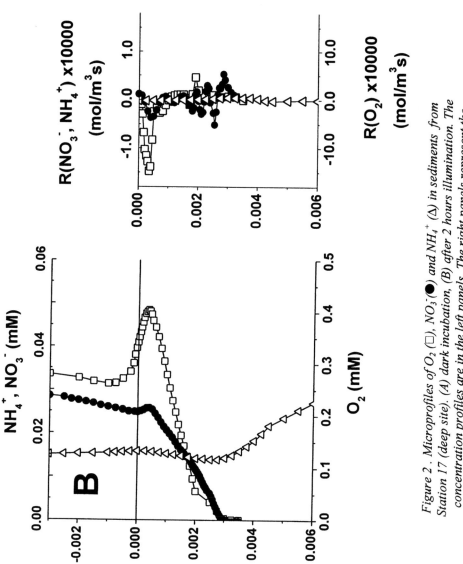

Figure 2 . Microprofiles of O_2 (□), NO_3^- (●) and NH_4^+ (△) in sediments from Station 17 (deep site). (A) dark incubation, (B) after 2 hours illumination. The concentration profiles are in the left panels. The right panels represent the volumetric conversion rates, calculated from the profiles.

- In the deeper station 17 some surfacial NO_3^- uptake was observed upon illumination, however less than 10% of the uptake at the shallow site (maximum volumetric rate 0.1×10^{-4} mol/m^{3}s, areal rate 0.25×10^{-8} mol/m^{2}s). The areal nitrification rate was lower than in the shallow station.

In both stations, NO_3^- and O_2 penetrated equally deep during illumination and the NO_3^- consumption in the deeper layer (2-3 mm) coincided with the O_2 consumption. During illumination of both sediments NH_4^+ was withdrawn gradually into the sediments. The intermediate states were recorded for ca. 2 hours, after which still no steady state was observed. The NH_4^+ consumption or production rates estimated from the profiles were very low, 1-2 orders of magnitude lower than the local NO_3^- production rates.

In summary, in both stations the O_2 profiles were similar and responded similarly to light. The deeper station had a higher NH_4^+ concentration probably due to mineralization at the surface. The NO_3^- profiles were similar in the dark, but responded different to light: only in the shallow station nitrate was consumed in the photic zone. NO_3^- production was lower in the deeper station, although NH_4^+ was higher.

Microbial mats

The O_2, pH and CO_2 profiles were profoundly influenced by light (Figure 3). In the dark O_2 penetrated 250 micron, and high CO_2 concentrations were observed. The pH showed a gradual decrease with depth. In the photosynthetic zone during illumination the O_2 concentration increased to supersaturation, the CO_2 concentration dropped to 0 and the pH went up to 9. However, below 0.5 mm, O_2 decreased and CO_2 appeared again.

Discussion

Sediments

The profiles allowed calculation of the local conversion rates with high spatial resolution, as given in Table I and in Figure 1 and 2. It may not be directly evident that a concentration profile is a resultant of consumption and production. For example the nitrate profile in Figure 1A rather gradually decreases with depth, however, the rather subtle changes in concentration gradients are caused by a consumption in the top of the sediment, a production at ca 1 mm depth and a consumption deeper in the sediment. The spatial resolution

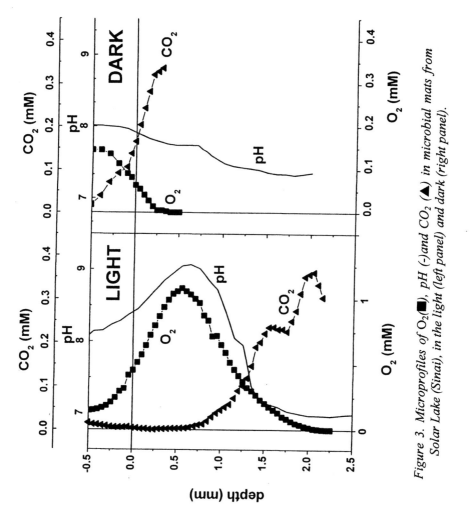

Figure 3. Microprofiles of O_2 (■), pH (-) and CO_2 (▲) in microbial mats from Solar Lake (Sinai), in the light (left panel) and dark (right panel).

is practically limited, besides by the measurement time, the tip size of the sensor. The spatial resolution is approximately 2 times the tip size of the sensor. It is obvious that porewater analysis using conventional extraction techniques has a far to low spatial resolution to distinguish concentration differences on the scale of less than 1 mm.

The microbial activities calculated from the microprofiles are of the same order of magnitude as reported values from eutrophic lakes in The Netherlands. From nitrogen budgets it was calculated that the nitrification in the oxic sediment layers was $0.6\text{-}1.2 \times 10^{-8}$ mol/m^{2}s, while nitrate diffusing in from the overlying water was 0.9×10^{-8} mol/m^{2}s, during the circulation period (60). Nitrate fluxes over the sediment-water interface in another lake near our sampling site ranged from $0.4 - 0.8 \times 10^{-8}$ mol/m^{2}s (61).

The dark profiles of NO_3^- and O_2 were similar in both stations: O_2 is respired within 1 mm, NO_3^- diffuses to 2 mm depth, where it is consumed by denitrification. NO_3^- sources for denitrification are the overlying water and nitrification inside the sediment. As the NO_3^- production in the oxic zone can only originate from the nitrification of NH_4^+, it is surprising that the NH_4^+ profiles did not clearly reflect this. The NH_4^+ consumption rates, calculated from the profiles, are orders of magnitude lower than the NO_3^- production rates. The discrepancy between NO_3^- and NH_4^+ profiles was observed in all measurements. Possibly, NH_4^+ formation by degradation of organic matter is balancing the NH_4^+ consumption by nitrification. Furthermore, it is known that NH_4^+ binds in large amounts to organic matter and can also be released (62). It is thus likely that the NH_4^+ concentration is well buffered in these organic rich sediments. Thus the measured NH_4^+ profiles reflect only a blurred image of the actual NH_4^+ distribution.

The NH_4^+ peak found in station 17 reflects a fast degradation of freshly deposited organic matter at the sediment surface. This organic material had obviously accumulated in the deeper parts of the lake. In the shallower parts of the lake most NH_4^+ was formed at a depth of 3 mm, but at much lower rate than in station 17. Only a little ammonification was observed at the surface. Probably sedimenting organic material preferentially accumulates into the deeper parts of the lake.

Photosynthesis occurred at similar rates in both stations. The effect of light on the NO_3^- profiles was complex and interesting. In the sediment from the shallow parts of the lake (Figure 1B) a rapid uptake of NO_3^- in the photic zone occurred, obviously for assimilation by photosynthetic organisms (maximum rate ca 2×10^{-4} mol/m^{3}s). In sediments from the deeper parts of the lake (Figure 2B) a much less pronounced response was observed in the NO_3^- profile. Only a small decrease was observed at the surface, resulting from a maximum uptake rate of 0.1×10^{-4} mol/m^{3}s. The explanation for this difference is that in the deeper sediments more NH_4^+ was available in the photic zone, which is preferred over

NO_3^- for assimilation. In the light the NH_4^+ peak disappeared. The NH_4^+ production at the top of the sediment continued and supplied sufficient N-source for assimilation. It should be noted that the NO_3^- assimilation rates in the light at station 14 are close to the NH_4^+ production rates at station 17. Thus at station 17 NH_4^+ was used for assimilation, at station 14 mainly NO_3^- was used.

As expected, in the presence of light the NH_4^+ concentrations decreased both by photostimulated assimilation and enhanced nitrification. The NH_4^+ profiles responded to light, but much slower than the O_2 and NO_3^- profiles. This can only be explained by strong buffering of the local NH_4^+ concentration by organic matter. Surprisingly, the NH_4^+ concentrations did not decrease below 10 μM, even in the zones where NH_4^+ was expected to be taken up at high rates (in the nitrifying zones and in the photo-assimilation zone). Again, this can be attributed to the NH_4^+ buffering capacity of the sediment.

Similar effects of light on NO_3^- profiles have also been found in more oligotrophic sediments (63). These authors also found that nitrification was absent in very top layers of the sediment, but rather could be found near the oxic anoxic interphase. They hypothesized that nitrifiers competed for NH_4^+ that diffused upwards from the anoxic zones. Unfortunately, they did not measure NH_4^+ profiles. In our experiments, NH_4^+ was present in all sediment layers and was seemingly not limiting nitrification. Actually, in the sediment from station 17, where the highest NH_4^+ concentrations were measured, nitrate formation by nitrification was lower (Table I and II). Nitrification was partially limited by O_2 availability, in the deeper parts of the sediment, which was alleviated by photosynthesis. In the top 0.5 mm of the sediment no nitrification occurred.

It should be noted that the local concentrations are a function of many processes: ammonification, nitrification, denitrification and aerobic mineralization are the processes considered. It was until recently assumed that nitrification is a typical aerobic process and denitrification exclusively anaerobic. In the last years it has become evident that nitrification, thought to be a purely aerobic process may be driven Mn-oxides in the absence of O_2 (64, 65). Also, ammonium can be anaerobically oxidised by nitrite (66, 67). Furthermore, aerobic denitrification was demonstrated (68-70). Thus the division between aerobic and anaerobic processes is blurred, and it is well possible that nitrification and denitrification occur simultaneous in both oxic and anoxic layers. This complicates the budgetting of the N-cycle from analysis of porewater profiles. From Table I and II it is clear that the oxygen fluxes and conversions are much higher than those of the nitrogen compounds. Thus the contribution of nitrate to mineralization of organic matter seems limited. In the dark the sum of nitrate influx and production in the sediment is ca 2-3 x 10^{-8} mol/m²s, which is ca 15% of the oxygen input. In the light, the nitrate turnover is ca 1-5 x 10^{-8} mol/m²s (depending on the station), which is less than 10 % of the oxygen turnover. As discussed previously, the porewater ammonium concentration cannot reflect the actual available ammonium.

Our experiments do not enable conclusions on the presence or absence of anaerobic nitrification by Mn oxides. The nitrate profiles do not show anaerobic nitrate production. If it occurred the denitrification exceeded the anaerobic nitrate production. Anaerobic ammonium consumption was not observed either. If it occurred, it was lower than the ammonification. The anaerobic oxidation of ammonium by nitrite may well be significant in gradient systems, thus in sediments where nitrate/nitrite and ammonium profiles overlap. However, in these sediments this process can be active only at night, as it is inhibited by oxygen that penetrates as deep as nitrate during illumination. The apparent discrepancies between the NH_4^+ and NO_3^- fluxes in our data-set cannot be explained by this process. The microsensor measurements demonstrate that denitrification can occur in the presence of oxygen, as in the light nitrate and oxygen have the same penetration depth (Figure 1B and 2B). In the dark the 'normal' nitrate distribution was found, where nitrate penetrates deeper than oxygen, and is only consumed in the anoxic zone.

Mats

The mats were photosynthetically highly active. O_2 levels reached over 5 times airsaturation, the pH went up with 1 unit and CO_2 was depleted. Although CO_2 is the preferred substrate, it is known that most cyanobacteria can also use HCO_3^- for carbon fixation, which is indeed needed in the absence of CO_2 to explain the observed O_2 production. Most likely, photosynthesis is not limited by the availability of inorganic carbon, but by the high O_2/CO_2 ratio which switches the microbial activity of RuBisCo to photorespiration.

Competition between autotrophs

Autotrophic growth involves the fixation of CO_2 to biomass. Both phototrophs and nitrifiers are autotrophs. Nitrifying organisms and phototrophs compete for at least 2 substrates: CO_2 and NH_4^+. It is probable that the affinity of the photosynthetic organisms for NH_4^+ is higher than that of nitrifying organisms. The lowest reported K_m (the Michealis-Menten saturation constant) for nitrifiers is 40 µM (71, 72), while a K_m of 2 µM was reported for phototrophs (73). Thus phototrophs can outcompete nitrifiers for NH_4^+. CO_2 is fixed by both phototrophs and nitrifiers by the Calvin-Bassham-cycle, i.e. by the action of RuBisCo (74). During illumination, the CO_2 concentration is probably zero in the photosynthetic zone (19), as most benthic algae possess an efficient CO_2

uptake system (75-78). Unfortunately, CO_2 measurements were not done in the lake sediment, but it can be assumed that CO_2 was depleted in the photic zone, as measured in microbial mats. No CO_2 accumulating mechanism or bicarbonate uptake has been reported for nitrifiers. It is, therefore, unlikely that nitrifiers can compete under light conditions with the phototrophs.

Conclusions

Phototrophs compete with aerobic nitrifiers for ammonium and CO_2. In dense communities, the nitrifiers will loose this competition because of lower substrate affinities. The competition for CO_2 and nutrients between different autotrophic populations should be investigated further.

References

1. Albanese, R. A. *J. Theor. Biol.* **1973**, *38*, 143-154.
2. Zhang, T. C.; Bishop, P. L. *Wat. Res.* **1994**, 2279-2287.
3. Muller, W.; Winnefeld, A.; Kohls, O.; Scheper, T.; Zimelda, W.; Baumgartl, H. *Biotechnol. Bioeng.* **1994**, *44*, 617-625.
4. Glud, R. N.; Gundersen, J. K.; Revsbech, N. P.; Jørgensen, B. B. *Limnol. Oceanogr.* **1994**, *39*, 462-467.
5. Bungay, H. R.; Whalen, W. J.; Sanders, W. M. *Biotechnol. Bioeng.* **1969**, *11*, 765-772.
6. Revsbech, N. P.; Ward, D. M. *Appl. Environ. Microbiol.* **1983**, *45*, 755-759.
7. Revsbech, N. P. *Limnol. Oceanogr.* **1989**, *55*, 1907-1910.
8. Revsbech, N. P.; Nielsen, L. P.; Christensen, P. B.; Sorensen, J. *Appl. Environ. Microbiol.* **1988**, *45*, 2245-2249.
9. Hinke, J. In *Glass microelectrodes*; Lavallee, M., Schanne, O. F., Herbert, N. C., Eds.: Wiley, 1969, pp 349-375.
10. De Beer, D.; Van den Heuvel, J. C. *Talanta* **1988**, *35*, 728-730.
11. De Beer, D.; Sweerts, J. P. R. A. *Anal. Chim. Acta* **1989**, *219*, 351-356.
12. Larsen, L. H.; Revsbech, N.; Binnerup, S. J. *Appl. Environ. Microbiol.* **1996**, *62*, 148-1251.
13. Larsen, L. H.; Kjaer, T.; Revsbech, N. P. *Anal. Chem.* **1997**, *69*, 3527-3531.
14. Revsbech, N. P.; Jørgensen, B. B.; Blackburn, T. H.; Cohen, Y. *Limnol. Oceanogr.* **1983**, *28*, 1062-1074.
15. Jeroschewski, P.; Steukart, C.; Kühl, M. *Anal.Chem.* **1996**, *68*, 4351-4357.
16. De Beer, D.; Schramm, A.; Santegoeds, C. M.; Kühl, M. *Appl. Environ. Microbiol.* **1997**, *63*, 973-977.
17. Damgaard, L. R.; Revsbech, N. P. *Anal. Chem.* **1997**, *69*, 2262-2267.

244

18. Ammann, D.; Bührer, T.; Schefer, U.; Müller, M.; Simon, W. *Pflügers Arch.* **1987**, *409*, 223-228.

19. De Beer, D.; Glud, A.; Epping, E.; Kühl, M. *Limnol. Oceanogr.* **1997**, *42*, 1590-1600.

20. Holst, G.; Klimant, I.; Kuehl, M.; Kohls, O. In *Chemical sensors in oceanography*; Varney, M. S., Ed.; OPA: Amsterdam, 2000, pp 143-188.

21. Brendel, P. J.; Luther, G. W. *Env.Sci. Technol* **1995**, *29*, 751-761.

22. Tercier, M. L.; Parthasarathy, N.; Buffle, J. *Electroanal.* **1995**, *7*, 55-69.

23. Buffle, J.; Wilkinson, K. J.; Tercier, M. L.; Parthasarathy, N. *Annali di Chimica* **1997**, *87*, 67-82.

24. De Beer, D.; Sweerts, J.-P. R. A.; van den Heuvel, J. C. *FEMS Microbiol. Ecol.* **1991**, *86*, 1-6.

25. Epping, E.; Jorgensen, B. B. *Mar.Ecol.Prog.Ser.* **1996**, *139*, 193-203.

26. Revsbech, N. P.; Jorgensen, B. B. *Limnol. Oceanogr.* **1983**, *28*, 749-756.

27. Revsbech, N. P.; Jørgensen, B. B. *Adv.Microbial Ecol.* **1986**, *9*, 293-352.

28. Revsbech, N. P.; Madsen, B.; Jørgensen, B. B. *Limnol. Oceanogr.* **1986**, *31*, 293-304.

29. Sweerts, J.-P. R. A.; Bar-Gillisen, M. J.; Cornelise, A. A.; Cappenberg, T. E. *Limnol. Oceanogr.* **1991**, *36*, 1124-1133.

30. Archer, D.; Emerson, S.; Reimers, C. *Geochim.Cosmochim Acta* **1989**, *53*, 2831-2845.

31. Smith, K. L. J.; Glatts, R. C.; Baldwin, R. J.; Beaulieu, S. E.; Uhlman, A. H.; Horn, R. C.; Reimers, C. E. *Limnol. Oceanogr.* **1997**, *42*, 1601-1612.

32. Wieland, A.; Kühl, M. *Mar. Ecol. Prog. Ser.* **2000**, *196*, 87-102.

33. Wieland, A.; Kuehl, M. *Mar. Biol.* **2000**, *137*, 71-85.

34. Kühl, M.; Jørgensen, B. B. *Appl. Environ. Microbiol.* **1992**, *58*, 1164-1174.

35. De Beer, D.; Stoodley, P.; Roe, F.; Lewandowski, Z. *Biotechnol. Bioeng.* **1994**, *43*, 1131-1138.

36. De Beer, D.; van den Heuvel, J. C.; Ottengraf, S. P. P. *Appl. Environm. Microbiol.* **1993**, *59*, 573-579.

37. De Beer, D. , University of Amsterdam, Amsterdam, 1990.

38. Schramm, A.; Santegoeds, C. M.; Nielsen, H. K.; Ploug, H.; Wagner, M.; Pribyl, M.; Wanner, J.; Amann, R.; de Beer, D. *Appl. Environ. Microbiol.* **1999**, *65*, 4189-4196.

39. Ploug, H.; Jorgensen, B. B. *Mar. Ecol. Prog. Ser.* **1998**, submitted.

40. Ploug, H.; Grossart, H. P. *Aquat.Microb.Ecol.* **1999**, *20*, 21-29.

41. Ploug, H.; Kühl, M.; Buchholz-Cleven, B.; Jørgensen, B. B. *Aquat. Microb. Ecol.* **1997**, *13*, 285-294.

42. Kühl, M.; Cohen, Y.; Dalsgaard, T.; Jørgensen, B. B.; Revsbech, N. P. *Mar. Ecol. Prog. Ser.* **1995**, *117*, 159-172.

43. de Beer, D.; Kühl, M.; Stambler, N.; Vaki, L. *Mar. Ecol. Prog.Ser.* **2000**, *194*, 75-85.

245

44. Rink, S.; Kühl, M.; Bijma, J.; Spero, H. J. *Mar. Biol.* **1998**, *131*, 583-595.
45. Revsbech, N. P. In *Polarographic oxygen sensors: Aquatic and physiological applications*; Gnaigner, E., Forstner, H., Eds.; Springer Verlag: Berlin, 1983, pp 265-273.
46. Visscher, P. T.; Beukema, J.; Van Gemerden, H. *Limnol. Oceanogr.* **1991**, *36*, 1476-1480.
47. Thomas, R. C. *Ion-sensitive intracellular microelectrodes, how to make and use them*; Academic Press: London, 1978.
48. Morf, W. E.; Simon, W. *Ion-selective electrodes based on neutral carriers*; Plenum Press: New York, 1978.
49. Khuri, R. N.; Bogharian, K. K.; Agulian, S. K. *Pflügers Arch.* **1974**, *349*, 285-294.
50. Damgaard, L. R.; Larsen, L. H.; Revsbech, N. P. *Trends Anal. Chem.* **1995**, *14*, 300-303.
51. Lee, W.; de Beer, D. *Biofouling* **1995**, *8*, 273-280.
52. Schulthess, P.; Shijo, Y.; Pham, H. V.; Pretsch, E.; Ammann, D.; Simon, W. *Anal. Chim. Acta* **1981**, *131*, 111-116.
53. Cai, W. J.; Reimers, C. E. *Limnol. Oceanogr.* **1993**, *38*, 1762-1773.
54. Komada, T.; Reimers, C. E.; Boehme, S. E. *Limnol. Oceanogr.* **1998**, *43*, 769-781.
55. Luther, I., G. W.; Bono, A.; Taillefert, M.; Cary, S. C. In *Electrochemical Methods for the Environmental Analyses of Trace Element Biogeochemistry*; Taillefert, M., Rozan, T., Eds.; American Chemical Society Symposium Series; American Chemical Society: Washington, D.C., 2001, Chapter 4.
56. Tercier, M. L.; Buffle, J.; Koudelka-Hep, M.; Graziottin, F. In *Electrochemical Methods for the Environmental Analyses of Trace Element Biogeochemistry*; Taillefert, M., Rozan, T., Eds.; American Chemical Society Symposium Series, American Chemical Society: Washington DC, 2001, Chapter 2.
57. Sweerts, J.-P. R. A.; de Beer, D. *Appl. Environ. Microbiol.* **1989**, *55*, 754-757.
58. Santegoeds, C. M.; Damgaard, L. R.; Hesselink, G.; Zopfi, J.; Lens, P.; Muyzer, G.; de Beer, D. *Appl. Environm. Microbiol.* **1999**, *65*, 4618-4629.
59. de Beer, D. In *Immobilized cells.*; Wijffels, R., Ed.; Springer Verlag: Heidelberg, Germany, 1999, pp in press.
60. Sweerts, J. P. R. A. , Ph.D. thesis, University of Amsterdam, Amsterdam, 1990.
61. van Luijn, F.; Boers, P. C. M.; Lijklema, L.; Sweerts, J. P. R. A. *Wat. Res.* **1999**, *33*, 33-42.
62. Nielsen, P. H. *Wat.Res.* **1996**, *30*, 762-764.

63. Lorenzen, J.; Larsen, L. H.; Kjaer, T.; Revsbech, N. P. *Appl. Environ. Microbiol.* **1998,** *64,* 3264-3269.

64. Hulth, S.; Aller, R.; Gilbert, F. *Geochim. Cosmochim. Acta* **1999,** *63,* 49-66.

65. Luther III, G. W.; Sunby, B.; Lewis, B. L.; Brendel, P.; Silverberg, N. *Geochim. Cosmochim. Acta* **1997,** *61,* 4043-4052.

66. Strous, M.; Van Gerven, E.; Kuenen, J. G.; Jetten, M. *Appl. Environ. Microbiol.* **1997,** *63,* 2446-2448.

67. Strous, M.; Kuenen, J. G.; Jetten, M. S. M. *Appl. Environ. Microbiol.* **1999,** *65,* 3248-3250.

68. Robertson, L. A.; Dalsgaard, T.; Revsbech, N. P.; Kuenen, J. G. *FEMS Microb. Ecol.* **1995,** *18,* 113-120.

69. Robertson, L. A.; van Niel, E. W. J.; Torremans, R. A. M.; Kuenen, J. G. *Appl. Environ. Microbiol.* **1988,** *54,* 2812-2818.

70. Kuenen, J. G.; Robertson, L. A. *FEMS Microbiol. Rev.* **1994,** *15,* 109-117.

71. Schramm, A. , Ph.D. thesis, Bremen University, Bremen, 1998.

72. Schramm, A.; de Beer, D.; van den Heuvel, J. C.; Ottengraf, S. P. P.; Amann, R. *Appl. Environ. Microbiol.* **1999,** *65,* 3690-3696.

73. Matyas, P.; Katalin, V.; Lajos, V.; Hesham, S. M. *Hydrobiologia* **1997,** *342-343,* 55-61.

74. Schlegel, H. G. *Allgemeine Mikrobiologie*; Thieme Verlag: Stuttgart, Germany, 1985.

75. Tortell, P. D.; Reinfelder, J. R.; Morel, F. M. M. *Nature* **1997,** *390,* 243-244.

76. Goiran, C.; Al-Moghrabi, S.; Allemand, D.; Jaubert, J. *J. Exp. Mar. Biol. Ecol.* **1996,** *199,* 207-225.

77. Badger, M. R.; Andrews, T. J. *Plant Physiol.* **1982,** *70,* 517-523.

78. Huertas, I. E.; Colman, B.; Espie, G. S.; Lubian, L. M. *J. Phycol.* **2000,** *36,* 314-320.

Chapter 13

Seasonal Variations of Soluble Organic-Fe(III) in Sediment Porewaters as Revealed by Voltammetric Microelectrodes

M. Taillefert[1], T. F. Rozan[2], B. T. Glazer[2], J. Herszage[3], R. E. Trouwborst[2], and G. W. Luther, III[2]

[1]School of Earth and Atmospheric Sciences, Georgia Institute of Technology, Atlanta, GA 30332
[2]College of Marine Studies, University of Delaware, Lewes, DE 19958
[3]Department of Inorganic, Analytical and Physical Chemistry, University of Buenos Aires, Buenos Aires, Argentina

The seasonal cycling of iron and sulfur in the sediment from Rehoboth Bay (Delaware) was examined with solid state Hg/Au voltammetric microelectrodes. High resolution vertical profiles of ΣH_2S, Fe(II), and soluble organic-Fe(III) in the porewaters indicate that dissolved sulfide controls the fate of soluble organic-Fe(III). In the spring and in the fall, dissolved sulfide concentrations are negligible, and soluble Fe(III) and Fe(II) are dominant. This may be explained if the reduction of Fe(III) occurs by dissolution of solid Fe(III), then reduction to Fe(II), or if Fe(II) is oxidized anaerobically in the presence of organic ligands. During the summer, intensive sulfate reduction precipitates iron as FeS and pyrite. Because of its high reactivity and mobility, soluble organic-Fe(III) produced in porewaters can be considered a new reactive electron acceptor.

Introduction

Solid and colloidal iron oxides are commonly involved in the remineralization of natural organic matter in sediments. The reactivity of ferric iron has been assessed on the premise that this species is in its precipitated or colloidal form and that the reduction step proceeds at the surface of the particle (*1-4*). Kinetic studies have shown that the rate of reduction depends on: (i) the reactivity of the Fe(III) phase, an amorphous species being more reactive than a well-crystallized form; and (ii) the strength of the reductant (*1-3*). Among reductants present in sediments, dissolved inorganic sulfide may reduce Fe(III) (*1, 2, 5, 6*), with the rate limiting step being apparently the rate of dissolution of the surface ferrous hydroxide (*1*). However, more readily available soluble Fe(III) could accelerate the remineralization rate of natural organic matter.

The behavior of soluble Fe(III) at a voltammetric electrode has been described previously (*7*). Experiments with synthetic solutions showed that soluble Fe(III) (i.e. < 50 nm diameter) reacts at a mercury voltammetric electrode at circumneutral pH if it is complexed by an organic ligand. Several organic-Fe(III) complexes can be detected simultaneously, but they have not been identified. Thus, it is not possible to quantify these electroactive species by voltammetry, and their concentrations are reported in current intensities. Experiments performed in seawater (*7*) showed that these organic-Fe(III) complexes aggregate with time: their molecular weight increases and their reduction potential simultaneously shifts negatively. Thus, generally "fresh" or less aggregated species measured in natural waters are found around -0.3 V (*7, 8*), while "aged" or more aggregated species are reduced around -0.6 V (*7, 9*). These reduction potentials may change slightly with pH and ionic strength.

The reactivity of soluble organic-Fe(III) with dissolved sulfide has been studied in the laboratory with an eight day-old solution of Fe(III) complexed by TRIS (*7*). Upon addition of bisulfide, Fe(III) is reduced to Fe(II) and the signal for Fe(II) increases drastically. This indicates that the reaction is extremely fast, with a half-time of seconds rather than minutes and hours previously reported with ferrihydrite (*6*), lepidocrocite (*6, 10*) and goethite (*1*). In addition, a non-reductive breakdown of "aged" Fe(III) is promoted when adding bisulfide (*7*). This breakdown could supplement the production of soluble Fe(III) in aquatic systems, usually promoted by proton dissolution (*4*), by oxidation of organically complexed ferrous iron with solid ferric oxides (*11*), or by non-reductive dissolution in the presence of organic ligands (*11, 12*), including siderophores (*13*). Fe(II) produced reacts simultaneously with bisulfide to form aqueous FeS (*14*), but FeS precipitates rapidly, and the voltammetric signals for FeS and Fe(II) usually decrease within minutes. Bisulfide is simultaneously oxidized to elemental sulfur by soluble organic-Fe(III) (*7*).

In a more recent study (*9*), the behavior of soluble organic-Fe(III) in two low-ionic-strength anoxic sediment cores from the same site was shown to be controlled by the occurrence of dissolved sulfide. Soluble organic-Fe(III) and Fe(II) were dominant in a sediment core collected from a large creek where dissolved sulfide was negligible. The pH in this system was below the the first acidity constant of the sulfide species, which favored the formation of pyrite by reaction of dissolved sulfide with $FeS_{(s)}$ (Wächterhäuser reaction, *15*). Pyrite formation by incorporation of S(0) from elemental sulfur (S_8) and polysulfides (S_x^{2-}) into $FeS_{(s)}$ was dismissed because this reaction, which probably occurs mainly at the sediment-water interface where S_8 and S_x^{2-} are produced by oxidation of H_2S, is slower than the Wächterhäuser reaction. In addition, the more recent proposed pathway involving the loss of Fe(II) by $FeS_{(s)}$ followed by incorporation of S(0) into another $FeS_{(s)}$ (*16*) was also not considered because this pathway was not consistent with the pH profile in this sediment. These data indicated that the production of dissolved sulfide was the rate limiting step in the Fe(III)-H_2S-pyrite system, thus preserving soluble organic-Fe(III). In contrast in another sediment core collected near a saltmarsh, dissolved sulfide produced by sulfate reduction occurred in significant concentration one centimeter below the sediment-water interface (SWI), and soluble organic-Fe(III) was observed only in the top centimeter of the porewaters. At the onset of dissolved sulfide, soluble organic-Fe(III) was reduced to Fe(II), forming $FeS_{(aq)}$ which further precipitated. The pH was higher in this sediment, indicating that pyritization rates by the Wächterhäuser mechanism were slower, thus favoring the removal of dissolved sulfide by reduction of soluble organic-Fe(III). These data indicated this system could not be considered at steady-state. The fact that sulfate reduction controls pyritization rates has already been suggested in a recent study comparing different sulfidic environments (*17*), but the role of pH in controlling pyritization rates has not been evidenced in this study.

In this paper, we report seasonal variations showing that the formation and disappearance of soluble organic-Fe(III) is seasonally controlled by dissolved sulfide and the pH.

Experimental Methods

Sediment cores were collected monthly at the same site (i.e., within a 20 m radius) in Rehoboth Bay (Delaware), a shallow inland bay (water depth ~ 1.5 m) characterized by seasonal eutrophication, and brought back to the laboratory for analysis. Within an hour after sampling, high resolution vertical profiles of O_2, ΣH_2S (= H_2S + HS^- + S^{2-} + S^0 + S_x^{2-}), Mn(II), Fe(II), $FeS_{(aq)}$, and soluble organic- Fe(III) were acquired by voltammetry with a Au/Hg microelectrode (*18-20*) coupled to a potentiostat (DLK-100A, Analytical Instrument Systems).

Minimum detection limits for O_2, ΣH_2S, Mn(II), and Fe(II), were 2 μM, <0.2 μM, 5 μM, and 10 μM, respectively. Results for $FeS_{(aq)}$ and soluble organic-Fe(III) measurements are reported in current intensities as the electrochemical sensitivity of these species has not ($FeS_{(aq)}$) or cannot (soluble organic-Fe(III)) be determined. For oxygen measurements, a linear-sweep method was used from -0.05 V to -1.85 V at a scan rate of 200 mV/s. For ΣH_2S, Mn(II), Fe(II), $FeS_{(aq)}$, and soluble organic-Fe(III) measurements, a square-wave method with a pulse amplitude of 24 mV and a scan rate of 200 mV/ s was used from -0.1 V to -1.85 V. A conditioning period was set at -0.1 V for 5 s when dissolved sulfide was not present and at -0.9 V for 10 s when dissolved sulfide was measured in the porewaters. When dissolved sulfide concentrations exceeded 20 μM, the voltammograms were collected anodically from -1.8 V to - 0.1 V.

Typical triplicate measurements of soluble organic-Fe(III) obtained by cathodic square wave voltammetry 4 mm below the SWI in the porewaters of Rehoboth Bay are shown in Figure 1a. These voltammograms show with good reproducibility a peak at -0.53 V for ferric iron species and a peak at -1.43 V for Fe(II). The Figure 1b displays, also with good reproducibility, a strong dissolved sulfide peak at -0.65 V measured by anodic stripping voltammetry 19 mm below the SWI.

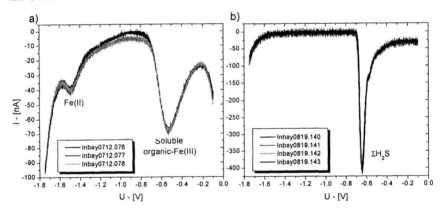

Figure 1. Replicates of square wave voltammograms collected in the sediment porewaters of Rehoboth Bay: a) soluble-organic Fe(III) 4 mm below the SWI in July 1999; b) dissolved sulfide 19 mm below the SWI in August 1999. The dissolved sulfide profile was measured anodically from -1.8 to -0.1 V.

In addition, pH profiles were simultaneously collected from July to October 1999 with a combination pH microelectrode (Diamond General Development Corp.) coupled to a pH-meter (Accumet). The tip of the pH microelectrode was positioned at the same height, but 7 mm away from the voltammetric microelectrode, and lowered with a micromanipulator with minimal depth increments of 0.5 mm. Prior to deployment, the voltammetric microelectrode was calibrated for Mn(II), Fe(II), and ΣH_2S using the pilot ion method (18). It was also calibrated for oxygen by calculation of its solubility at the temperature and salinity of the overlying waters (21). The pH microelectrode was calibrated by measuring the potential and temperature of buffers at the ionic strength of seawater. The pH in the porewaters was recalculated from the potentials measured and temperature of the overlying waters, assuming the temperature of the core was constant with depth. All the potentials are reported with respect to the Ag/AgCl reference electrode.

Results and Discussion

Electrochemical porewater profiles were collected in April and monthly from June to October 1999. High-resolution vertical profiles of oxygen, dissolved sulfide (ΣH_2S), and pH obtained are displayed in Figure 2. Except for the September data set, oxygen (filled squares) consistently decreases from saturation in the overlying waters a few millimeters above the SWI to insignificant values at the SWI. This indicates that respiration is intense in these sediments during most of the year. The data set of September indicates that oxygen penetrates 2 mm in the sediment, but this core was collected right after the passage of Hurricane Floyd over Rehoboth Bay, suggesting that the surface of the sediment may have been mixed.

The pH is constant around 7 in the overlying waters. Generally, a sharp decrease in pH is observed right below the SWI to a minimum value around the first pK_a of the carbonate system (i.e., 6.28 to 6.5) at a depth varying between 7 and 9 mm. This sharp decrease is followed by a progressive increase in pH deeper in the sediment to ca. 6.9, 6.45, and 6.8 in July, August, and October, respectively. This behavior is observed monthly except in September. After the hurricane, the pH also decreases sharply below the SWI but to ca. 7.2 at a depth of 5 mm only. In addition, this sharp decrease is followed by a constant diminution to a pH of 7.05 at 40 mm. The pH profiles in July, August, and October indicate that remineralization of natural organic matter by oxygen and nitrate occurs within the first 9 mm of the sediment because these reactions release acid equivalents (20, 22). The pH minima indicate the minimum penetration depth of oxygen and/or nitrate and the upper limit of manganese and iron utilization during remineralization (23) since manganese and iron reduction consumes protons (22). Interestingly, the pH rise in August is much smaller that in July and October. This may be explained if the pH is buffered, possibly by the release of protons during precipitation of $FeS_{(s)}$ (see below).

252

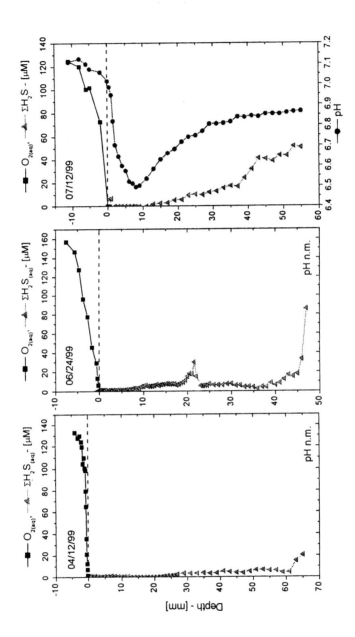

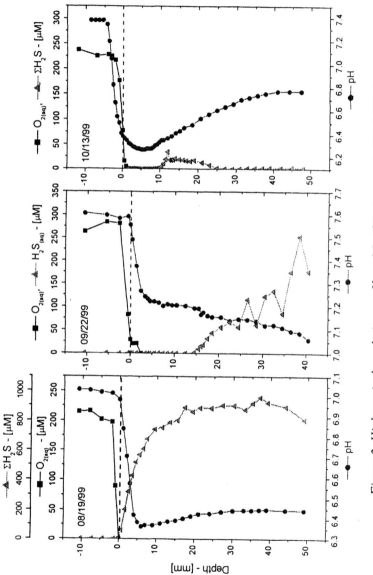

Figure 2. High vertical resolution profiles of O_2 (filled squares), ΣH_2S (filled triangles), and pH (filled circles) in the sediment porewaters of Rehoboth Bay (Delaware) from April to October 1999 (the pH was not measured (n.m.) in May and June). The dashed line represents the SWI.

The fact that the pH measured in September decreases to a depth of 40 mm indicates that the penetration depth of oxygen and nitrate has reached 40 mm, and thus the sediment has been partially mixed during the hurricane, or that the reduction of manganese and iron oxides has been shut down during this event. The pH profile of October shows that the sediment had returned to normal conditions for Rehoboth Bay.

In April, dissolved sulfide is close to the detection limit over the first twenty millimeters and reaches ca. 6 µM between 30 and 60 mm. It is only below 60 mm that the dissolved sulfide concentration increases significantly. In June, the concentration of dissolved sulfide is close to detection limit above 10 mm. It increases regularly below that depth to reach a maximum of 30 µM at 21.5 mm, then decreases to close to detection limit at 35 mm, where ΣH_2S raises again to 80 µM at 47 mm. In July, small concentrations of dissolved sulfide are measured at the SWI, probably in the form of elemental sulfur (7) or polysulfides (24). Below the SWI, ΣH_2S stays below detection limit until 10 mm, the depth below which sulfate reduction produces ΣH_2S to concentrations up to 50 µM at 50 mm. In August, sulfate reduction is extremely intense and dissolved sulfide reaches the SWI with values exceeding 900 µM at depth, the greatest concentrations reached during the studied year. In September, right after Hurricane Floyd, ΣH_2S is not found until 15 mm below the SWI, but then increases systematically to reach ca. 200 µM at 40 mm. Finally, ΣH_2S was only detected between 10 and 25 mm in October, with a maximum of 35 µM at 12 mm. A close examination of the sulfide profiles in Figure 2 reveals that the production of dissolved sulfide increases progressively during the spring and early summer and that dissolved sulfide diffuses slowly towards the SWI. In August, the porewaters seem to be saturated with sulfide, and dissolved sulfide probably diffuses to the overlying waters. However, the pH is unusually low for such a high concentration of dissolved sulfide. This may be explained if the system is buffered either by the carbonate system (pH ~ 6.5 close to the pK_a of bicarbonates) or that $FeS_{(s)}$ precipitation (eq. 1) releases enough acidity to buffer the pH around 6.5:

$$Fe^{2+} + H_2S \leftrightarrow FeS_{(s)} + 2H^+ \tag{1}$$

The profiles of soluble Mn(II), organic-Fe(III), Fe(II), and $FeS_{(aq)}$ from April to October 1999 are displayed in Figure 3, together with the profiles of O_2 and ΣH_2S already described in Figure 2. Mn(II) is only above detection limits in July, September and October. This indicates that manganese oxides (MnO_x) are not reduced during remineralization of organic matter in the spring and that Mn(II) has probably fluxed out of the sediment in August.

In April, soluble organic-Fe(III) builds up right below the SWI. It reaches a maximum current intensity of 17 nA at 4 mm but decreases to non-detectable values with the onset of dissolved sulfide. Fe(II) and $FeS_{(aq)}$ are also produced right below the SWI but remain relatively constant and low in the porewaters (ca. 10-30 μM for Fe(II) and 1 nA for $FeS_{(aq)}$).

In June, soluble organic-Fe(III) is minimal over the first 10 mm. Deeper, it forms a broad peak with a maximum current intensity of 80 nA at 20 mm. Soluble organic-Fe(III) progressively decreases below 20 mm to a minimum of 10 nA, when the concentration of dissolved sulfide increases in the porewaters. Fe(II) and $FeS_{(aq)}$ are generally low throughout the profile (ca. 10-30 μM for Fe(II) and 5-10 nA for $FeS_{(aq)}$), but their maximum concentrations coincide with the peak maximum of soluble organic-Fe(III).

In July, Mn(II) is produced 2 mm below the SWI and reaches ca. 150 μM at 4 mm. It decreases slightly to ca. 90 μM between 4 and 7 mm and remains constant to 38 mm. At this depth, Mn(II) decreases in the porewaters to concentrations below the detection limit at 41 mm. Although, dissolved sulfide can reduce manganese oxides with a 1:1 stoichiometric ratio (*25, 26*), the profiles of Mn(II) and dissolved sulfide show that, assuming a steady-steate is reached, manganese oxides are also reduced during remineralization of organic matter and/or Fe(II) oxidation (see below) because the maximum concentration of dissolved sulfide found in July is approximately half of that of Mn(II). A peak of soluble organic-Fe(III) with a maximum current intensity of 55 nA is found 4 mm below the SWI and is coincident with the Mn(II) maximum. In contrast to Mn(II), soluble organic-Fe(III) slowly decreases at the onset of dissolved sulfide to concentrations below the detection limit at 42 mm. Fe(II) forms a very sharp peak with a maximum of ca. 1450 μM 1.5 mm below the interface but decreases to concentrations below detection limit at 6 mm. $FeS_{(aq)}$ was not detected at this time of the year.

In August, the dissolved sulfide concentration reaches 900 μM and soluble organic-Fe(III) is only detected at the SWI with a small current intensity of 3 nA. In addition, Mn(II) is not detected, and Fe(II) is measured at only one location in the porewaters, while $FeS_{(aq)}$ forms a peak 5 mm below the SWI with a maximum current intensity of 15 nA.

In September, a peak of Mn(II) is found below the SWI with a concentration reaching 92 μM at 5 mm. At 10 mm, this peak has decreased to ca. 10 μM and remains constant until 25 mm, where Mn(II) is removed from the porewaters. This peak of Mn(II) could result from the input of manganese oxides from the surface during the hurricane followed by rapid reduction. The profiles of Mn(II) and dissolved sulfide in September show that reactive manganese oxides are not present below 25 mm because Mn(II) is not produced with the onset of dissolved sulfide. Soluble organic-Fe(III) is only detected between 26 and 40 mm and remains low as compared to the other months, with current intensities ca. 5 nA.

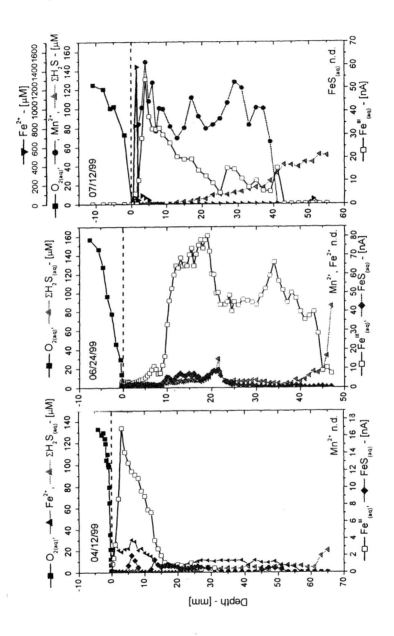

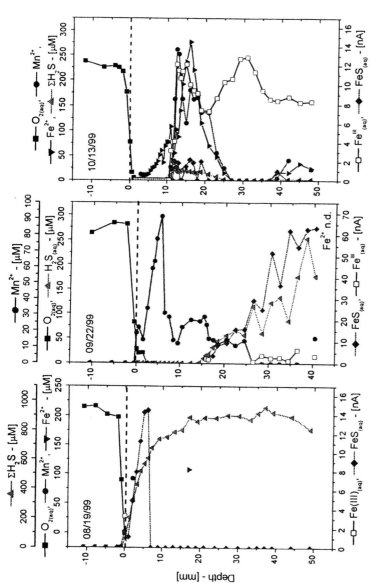

Figure 3. High vertical resolution profiles of soluble organic-Fe(III) (open squares), FeS$_{(aq)}$ (filled diamonds), Mn(II) (filled circles), Fe(II) (downward triangles), O$_2$ (filled squares), and ΣH$_2$S (upward triangles) in the sediment porewaters of Rehoboth Bay from April to October 1999. Species not detected (n.d.) are identified in each profile. The dashed line represents the SWI.

Fe(II) was not detected in the porewaters at this time of the year, but $FeS_{(aq)}$ increased regularly at the onset of dissolved sulfide to reach a maximum current intensity of 65 nA at the maximum depth of observation.

Finally, in October Mn(II) is observed 3 mm below the SWI and increases with depth to reach a maximum of ca. 280 μM at 16 mm. Its concentration decreases regularly below this maximum to values under detection limit at 28 mm. The peak of Mn(II) and the dissolved sulfide maximum are found at the same location, indicating that Mn(II) could be produced by dissolved sulfide reduction of manganese oxide between 10 and 25 mm. In fact, oxidation of bisulfide by manganese oxides produces elemental sulfur (25, 26), which should be detected as dissolved sulfide by voltammetry (24, 27). These measurements performed after porewater extraction and acidification (not shown) confirmed that dissolved sulfide measured between 10 and 25 mm exclusively consists of elemental sulfur. Soluble organic-Fe(III) is only found below 10 mm, where it increases sharply to reach a maximum current intensity of 12 nA. It remains relatively constant with greater depth because the concentration of dissolved sulfide is low, with current intensities varying between 7 and 13 nA. Fe(II) forms a peak with a maximum concentration of 260 μM between 10 and 26 mm coincident with the peak of Mn(II), ΣH_2S, and soluble organic-Fe(III). It is below detection limit above 10 mm and below 26 mm. $FeS_{(aq)}$ is low (ca. 1 nA) and only measured where Fe(II) and ΣH_2S are found simultaneously (i.e., between 10 and 20 mm).

The mechanism of formation of soluble organic-Fe(III) can be inferred from the chemical profiles in the porewaters. Oxidation of Fe(II) in the presence of oxygen and dissolved organic ligands may well produce soluble organic-Fe(III) (7, 25), but this is unlikely in these sediments because the penetration depth of oxygen does not exceed few millimeters below the SWI (Figure 2). Proton promoted non- reductive dissolution is also known to occur but generally below a pH of 6 (4). Manganese oxides may also oxidize Fe(II) (29) and form soluble Fe(III) in the presence of dissolved organic ligands at depth in the sediment. This mechanism, which produces Mn(II), may be significant in July and October at locations where the profiles of soluble organic-Fe(III) and Mn(II) coincide (i.e., between 2 and 40 mm in July and between 10 and 20 mm in October). However, the presence of soluble organic-Fe(III) in May and June as well as below 20 mm in October, when Mn(II) is not observed, indicates that MnO_x reduction is not the sole process producing soluble organic-Fe(III) in these porewaters. Oxidation of Fe(II) by hydrous iron oxides in the presence of organic ligands could be responsible for the formation of soluble organic-Fe(III) complexes. It has been proposed (11, 30, 31) that a Fe(II) complex with a weak-field ligand (i.e., oxygen donor ligand) can facilitate the electron transfer from the Fe(II) to the hydrous iron oxide and thus promote the reductive dissolution of the oxide and simultaneously produce a soluble Fe(III)-ligand complex (eq. 2). In this equation, one atom of iron is labeled by a star to track its reactive pathway.

$$Fe(OH)_{3(s)} + {}^*FeL^{2+} + 3H^+ \rightarrow Fe^{2+} + {}^*FeL^{3+} + 3H_2O \tag{2}$$

The mechanism of electron transfer from the reduced complexed iron to the solid hydrous iron oxide is still open to discussion: on one hand, it is proposed that the weak-field ligand bridges Fe(II) with the hydrous iron oxide through an innersphere complex (30, 31; on the other hand, it is possible that the weak-field ligand is only complexed to Fe(II) and pushes the electron density of Fe(II) directly to Fe(III) without ligand bridging and without formation of a bond between the two iron atoms (11). The latter mechanism involves an outersphere complex and may be faster producing soluble organic-Fe(III). It is important to mention that the rate of oxidation of Fe(II) by hydrous iron oxides in the presence of an organic ligand increases with pH, but the reaction has only been observed in the laboratory below pH ~ 6 (11, 32). Anaerobic oxidation of Fe(II) has been shown to occur biotically in acidic conditions (33, 34) and mediated by nitrate reduction (35). However, the pH is too high in the sediments of the present study to favor such microbial production of soluble organic-Fe(III).

Finally, it has been shown that non-reductive dissolution of amorphous iron oxides by organic ligands can occur at circumneutral pH (4, 11, 36, 37) and that microbial reduction to Fe(II) can be successively faster if solid iron oxides are first non-reductively dissolved in the presence of an organic ligand (12). The composition of the organic ligand promoting the non-reductive dissolution of amorphous iron oxides is not known, but ligands that form mononuclear complexes with iron oxides will facilitate their dissolution (36, 37). In addition, multidentate ligands will further increase the reactivity of iron oxides (38). The proposed reaction (11) is presented in eq. 3 for a mononuclear complex formed with a bidentate ligand.

$$Fe(OH)_{3(s)} + LH_2 \rightarrow FeL^+ + OH^- + 2H_2O \tag{3}$$

The porewater profiles of the present study indicate that dissolved sulfide concentrations control the fate of soluble organic-Fe(III). In the spring and the fall, dissolved sulfide concentrations are negligible, and soluble organic-Fe(III) is dominant. During the summer, intensive sulfate reduction produces H_2S which reduces soluble organic-Fe(III) to Fe(II). According to our previous laboratory study (7), the reaction should produce elemental sulfur (eq. 4 in the case of a bidentate ligand).

$$2FeL^+ + H_2S \rightarrow 2FeL + S^0 + 2H^+ \tag{4}$$

As a result, soluble organic-Fe(III) always decreases in the porewaters with the onset of dissolved sulfide, and the product of this reaction can form aqueous FeS (eq. 5), which eventually precipitates into solid FeS (eq. 6-1).

$$FeL + H_2S \rightarrow FeS_{(aq)} + 2H^+ + L^{2-} \tag{5}$$

$$FeS_{(aq)} \xrightarrow[2]{1} FeS_{(s)} \tag{6}$$

Finally, pyrite can be formed by several mechanisms including the Wächterhäuser reaction where FeS reacts rapidly with an excess of dissolved sulfide according to eq. 6-2 and eq. 7 (*15, 39*):

$$FeS_{(aq)} + H_2S \rightarrow FeS_{2(s)} + H_{2(g)} \tag{7}$$

In July, dissolved sulfide is only produced below one centimeter, and soluble organic-Fe(III) is reduced, but Fe(II) and $FeS_{(aq)}$ are not observed. In our previous study (*9*), we showed that the pH was correlated to the amount of $FeS_{(s)}$ and inversely related to the amount of pyrite found in anoxic sediments. This was explained by the fact that the rate of pyritization by the Wächterhäuser mechanism (eq. 7) increases when the pH decreases below the first acidity constant of dissolved sulfide (pH < 7) because H_2S is the most reactive dissolved sulfide species to form pyrite according to eq. 7 in anoxic conditions. In July, the pH is below 6.8, indicating that, if it occurs, eq. 7 is not the rate limiting step in the formation of pyrite at depth and that pyritization is limited by the supply of Fe(II) and thus $FeS_{(aq)}$. In August, the excess of dissolved sulfide has completely reduced soluble organic-Fe(III) to Fe(II), and $FeS_{(aq)}$ is measured within the first centimeter of the porewaters. This indicates that pyritization by the Wächterhäuser reaction is only limited below 8 mm by the production of Fe(II) and $FeS_{(aq)}$, which depends on the rate of reduction of Fe(III). This is not surprising because the concentration of dissolved sulfide is sufficiently high at this time of the year that pyritization is expected to be limited by the availability of Fe(III). Previous studies, which did not include the reduction of Fe(III) prior to pyritization, have shown that the formations of H_2S and $FeS_{(aq)}$ are the limiting steps in the formation of pyrite by the Wächterhäuser reaction (*15, 39*). However, the extreme reactivity of soluble organic-Fe(III) in the presence of dissolved sulfide indicates that the limiting rate in the formation of pyrite can be the formation of $FeS_{(aq)}$ (eq. 6-2). In September, the pH reached the highest values found in 1999 in the porewaters (i.e., $7.1 < pH < 7.5$). This indicates that the rate of pyritization by the Wächterhäuser reaction is relatively slow as this time of the year, which allows $FeS_{(aq)}$ to accumulate and reach the highest currents measured in 1999 in Rehoboth Bay sediments. Finally, in October the pH decreased to values between 6.3 and 6.8, suggesting that pyritization should be efficient again. However, it has been shown that pyritization can be limited by the rate of sulfate reduction (17). Indeed, the availability of dissolved sulfide in these sediments seems to limit the production of pyrite at this time of the year.

Conclusions

State of the art electrochemical microprofiles were collected from April to October 1999 in the same area in Rehoboth Bay (Delaware). These microprofiles show that electrochemically reactive soluble organic-Fe(III) can be produced in the porewaters as long as dissolved sulfide is not present. This study confirms previous work performed with synthetic solutions (7) that showed soluble organic-Fe(III) complexes are reduced extremely rapidly by dissolved sulfide, as well as other field measurements that demonstrated the incompatibility of reduced sulfur and soluble organic-Fe(III) in sediment porewaters (7, 9).

These new measurements show the seasonal interaction between dissolved sulfide and soluble organic-Fe(III). In the spring, sulfide produced by sulfate reduction is found in low concentrations, suggesting either that sulfate reduction is less intense in the spring season or that sulfide is readily removed from the porewaters. In the summer, sulfate reduction produces millimolar concentrations of dissolved sulfide, preventing the existence of soluble organic-Fe(III) in the porewaters. Finally, in the fall dissolved sulfide concentrations decrease and soluble organic-Fe(III) can build up in the porewaters.

Soluble organic-Fe(III) can be produced by several processes, and its exact source in sediment porewaters remains to be identified. However, three anoxic processes responsible for the production of these complexes in the sediment of Rehoboth Bay can be inferred from the microprofiles obtained by voltammetry. First, oxidation of Fe(II) by reactive manganese oxides followed by complexation by natural organic ligands may partially contribute to the production of soluble organic-Fe(III) in these sediments. Generally, the reduction of reactive manganese oxides with dissolved sulfide is fast, suggesting that competition between soluble organic Fe(III) and manganese oxides may occur in these sediments; Second, the oxidation of Fe(II) by hydrous iron oxides in the presence of organic ligand could be responsible for the formation of soluble organic-Fe(III) complexes. This reaction, usually observed at pH values

262

very close to those measured in these porewaters, cannot be excluded in these sediments. Finally, the non-reductive dissolution of amorphous iron oxides by organic ligands could also produce soluble organic-Fe(III), which might be an intermediate in the microbial reduction of solid ferric iron to Fe(II).

These measurements seem to indicate that soluble organic-Fe(III) may play an important role in diagenetic processes. Because of its high reactivity and mobility, soluble organic-Fe(III) produced in porewaters may provide an electron acceptor at locations in the sediment where metal oxides are already consumed, potentially shifting the activity of local microbial communities. In addition, the high reactivity of soluble organic-Fe(III) may accelerate the formation of $FeS_{(aq)}$ and FeS_2. Finally, these soluble organic-Fe(III) complexes may influence phytoplankton growth in coastal areas. Iron is a very well known limiting nutrient for phytoplankton growth in surface waters, but its source and composition have not been clearly identified. If dissolved sulfide is not produced in the porewaters, soluble organic-Fe(III) may accumulate and eventually diffuse out of the sediment to supply readily bioavailable soluble Fe(III) to the surface waters of coastal marine settings.

Acknowledgements

This research was supported by a NOAA Sea Grant (NA16RG0162-03). We would like to thank two anonymous reviewers for their helpful comments on an earlier version of this paper.

References

1. Pyzik, A. J.; Sommer, S. E. *Geochim. Cosmochim. Acta.* **1981**, *45*, 687-698.
2. Dos Santos Afonso, M.; Stumm, W. *Langmuir.* **1992**, *8*, 1671-1675.
3. Deng, Y.; Stumm, W. *App. Geochem.* **1994**, *9*, 23-36.
4. Zinder, B.; Furrer, G.; Stumm, W. *Geochim. Cosmochim. Acta.* **1986**, *50*, 1861-1869.
5. Canfield, D. E. *Geochim. Cosmochim. Acta.* **1989**, *53*, 619-632.
6. Canfield, D.E.; Raiswell, R.; Bottrell, S. *Am. J. Sci.* **1992**, *292*, 659-683.
7. Taillefert, M.; Bono, A. B.; Luther III, G. W. *Environ. Sci. Technol.* **2000**, *34*, 2169-2177.
8. Rickard, D.; Oldroyd, A.; Cramp, A. *Estuaries.* **1999**, *22(3A)*, 693-701.
9. Taillefert, M; Hover, V. C.; Rozan, T. F.; Theberge, S. M.; Luther III, G. W. *Estuaries.* **2000**, submitted.
10. Peiffer, S.; Dos Santos Afonso, M.; Wehrli, B.; Gächter, R. *Environ. Sci. Technol.* **1992**, *26*, 2408-2413.

11. Luther III, G. W.; Kostka, J. E.; Church, T. M.; Sulzberger, B.; Stumm, W. *Mar. Chem.* **1992**, *40*, 81-103.
12. Lovley, D. R.; Woodward. J. C. *Chem. Geol.* **1996**, *132*, 19-24.
13. Butler, A. *Science.* **1998**, *281*, 207-210.
14. Theberge, S. M.; Luther III, G. W. *Aquat. Geochem.* **1997**, *3*, 191-211.
15. Rickard, D. *Geochim. Cosmochim. Acta.* **1997**, *61*, 115-134.
16. Wilkin, R. T.; Barnes, H. L. *Geochim. Cosmochim. Acta.* **1996**, *60*, 4167-4179.
17. Hurtgen, M. T.; Lyons, T. W.; Ingall, E. D.; Cruse, A. M. *Am. J. Sci.*, **1999**, *299*, 556-588.
18. Brendel, P. J.; Luther III, G. W. *Environ. Sci. Technol.*, **1995**, *29*, 751-761.
19. Luther III, G. W.; Brendel, P. J.; Lewis, B. L.; Sundby, B.; Lefrançois, L.; Silverberg, N.; Nuzzio, D. B. *Limnol. Oceanogr.* **1998**, *43(2)*, 325-333.
20. Luther III, G. W.; Reimers, C. E.; Nuzzio, D. B.; Lovalvo, D. Environ. Sci. Technol. **1999**, *33(23)*, 4352-4356.
21. Kester, D. R. In *Chemical Oceanography*; Editors: Riley & Skirrow, 1975; 2nd Ed., Volume. 1, Chapter 8, p 498.
22. Froelich, P. N.; Klinkhammer, G. P.; Bender, M. L.; Luedtke, N. A.; Heath, G. R.; Cullen, D.; Dauphin, P. *Geochim. Cosmochim. Acta.* **1979**, *43*, 1075-1090.
23. Van Cappellen, P.; Wang, Y. *Am. J. Sci.* **1996**, *296*, 197-243.
24. Rozan, T. F.; Theberge, S. M.; Luther III, G. W. *Anal. Chim. Acta.* **2000**, *415*, 175-184.
25. Burdige, D. J.; Nealson, K. H. *Geomicrobiol. J.* **1986**, *4*, 361-387.
26. Yao, W.; Millero, F. J. *Geochim. Cosmochim. Acta.* **1993**, *57*, 3359-3365.
27. Wang, F.; Tessier, A.; Buffle, J. *Limnol. Oceanogr.* **1998**, *43*, 1353-1361.
28. Liang, L.; McCarthy, J. F.; Jolley, L. W.; McNabb, J. A.; Melhorn T. L. *Geochim. Cosmochim. Acta.* **1993**, *57*, 1987-1999.
29. Myers, C. R.; Nealson, K. H. *Geochim. Cosmochim. Acta.* **1988**, *52*, 2727-2732.
30. Sulzberger, B.; Suter, D.; Siffert, C.; Banwart, S.; Stumm, W. *Mar. Chem.* **1989**, *28*, 127-144.
31. Stumm, W. *Coll. Surf. A: Physicochem. Eng. Aspects* **1997**, *120*, 143-166.
32. Suter, D.; Banwart, S.; Stumm, W. *Langmuir.* **1991**, *7*, 809-813.
33. Razzell, W.; Trussell, P. C. *J. Bacteriol.* **1963**, *85*, 595-603.
34. Pronk, J. T.; Johnson, D. B. *Geomicrobiol. J.* **1992**, *10*, 153-171.
35. Straub, K. L.; Benz, M.; Shink, B.; Widdel F. *Appl. Environ. Microbiol.* **1996**, *62*, 1458-1460.
36. Bondietti, G.; Sinniger, J.; Stumm, W. *Coll. Surf.* **1993**, *79*, 157-174.
37. Nowack, B.; Sigg, L. *J. Coll. Interf. Sci.* **1996**, *177*, 106-121.

38. Huheey, J. E.; Keiter, E. A.; Keiter, R. L. **1993**. Inorganic chemistry. Principals of structure and reactivity, 4th Ed. HarperCollins College Publishers. New York, NY.
39. Rickard, D.; Luther III, G. W. *Geochim. Cosmochim. Acta.* **1997**, *61(1)*, 135-147.

Chapter 14

Microelectrode Measurements in Stromatolites: Unraveling the Earth's Past?

Pieter T. Visscher[1,*], Shelley E. Hoeft[1], Tonna-Marie L. Surgeon[1],
Daniel R. Rogers[1], Brad M. Bebout[2], John S. Thompson, Jr.[3],
and R. Pamela Reid[4]

[1]Department of Marine Sciences, University of Connecticut, Groton, CT 06340
[2]Exobiology Branch, NASA Ames Research Center, Mail Stop 239–4,
Moffett Field, CA 94035
[3]University of Miami, University Drive, Coral Gables, FL 33124
[4]Marine Geology and Geophysics, RSMAS – University of Miami,
4600 Rickenbacker Causeway, Miami FL 33149
*Corresponding author: email: Pieter.Visscher@uconn.edu

Oxygen, sulfide and pH microelectrodes were used to study the biogeochemistry of modern marine stromatolites in the Exuma Cays, Bahamas. Measurements included chemical characterization of the stromatolite mats, both in stagnant water on the beach during a diel cycle as well as short-term *in situ* measurements, and determination of O_2 production and consumption. Experiments with slurried stromatolite mats, in combination with sulfide and/or oxygen electrodes, were used to estimate potential rates of sulfate reduction and sulfide oxidation. Combining these measurements, which are instrumental in the understanding of the microbiology associated with stromatolite formation, facilitate a better biogeochemical interpretation of the formation of these sedimentary structures as well as their ancient ancestors.

Introduction

Stromatolites are lithified laminated sedimentary structures that date back as far as 3.2 Ga (*1,2*) and reached their maximum abundance in the Proterozoic (*3*). Stromatolites are used as evidence for Archean life (*4*) and understanding their biogenic origin may facilitate the search for extraterrestrial life (*5*). The most characteristic feature of stromatolites is the presence of continuous layers of calcium carbonate precipitate that form the lithified layers in these geomicrobiological structures. The presence of alternating lithified and unlithified layers differentiate stromatolites from other microbial mats, the latter which have been studied under the premise that they are analogues of these ancient structures (*6*). Although continuous lithified layers are absent, occasional $CaCO_3$ precipitates are found in these "non-lithifying" analogues (*7,-9*). Studies in these analogues have implied that cyanobacterial sheath material (*10,11*) and sulfate reducers (*7*) may be involved in the lithification process in stromatolites. Recently, in a study of modern marine stromatolites, Reid et al. (*12*) demonstrated that succession of microbial communities dominated by the cyanobacterium *Schizothrix* sp. are responsible for the observed lithification which ultimately results in the characteristic lamination. Three key communities were recognized in development of lithified and unlithified layers, ranging from a cyanobacterial pioneer community to a climax community with aerobic heterotrophs and sulfate reducers. Copious amounts of amorphous microbial exopolymer, presumably of cyanobacterial origin, and not the more structured sheath material of these organisms, is the key to formation of microcrytalline $CaCO_3$ crusts.

Several key functional groups of microbes that form regular marine and hypersaline mats and modern marine stromatolites are identical and include oxygenic phototrophs, aerobic heterotrophs, sulfate-reducing bacteria (SRB) and sulfide-oxidizing bacteria (SOB) (*12 -15*). Compared to other benthic ecosystems, microbial mats have extremely high rates of oxygenic photosynthesis, aerobic respiration, sulfate reduction and sulfide oxidation (*13,16,17*). The functional groups mentioned above play different roles in the formation of lithified layers (Figure 1): Some contribute to microcrystalline $CaCO_3$ precipitation (oxygenic photosynthesis and sulfate reduction) while others contribute to the dissolution of $CaCO_3$ (aerobic respiration and sulfide oxidation), and may do so in both regular microbial mats and stromatolites.

Microelectrode applications in benthic ecology were pioneered by Teal and Kanwisher (*18*), but did not find a wide application until after Revsbech, Jørgensen and coworkers deployed polarographic oxygen microelectrodes in their studies of laminated microbial mats in Solar Lake, Egypt (*19*), and other locations (*20-22*). Revsbech and coworkers made many improvements in the early electrode design, including development of extremely small sensing tip

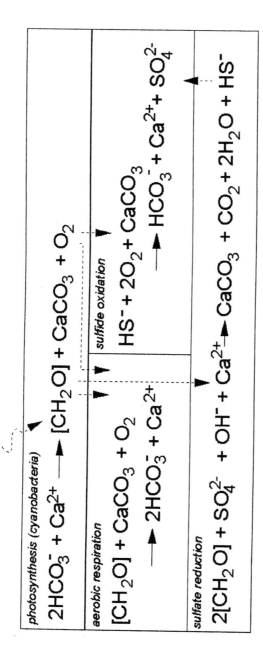

Figure 1. Representation of key microbial processes combined with the geochemical signature as it affects CaCO₃ precipitation. Four major functional groups of bacteria in the stromatolite mat are presented. Dashed arrows represent inputs (hv is photon energy which drives photosynthesis) and mass transport of key metabolic products ([CH₂O] is the generic formula for photosynthate) as it takes place between the different microbial populations.

dimensions (<10μm) and a guard cathode to allow for a very rapid response time (<0.2 s) minimizing O_2 interferences and reducing the stirring effect (23). Electric noise, fluctuating environmental conditions and other technical issues (including fragility of the thin glass sensing tips) in earlier designs resulted in laboratory studies rather than field deployment. Development of stainless steel needle electrodes (outer diameter 0.8 mm) was one of several modifications that enabled measurements *in situ* in a variety of sediment types. In these needle electrodes, the glass has been replaced by epoxy, while the rugged steel casing enables deployment in environments that are hostile to glass electrodes (24). There are inherent issues associated with needle electrodes, including dimensions and stirring sensitivity, but the combination of sensing tips for O_2 and S^{2-} in a single needle in combination with the rugged design (24) makes these needle electrodes the only suitable tool to obtain depth profiles of O_2 and S^{2-} in lithified structures such as stromatolites.

In this paper, we describe a microelectrode study of modern marine stromatolites using needle and glass microelectrodes for O_2, S^{2-} and pH. In addition to chemical characterization of the stromatolite mat, measurements were carried out to estimate rates of aerobic respiration, sulfate reduction and sulfide oxidation in intact stromatolites and slurried material. Finally, measurements were made *in situ* submerged in approximately 50 cm of water.

Materials and Methods

Site description

Our study was conducted at Highborne Cay, Bahamas, located at the northern end of the Exuma Cays (24°43'N, 76°49'W). Highborne Cay is flanked by Great Bahama Bank on the west and by Exuma Sound on the east. Stromatolites are found on the Exuma Sound side of the island where they occur along a 2 km long beach in between the low tide mark and the fringing reef, which defines the boundary between the shelf and Exuma Sound. Most stromatolites, ranging from 10 cm to several meters, are found in the subtidal area of the shelf. Stromatolite samples were collected from sites NS8 and NS11 (12,15). These stromatolites had developed lithified white crusts at their surface, indicative of a "climax" community, that comprises of cynaobacteria (Schizothrix spp, Solentia), aerobic heterotrophic bacteria, SRB and SOB (12, 15).

Electrode measurements

The construction and functioning of a variety of microelectrodes has reviewed recently by Glud et al. (*24*). In this study, microelectrode measurements were made with commercially available needle electrodes (Microscale Measurements, Haren, The Netherlands). These probes are built in stainless-steel needles in which the glass has been replaced by epoxy (*25*) as for reason outlined above.

The O_2 needle electrodes consist of an etched Pt-anode with an Au-plated tip (10-20 µm diameter) covered with a cellulose-nitrate membrane, and is polarized at 750 mV negative with respect to a calomel reference electrode. This electrode typically has a 90%-response time of 4-10 s and a drift of approximately 6%-signal min^{-1}. Since the O_2 saturation concentration changes with temperature and salinity, the electrode was calibrated in seawater at the appropriate temperature and salinity (or a range of values of these). Air saturated (100%) values are measured before and after taking readings in the field in seawater bubbled with air for approximately 10 min. The "0%" value is obtained by bubbling with O_2-free N_2 for at least 10 min, the actual concentration checked by a Winkler titration (*26*). The O_2 probe displays a linear response with respect to the ambient O_2 concentration over a wide range (0 to >700% air saturation) (*24,25*), which was tested for 0 and 100% in this study (data not shown). The glass O_2 electrodes with internal reference and guard cathode which were used in this study were obtained commerically (Diamond General, Ann Arbor, MI) with a variety of sensing tip diameters.

The sulfide sensor consists of a silver wire coated with Ag_2S (diameter 100 µm). This sulfide sensor is only sensitive to the fully dissociated form of hydrogen sulfide, S^{2-}, and thus sensitive to the pH of the measuring environment. Under typical seawater conditions, the detection limit is approximately 3-5 µM total sulfide and slightly higher when the pH increases. Calibration was carried out in an airtight vessel containing an oxygen-free buffer of the appropriate pH (7.5, 8.5 and 9.0; see (*25*)), to which incremental amounts of a sulfide stock solution are added. The actual concentration in this calibration solution is checked by the colorimetric methylene-blue method (*27*). The mV decreases 30 mV per decade over a concentration range of 3-3000 µM.

The needle probes deployed in this study included O_2-electrodes, ion-selective sulfide sensors and combined O_2/S^{2-} electrodes (0.8-1 mm outer diameter). Glass Clark-type oxygen microelectrodes with guard cathodes (sensing tip diameter 50 µm) and combined-glass O_2/pH electrodes, all with internal reference electrodes, were obtained from Diamond General (Ann Arbor, MI). Motor-driven stages were used as micromanipulators (MM-3M series, National Aperture, Salem, NH) and were mounted on heavy custom-made PVC adjustable stands. This allows for deployment in up to 15 cm of water and manipulation of 50-100 μm increments with the aid of a controller box (MC-3B-II, National Aperture, Salem, NH).

In situ underwater measurements in the surf zone were carried out with a hand-driven micromanipulator (model MM33, Märzhäuser-Wetzlar GmbH, Wetzlar, Germany) attached to an Al-stand with additional lead weights to prevent the electrodes from moving in the strong currents. A Keithley model 385 picoammeter with customized variable polarization (Keithley Instruments, Cleveland, OH) or a MasCom ME 01000 meter (MasCom, Bremen, Germany) were used to register the signal of the O_2 electrode and a Microscale Measurements (Haren, The Netherlands) multi-channel high-impedance meter or Beckman F-240 portable pH meter (Fisher Scientific, Pittsburgh, PA) were used to record the signal of S^{2-} and pH electrodes. All meters and micromanipulators were powered by a 12 V deep-cycle marine battery, which allowed for 24-36 h of continuous measurements.

Field Measurements

Diel fluctuations of O_2, pH and S^{2-} were monitored in the upper 25 mm of the stromatolite mat. Profiles were measured every 30 min during daytime and every 60-90 min during nighttime, covering 25 hr in total. In order to avoid submersion of the micromaniopulator, the intact stromatolite head (approximately 20 cm in diamter) was placed in a plastic container, in which void spaces were filled with sand from the location. This restored the stromatolite to an identical condition as experienced prior to sampling. The container with the stromatolite head was kept in *circa* 30 cm of flowing seawater, except when measurements were taken in which case the container was moved to the beach. Measurements were carried out while covered with 2-4 cm of stagnant seawater. The light intensity was registered every 10 min at the beach with a LiCor LI-1000 equipped with a quantum sensor (LiCor, Lincoln, NE). In a separate set of observations *in-situ* was made in a submersed, undisturbed stromatolite. These measurements were carried out during the light period only. The estimate for the flow rate of the water during these measurements was 20-25 cm s^{-1} (*28*). Estimates of oxygenic photosynthesis and aerobic respiration were made by measuring O_2 profiles in the light (1650 μE m^{-2} s^{-1}) followed by measurements taken after the sample and electrode stand were darkened (*20,22,29,30*). The dark profiles were taken for 10 min at 35-68 s intervals. Upon exposure to the light after darkening, O_2 profiles were measured every 45-60 sec for 10 min. The initial changes in the O_2 concentration over time for each depth were calculated from the individual depth profiles.

Shipboard measurements

Measurements under controlled conditions were carried out in a clear PVC flow box in which a water flow of about 5 cm s^{-1} was maintained using a submersible pump. Transient microprofiles were measured upon switching the light intensity from 100 µE m^{-2} s^{-1} to zero using a glass Clark-type electrode with a guard cathode. Profiles of O$_2$ were measured at 100 µm intervals every 60 s for 23 min. Calculations of aerobic respiration were carried out as outlined above.

Slurry experiments were carried out using either the upper 8-10 mm of the stromatolite or with isolated (individual) lithified layers. This latter approach was chosen since it was demonstrated that the highest microbial metabolic activity was associated with lithified layers (15). However, the amount of material required for slurry preparation limited the number of substrates taested for experiments using these individual layers. The stromatolite was dissected and material from lithified layers was homogenized in an equal amount of seawater (1:1 v/v) collected from the site. The 20-ml slurries were incubated in a gastight vessel (no headspace) in which an O$_2$ and/or S^{2-} electrode was inserted and kept in the dark. The potential for aerobic respiration was determined with a glass oxygen electrode having a guard cathode, and sulfate reduction with a S^{2-} electrode. After obtaining the endogenous O$_2$ uptake rate for each of the individual slurries, changes in the basal rate of aerobic respiration was calculated from the decrease in O$_2$ concentration over time upon addition of exogenous organic carbon. Organic carbon sources used included *Schizothrix* exopolymer (EPS), sugars, which are component of the EPS matrix, fatty acids, simple amino acids and alcohols that may result from fermentation and incomplete oxidation and sulfonates which may be part of EPS as well (31). Carbon additions were 2-50 µM and O$_2$ uptake was measured 2-3 times during 15-45 min each and were repeated for at least two concentrations of each exogenous carbon source. The potential for sulfide oxidation was determined by adding sulfide (100-200 µM) and following the decrease in O$_2$ and S^{2-} concentrations (after 30-60 min preincubation to sequester any free iron), or by adding S$_2$O$_3{}^{2-}$ (100 µM) and measuring the decrease in O$_2$ concentration over time. Molybdate was added to prevent potential microbial sulfide production. The potential rates were calculated from the initial slope of O$_2$ decrease and S^{2-} increase, for aerobic respiration and sulfate reduction, respectively. Controls for chemical oxidation of carbon and sulfur compounds were carried out for all experiments.

Results

Stromatolite Mat Characterization

Diel measurements of O$_2$, S^{2-} and pH including two tidal cycles and a diurnal/nocturnal cycle revealed a pattern that is typical for microbial mats: During nighttime, lack of photosynthesis and continued sulfate reduction resulted in anoxic conditions at or close to the surface of the stromatolite (Figure 2). Late in the dark period (2:00-6:00 AM) free sulfide ($\geq$10 µM) was found

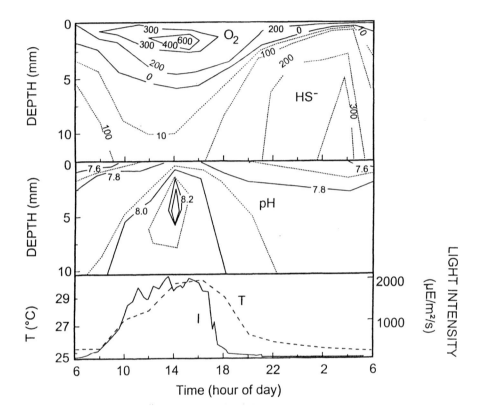

Figure 2. Isopleths of (top panel) the concentration of O₂ (solid lines) and HS⁻ (dashed lines) in the upper 12 mm of the stromatolite mat; (middle panel) pH (solid lines indicate even decimal pH values, dotted lines odd decimal values); (bottom panel) light intensity (solid lines) and temperature at the surface of the stromatolite (dashed line). Measurements were made on Jun 14-15, 1997, and cover an entire diel cycle.

near the surface at 0.8 mm depth. During the daytime, O_2 production by the cyanobacterial community reestablished an oxic zone in the upper few mm of the stromatolite mat, and appreciable concentrations of sulfide were found only in deeper layers due to chemical and microbial oxidation in the surface of the mats. During the early afternoon, between 8:00 and 4:00 PM, O_2 reaches supersaturation, with peak values of 600-650 μM at 1.5-2 mm depth between 11:30am and 2:45pm. Although the measurements shown in Figure 2 were carried out at the site, thus exposing the sample to natural illumination, they were performed in stagnant water and therefore deviated from true *in situ* conditions. An additional serie of measurements was taken *in situ* in ca. 50 cm of water under a relatively high hydrodynamic flow regime. The large number of particles (ooids) in the water column contributed to extensive light scatter, which may have lead to lower light intensity at the surface of the stromatolite mat. *In situ* measurements were carried out under high tidal flow conditions, which contributed to the observed deeper oxygen penetration and also a lower maximum O_2 concentration (Figure 3). The somewhat rougher shape of the profiles measured *in situ* as opposed to those measured in quiescent water may be caused by stress on the needle probe due to the high flow velocity. Alternatively, this could be caused by the pumping effect of the periodically breaking waves and accompanied entrainment of ooids in the water column resulting in fluctuating light.

Production and Consumption of Oxygen

Rates of photosynthesis and aerobic respiration, estimated from transient O_2 profiles, displayed an increase with depth in the upper mm of the mat, followed by a decrease (Figure 4). Photosynthetic production peaked at a rate of 56 μM O_2 min^{-1} at the 0.5-0.75 mm depth interval. Aerobic respiration rates were lower and peaked at greater depth: 32 μM O_2 min^{-1} at 1-1.25 mm depth. Measurements with glass electrodes resulted in a comparable profile although with a much higher spatial resolution. Also, the rates determined with the thinner glass electrodes were 33-200% higher than those calculated from needle electrode measurements. The needle electrode measurements were carried out under natural light conditions, which were higher than those used during shipboard measurement with glass electrodes. Needle electrodes are known to have a higher O_2 consumption than glass electrodes due to a thicker sensing tip. Despite this, estimations of aerobic respiration were very similar for needle and glass electrodes (Figure 4).

Slurry Experiments

The addition of organic carbon to a slurried stromatolite mat stimulated microbial activity resulting in increase of up to 25 and 15 times for aerobic

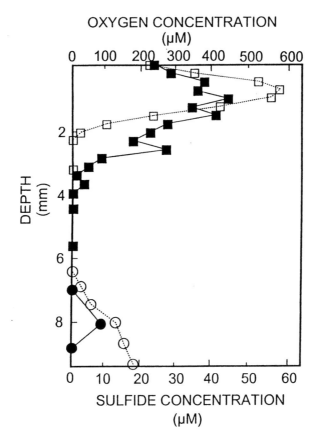

Figure 3. Depth profiles of O_2 and HS⁻ measured in a stromatolite mat. Squares represent [O_2] and circles [HS⁻]. Open symbols depict measurements in stagnant water on the beach, closed symbols are in situ *measurements made in approximately 50 cm of water. Dashed lines indicate the approximate depths of lithified (0-0.5 mm and 3-4.5 mm) and unlithified (0.5-3 mm) layers.*

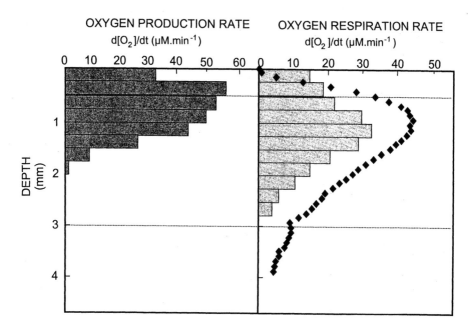

Figure 4. Rates of photosynthesis (left panel) and aerobic respiration (right panel) determined from transient O_2 profiles. Right panel shows a comparison of calculations using needle and glass electrodes, respectively. Bars indicate measurements at the beach in stagnant water under natural illumination (ca. 1600-1800 $\mu M\ m^{-2}\ s^{-1}$) using needle electrodes, diamonds represent measurements in the laboratory in flowing water using a glass microelectrode.

Table I. Potential rates of aerobic respiration (i.e., O_2 uptake) and sulfate reduction (i.e., S^{2-} production) by bacteria in stromatolite homogenate. Changes in O_2 and HS^- concentration over time were measured with a glass microelectrode with a guard cathode and needle electrode, respectively. EPS = Schizothrix exopolymer (1.67 $\mu g.ml^{-1}$). Listed are the mean and in parenthesis the standard deviation.

Carbon source	Aerobic respiration potential $d[O_2]/dt$ ($\mu mol.ml^{-1}.h^{-1}$)	Sulfate reduction potential $d[S^{2-}]/dt$ ($\mu mol.ml^{-1}.h^{-1}$)
Acetate	158 (14)	127 (8)
Lactate	142 (9)	79 (11)
Ethanol	119 (6)	64 (8)
Glycolate	97 (7)	38 (5)
EPS	48 (11)	32 (6)
Glucose	93 (2)	34 (3)
Mannose	NT*	22 (4)
Xylose	53 (5)	14 (3)
Arabinose	41 (8)	16 (6)
Taurine	38 (8)	35 (2)
Cysteate	NT	24(3)
2-Sulfobenzoate	NT	13 (2)
Glutamate	56 (4)	30 (3)
Endogenous rate	6(1)	8 (1)

*NT = not tested

respiration and sulfate reduction, respectively (Table I). Acetate stimulated both respiratory modes by approximately a factor of 10-20, but more complex molecules, such as sulfonates and the C-6 sugar mannose and C-5 sugar xylose, had a much smaller effect. Interestingly, exopolymer isolated from laboratory cultures of the cyanobacterium *Schizothrix* stimulated respiration more than most of the sugars that could be part of the EPS backbone structure (*31*) and had a similar stimulatory effect on sulfate reduction and on aerobic respiration. Sulfonates stimulated both aerobic and anaerobic respiration, similar to earlier findings (*32*). In an experimental comparison, addition of substrate resulted in a greater stimulation of both aerobic respiration and sulfate reduction in the surface lithified layer, relative to the first sub-surface lithified layer. (Table II). Thiosulfate oxidation rates were higher than rates of sulfide oxidation and both

Table II. Comparison of microbial activity in surface lithified layer (0-1 mm) and next lithified layer (approximately from 4 to 7 mm depth). Potential activities were determined in slurried stromatolite samples.

Carbon source	Surface lithified layer		Sub-surface lithified layer	
	Aerobic respiration potential $d[O_2]/dt$	Sulfate Reduction potential $d[S^{2-}]/dt$	Aerobic respiration potential $d[O_2]/dt$	Sulfate reduction potential $d[S^{2-}]/dt$
Acetate	168 (22)	94 (12)	76 (10)	114 (9)
Ethanol	147 (14)	51 (7)	24 (5)	81 (8)
Glycolate	73 (7)	48 (12)	23 (5)	17 (4)
Endogenous	6 (1)	6 (2)	2 (1)	9 (3)

NOTE: Units are $\mu mol.ml^{-1}.h^{-1}$

were unaffected by the addition of molybdate, an inhibitor of sulfate reduction (Table III). When comparing decreases in O_2 and S^{2-}, sulfide oxidation was incomplete, i.e. there was not a stoichiometric conversion from sulfide to sulfate.

Discussion

Microelectrodes have been applied in studies of microbial mats to understand the dynamics of O_2 production and consumption (22,24,30,33). The diel pattern of fluctuating O_2 observed in modern marine stromatolite mats (Figure 2) is not different from similar studies in unlithified microbial mats (34,35), although the stromatolite mat differed in that the maximum O_2 concentration was lower: 600 μM versus >1400μM for Solar Lake mats (34), and was reached later: between 12:00 noon and 2:00pm versus mid morning in other mats (34,35). Likewise, the pH values in most microbial mat studies, including the present, follow the O_2 concentrations and peak when oxygen concentrations are the highest (14,34). It is noteworthy that pH values higher than 10 have been reported for unlithifying mats (37), in contrast to values of 8.9 in the current study.

Table III. Potential rate of aerobic sulfide and thiosulfate oxidation by sulfide-oxidizing bacteria in slurried stromatolite samples.

Treatment	*HS⁻ oxidation rate*		*$S_2O_3^{2-}$ oxidation rate*
	d[O_2]/dt (μmol.ml^{-1}.h^{-1})	d[S^{2-}]/dt (μmol.ml^{-1}.h^{-1})	d[O_2]/dt (μmol.ml^{-1}.h^{-1})
Rate upon electron donor addition	45 (6)	41 (4)	87 (12)
Molybdate	43 (8)	48 (6)	81 (8)
Exogenous rate	12 (3)	5(2)	2(1)

Coinciding maximum values of pH and $[O_2]$ are contributed to the extremely high rates of photosynthesis and hence CO_2 consumption, which result in a shift of the carbonate equilibrium and a subsequent increase of the pH. The pattern of S^{2-} distribution followed the typical image with respect to the O_2 isopleths: low ($<10\mu M$) sulfide concentrations are found near the surface and increase with depth during the night when O_2 penetration is shallow but the presence of S^{2-} is confined to greater depth during daytime when O_2 reaches higher values and penetrates deeper.

Whereas diel fluctuations are nearly identical in unlithified benthic mats and the modern marine stromatolite mats in the present study, the maximum values of $[O_2]$, pH, $[S^{2-}]$ are lower in the latter. This is also reflected in the activities of key functional groups: photosynthetic rates as calculated from light and dark profiles were at least one order of magnitude lower in the stromatolite when compared hotspring mats (22) or temperate marine microbial mats (30,37). This lower photosynthetic activity is further confirmed by the lower maximum pH values observed in stromatolite mats compared to non-lithifying mats. Similarly, aerobic respiration was one order of magnitude lower. It should be noted that the stromatolite mats required the use of more rugged needle electrodes versus the unlithified mats. The needle electrodes have less favorable characteristics for applying the transient oxygen microprofiles technique (30) (a larger outer dimension, a longer response time and greater sensitivity to stirring). The measurements obtained with Clark-type glass microelectrodes with a guard cathode resulted in higher aerobic respiration rates (Figure 4), which suggest a less pronounced difference between the Bahamian stromatolite mats and other non-lithifying mats (20,22,30). However, the difference in microbial activity was shown by traditional measurements of sulfate reduction rates using ^{35}S radiolabel: stromatolite mats sustained maximum rates of 20-35 $\mu M\ h^{-1}$ (32,36) whereas rates ranging from 27 to 588 $\mu M\ h^{-1}$ were reported for unlithified mats (16,17,38). It should be noted that the hydrodynamic regime affects the depth profiles (Figure 3) (39), and thus the actual *in situ* microbial activities may differ from values calculated in this study.

Interestingly, our measurements of potential microbial metabolic activities in general and sulfate reduction in particular show considerably less of a difference in rates. Potential rates of aerobic respiration and sulfate reduction measured in stromatolite mat slurries are similar to those measured in temperate mats (Tables I, II, III) (15,32). It is tempting to speculate, therefore, that actual rates in the stromatolite mat are limited by the availability of organic carbon produced by oxygenic photosynthesis. This notion is consistent with the observed lower rates of photosynthesis discussed above. Unfortunately, there are few published studies in which electrodes are used to assess the potential of aerobic respiration, sulfate reduction and sulfide oxidation. The low concentrations of Fe in Bahamian waters (40) allow for the use of S^{2-} electrodes to measure potential rates of sulfate reduction and sulfide oxidation. However, the Fe-buffering potential of typical shallow benthic sediments (41) makes this technique of limited use.

As mentioned above, although $CaCO_3$ precipitates are occasionally observed in non lithifying mats, they only form continuous layers in stromatolites. In a conceptual model for stromatolite mat functioning, decoupling of the four key metabolic reactions outlined in Figure 1, in both time and space would ultimately determine the location and thickness of micritic $CaCO_3$ precipitates (15). Some of these micritic layers are further lithified through the boring activity of the endolithic cyanobacterium Solentia sp. (42). Lack of biomass and hence organic matter in the unlithified layers of stromatolites could possibly result in slightly higher mass transport than typical microbial mat sediments, but more so, both the thin micritic and the thicker lithified layers may form mass transport barriers. Therefore, it can be hypothesized that the physical sedimentological structure of stromatolites allow for a more pronounced decoupling of sulfate reduction and sulfide oxidation, which ultimately could result in the observed continuous $CaCO_3$ precipitates in stromatolites. These observations in modern marine stromatolites may aid in understanding of biotic and abiotic processes that formed Archean stromatolites. Clearly, needle microelectrodes are a crucial tool in the interpretation of the role various microbes play in geomicrobial processes in these lthified microbial mats.

Acknowledgements

This study was supported by NSF grant OCE 9619314 to PTV, Grants from NASA's Exobiology and Astrobiology Missions and Technology Programs to BMB and NSF grant OCE 9530215 to RPR. We thank Dr. R.N. Glud for valuable comments. This is RIBS contribution number 14.

References

1. Walter, M.R.; Buick, R.; Dunlop, J.S.R. *Nature* **1980,** *284,* 443-445.
2. Bertrand-Sarfati, J.; Awramik, S.M. *Bull. Geol. Soc. Am.* **1992,** *104(9),* 1138-1155.

3. Semikhatov, M.A.; Raaben, M.E. In *Microbial Sediments*; Riding, R.E.; Awramik, S.M., Eds.; Springer Verlag: Berlin, Germany, **2000**, pp. 295-314.

4. Hofmann, H.J. In *Microbial Sediments*; Riding, R.E.; Awramik, S.M., Eds.; Springer Verlag: Berlin, Germany, **2000**, pp. 315-327.

5. Farmer, J.D.; DesMarais, D.J. *Lunar Planet Sci. Conf.* **1994**, *25*, 367-368.

6. Awramik, S.M. In *Microbial mats: Stromatolites*; Y. Cohen, Y.; Castenholz, R.W.; Halvorson, H.O. Eds. Alan Liss, New York, NY, **1984**, pp.1-22.

7. Lyons, W.B.; Long, D.T.; Hines, M.E.; Gaudette, H.E.; Armstrong, P.B. *Geology* **1984**, *12*, 623-626.

8. Farmer, J.D. Arizona State University, personal communication, **2000**.

9. Rogers, D.R. University of Connecticut, personal observation, **1999**.

10. Monty, C.L.V. In *Stromatolites*; Walter, M.R. Ed.; Elsevier, Amsterdam, **1976**, pp. 193-249.

11. Chafetz, H.S.; Buczynski, C. *Palaios* **1992**, *7*, 277-293.

12. Reid, R.P.; Visscher, P.T.; Decho, A.W.; Stolz, J.; Bebout, B.M.; Macintyre, I.G.; Paerl, H.W.; Pinckney, J.L.; Prufert-Bebout, L.; Steppe, T.F.; DesMarais, D.J. *Nature* **2000**, *406*, 989-992.

13. Van Gemerden, H. *Mar. Geol.* **1993**, *113*, 3-25.

14. Visscher, P. T.; Van Gemerden, H. In *Biogeochemistry of Global Change: Radiatively Active Trace Gases*. Oremland, R.S. Ed.; Chapman and Hall, New York, NY, **1993**, pp. 672-693.

15. Visscher, P.T.; Reid, R.P.; Bebout, B.M.; Hoeft, S.E.; Macintyre, I.G.; Thompson, J.A. Jr. *Am. Mineral.* **1998**, *83*, 1482-1494.

16. Canfield, D.E.; DesMarais, D.J. *Science* **1991**, *251*, 1471-1473.

17. Visscher, P. T.; Prins, R. A.; Van Gemerden, H. *FEMS Microbiol. Ecol.* **1992**, *86*, 283-294.

18. Teal, J.M.; Kanwisher, J. *Limnol. Oceanogr.* **1961**, *6*, 388-399.

19. Jørgensen, B.B.; Cohen, Y. *Limnol. Oceanogr.* **1977**, *22*, 657-666.

20. Jørgensen, B.B.; Revsbech, N.P.; Blackburn, T.H.; Cohen, Y. *Appl. Environ. Microbiol.* **1979**, *38*, 46-58.

21. Jørgensen, B.B.; Revsbech, N.P.; Cohen, Y. *Limnol. Oceanogr.* **1983**, *28*, 1075-1093.

22. Revsbech, N.P.; Ward, D.M. *Appl. Environ. Microbiol.* **1984**, *48*, 270-278.

23. Revsbech, N.P. *Limnol. Oceanogr.* **1989**, *34*, 472-476.

24. Glud, R.N.: Gundersen, J.K.; Ramsing, N.B. In *In situ monitoring of Aquatic Systems: Chemical Analyis and Speciation;* Buffle, J.; Horvai, G. Eds.; John Wiley & Sons, New York, **2000**, pp.19-73

25. Visscher, P.T.; Beukema, J.; Van Gemerden, H. *Limnol. Oceanogr.* **1991**, *36*, 1476-1480.

26. *A manual of Chemical and Biological Methods for Seawater Analysis*; Parsons, T.R.; Maita, Y.; Lalli, C.M., Eds.; Pergamon, Oxford, England, **1984**, pp.135-141.

27. Cline, J.D. *Limnol. Oceanogr.* **1969**, *14*, 454-458.

28. Huettel, M., Max Planck Inst Marine Microbiol, personal communication, **2000**.

29. Glud, R.N.; Ramsing, N.B.; Revsbech. N.P. *J. Phycol.* **1992**, *28*, 51-60.

30. Epping, E.H.G.; Khalili, A.; Thar, R. *Limnol. Oceanogr.* **1999**, *44*, 1936-1948.

31. Decho, A.W. Oceanogr. Mar. Biol. Ann. Rev. **1990**, 28, 73-154.

32. Visscher, P.T.; Gritzer, R.F.; Leadbetter, E.R. *Appl. Environ. Microbiol.* **1999**, *65*, 3272-3278.

33. Revsbech, N.P.; Jørgensen, B.B. *Adv. Microb. Ecol.* **1986**, *9*, 293-352.

34. Revsbech, N.P.; Jørgensen, B.B.; Blackburn, T.H.; Cohen, Y. *Limnol. Oceanogr.* **1983**, *28*, 1062-1074.

35. Visscher, P.T.; Van den Ende, F.P. In *Microbial Mats. Structure, Development and Environmental Significance;* Stal, L.J.; Caumette, P. Eds.; Springer Verlag, Berlin, Germany, **1994**, pp. 353-360.

36. Visscher, P.T.; Reid, R.P.; Bebout, B.M. *Geology* **2000**, *28*, 919-922.

37. Visscher, P.T.; Quist, P.;Van Gemerden, H. *Appl. Environ. Microbiol.* **1991**, *57*,1758-1763

38. Fründ, C.; Cohen, Y. *Appl. Environ. Microbiol.* **1992**, *58*, 70-77.

39. Ziebis, W.; Huettel, M.; Foster, S. *Mar. Ecol. Progres. Ser.* **1996**, *140*, 227-237.

40. Till, R. *J. Sed. Petrol.* **1970**, *40*(1), 367-385.

41. Heijs, S.K.; Jonkers, H.M.; Van Gemerden, H.; Schaub, B.E.M.; Stal, L.J. *Estuar. Coast. Shelf Sci.* **1999**, *49*, 21-35.

42. MacIntyre, I.G.; Prufert-Bebout, L.; Reid, R.P. *Sediment.* **2000**, *47*, 915-921.

Chapter 15

In Situ Sulfur Speciation Using Au/Hg Microelectrodes as an Aid to Microbial Characterization of an Intertidal Salt Marsh Microbial Mat

Brian T. Glazer, S. Craig Cary, Laura Hohmann, and George W. Luther, III

College of Marine Studies, University of Delaware, 700 Pilottown Road, Lewes, DE 19958

Sulfur speciation was determined *in situ* in a mid-Atlantic salt marsh microbial mat using a solid-state gold-amalgam voltammetric microelectrode. Chemical constituents were measured in real time with no sample manipulation or processing. A transition from O_2 to partially oxidized sulfur species (polysulfides, thiosulfate, and elemental sulfur) to H_2S was detected through the mat. Metal oxidation (Fe and Mn) of hydrogen sulfide did not occur in the mat, where microbially mediated processes are responsible for H_2S oxidation. The ~7 mm thick mat was frozen *in situ* and cyromicrotome-sectioned into 20-micron sections for visual and molecular biological analyses of the microbial community. The upper 3.16 mm of the mat was dominated by a filamentous morphotype while the lower 3.59 mm was dominated by a rod morphotype. The shift between the two morphologies corresponded to a zone of transition between $S_8/S_2O_3^{2-}$ and S_x^{2-}.

Introduction

It has been the aim of many researchers to attempt to identify bacterial processes and associated geochemical gradients in microbial environments. Dense microbial mats provide one of the best environments for such studies, as they exhibit sharp gradients of redox species and they are relatively easy to sample in the field and manipulate in the lab (*1*). Specifically, much work has been dedicated to cyanobacterial mat communities living in intertidal coastal sediments, marine salterns, hypersaline ponds and lakes, thermal springs, dry and hot deserts and Antarctic lakes (*1-3*). These benthic microbial communities often develop a stratification of diverse chemical and biological species in response to periodic stressors and environmental variations including irradiance, tidal inundation, desiccation, temperature fluctuation, and salinity variations (*1,2*).

The use of microelectrodes in these complex microbial communities has allowed for accurate, fine-scale determination of chemical redox gradients for parameters such as pH, O_2, CO_2, and H_2S (*4, 6-8, 48, 49*). However, until recently advances have been limited by the inability to produce simultaneous *in situ* measurements and differentiations between varying chemical species (*4-8*). The solid-state gold-amalgam (Au/Hg) electrode described by Brendel and Luther (*9*) enables simultaneous measurements of multiple redox species *in situ*, and has recently been shown to differentiate between a variety of sulfur species (*10, 11*).

Here we examined a microbial mat found in a salt marsh environment that produces reduced sulfur species from sediments below the mat. Such sulfur species can then be oxidized back to sulfate as they diffuse upward, by oxygen, iron (III), and manganese (III, IV) compounds (*12-16*), with and without microbial intervention. Oxidation of sulfide can frequently result in the formation of intermediate compounds such as elemental sulfur, polysulfides and thiosulfate, which can be utilized by bacteria (*17*) or may react with metals (*18*) and organic compounds (*19, 20*). Since the sulfur cycle is so dominant in marine systems, with sulfate being the dominant electron acceptor in marine sediments, it is essential to attempt to better understand the dynamics of the complex reactions involved and couple them to the organisms living in those gradients.

In this work we describe the *in situ* application of voltammetric techniques using gold-amalgam microelectrodes to measure microscale vertical variation of chemical species in a seasonal salt marsh microbial mat. We quantify the presence or report the absence of the important chemical redox species, and characterize the microbial consortia using fluorescent microscopy and molecular techniques, with sub-millimeter resolution.

Experimental methods

Electrochemical techniques

A standard three-electrode cell was used in all electrochemical measurements. The working electrode was a gold amalgam (Au/Hg) electrode of 0.1 mm diameter made in 5 mm glass drawn to 0.2-0.3 mm and sealed with epoxy as described by Brendel and Luther (9). Counter (Pt) and reference (Ag/AgCl) electrodes, each of 0.5 mm diameter, were made in commercially available PEEK (polyethyl ether ketone) tubing.

For the voltammetric measurements, the voltage range scanned was from -0.1 V to –2.0 V. In linear sweep voltammetry (LSV) and cyclic voltammetry (CV) the scan rate was 200, 500, or 1000 mVs^{-1}, depending on targeted chemical species. The parameters for square wave voltammetry (SWV) were as follows: pulse height, 24 mV; step increment, 1 mV; frequency, 100 Hz; scan rate, 200 mVs^{-1}. LSV and CV were used to measure oxygen and sulfur species, while SWV was employed for detection of metal redox species. Electrochemically conditioning the electrode between scans removed any chemical species from the surface of the electrode, restoring it for the next measurement. To remove any deposited Fe or Mn the working electrode was conditioned at a potential of – 0.1V for 10s (9). Prior to field excursions, standard curves were produced for O_2, Mn, and sulfur species, as described previously (9, 10).

In July of 2000, a portable Analytical Instruments System (AIS) DLK 100(a) voltammetric analyzer was taken to a large panne (greater than 5 m^2) in Great Marsh, Delaware for *in situ* chemical analyses of a microbial mat. Mats in this location develop on a temporary basis, and can be destroyed by rain, tidal cycles, and dessication. An attempt was made to select a relatively flat, level section of mat with a few centimeters of overlying water for sampling. Measurements were made at approximately noon on a sunny day. The gold amalgam electrode was inserted into the water and into the mat with a manually controlled three-axis Narishige micromanipulator in 0.5 mm increments, and voltammetric scans were produced at each depth in triplicate. To prevent sulfidic fouling, counter and reference electrodes were placed in the water overlying the mat so they would not be in contact with the mat, sediment, or sulfide zone. Salinity and temperature of the overlying water were measured in order to calculate O_2 saturation.

Microbial mat samples

Immediately following voltammetric measurements from the deepest point of the profile, the electrochemical equipment was carefully extracted and a core of the mat was taken using a copper tube with a diameter of 1.6 cm. The mat sample was then immediately pushed through the core tube and plunged into liquid propane. A wide-mouth dewar containing dry ice and a copper receptacle was transported to the field in a small cooler packed with insulating closed-cell foam. Prior to removing the core from the mat, approximately 15 ml of liquid propane was allowed to accumulate in the copper reservoir by inverting a portable, disposable propane cartridge with a modified torch head attachment. This proved to be an inexpensive, efficient method to immediately freeze the mat in the field. Both freezing the mat sample in the field and the use of liquid propane to ensure rapid freezing and small ice crystal formation were expected to minimize distortion and accurately preserve the original spatial orientation of the microbial community (21). The mat sample was kept frozen and sectioned into 20 μm slices with a Spencer 830-C Cryo-Cut cryostat microtome. Three of these 20 μm sections were pooled for molecular analyses at 0.25 mm increments throughout the 7 mm mat. Twenty-μm sections were also removed at 1.75 mm increments and mounted on glass microscope slides to provide for fluorescent microscopic analysis. In addition to the single mat core described above, a replicate core was taken for a preliminary DNA extraction comparison experiment.

Fluorescent microscopy

Sections mounted on microscope slides were stained with DAPI (22) to visualize section constituents and minimize non-specific background staining (23, 24). Images were resolved on an Olympus Provis AX70 microscope using a 100x objective lens and captured with a CTD camera with effective pixel size of 0.074 microns. The filter used to visualize the DAPI stain was a Chroma 3100 band pass filter.

DNA extraction

Typically marsh sediments contain humics and tannins that can inhibit enzymes used in molecular biological analyses. In an effort to optimize methodology and minimize effects of sediment contamination we compared two differing methods of DNA extraction. Hexadecyltrimethyl ammonium bromide (CTAB) and an Isoquick DNA extraction kit (Isoquick, Inc.), were compared using mat material from an area of mat adjacent to where *in situ* electrochemical analyses took place. Each tube of thawed sample sections received 500 µL of CTAB extraction buffer (100 mM tris-HCl, 1.4 M NaCl, 0.4% B-ME, 1% PVP, 2% CTAB, and 20 mM EDTA) preheated to 60°C. Tubes were vortexed and heated to 60°C for 15 minutes. Upon cooling, 500 µL of chloroform:isoamyl alcohol (25:1) was added to the mixture, mixed, and rocked gently for 20 minutes. Following 15 minutes of centrifugation at 12,500 rpm with a Kompspin PC-18 micro table top centrifuge (Composite Rotor, Inc.), the extraction buffer phase was transferred to a clean tube. 5 M sodium chloride (0.5x of recovered volume) and 100% isopropyl alcohol (1x of recovered volume) was added to the tubes, followed by 1 hour freezing at –80°C to precipitate DNA. Tubes were thawed, centrifuged at 12,500 rpm for 15 minutes, and the DNA pellet was washed with cold 70% ethanol. The isolated pellet was then resuspended in distilled water and stored at –20°C for analyses. Replicate samples were subjected to the Isoquick DNA extraction kit (Isoquick, Inc.) according to the manufacturer's instructions.

DNA obtained from both methods was analyzed on a spectrophotometer to assess DNA purity and concentration (*25*) and resolved on a 1% Seakem agarose gel. These preparations were subsequently used as template DNA in the polymerase chain reaction (PCR) (*26*).

PCR amplification

The variable V3 region of 16S rDNA was amplified in the PCR (*26*) with primers designed for conserved regions of the 16S rRNA genes (*27*). The primer nucleotide names and sequences are as follows: 338FGC, 5'-CCTACGGGAGGCAGCAG-3', with an additional 40-nucleotide GC-rich sequence (GC clamp) at its 5' end; and 519RC, 5'-ATTACCGCGGCTGCTGG-3', as described in Muyzer et al. (*28*). Polymerase chain reaction amplification followed the procedure of Muyzer et al. (*28*). Briefly, amplification was performed on an MJ Research PTC-150 Minicycler™ (MJ Research, Inc., Watertown, MA) with a Hot Bonnett™ heated lid in 50 µL reactions with final concentrations as follows: approximately 50 ng of template genomic DNA, 0.4 µM of each of the primers, 0.2 mM (each) deoxyribonucleoside triphosphate, 1x Promega PCR buffer, and 1.5 mM $MgCl_2$. *Taq* DNA polymerase addition followed the hot start technique described by D'Aquila et al., (*45*) to minimize nonspecific annealing of primers to non-target DNA template. The PCR also encorporated the touchdown technique described by Muyzer et al., (28), lowering annealing temperature by 1 °C every second cycle until 55 °C to reduce the formation of spurious by-products during the amplification process. Primer extension was carried out at 72 °C for 1 minute. Amplification products were analyzed in 1% Seakem gels stained with ethidium bromide (EtBr) (1 mg/mL) and resolved with UV light.

DGGE analysis

Denaturing gradient gel electrophoresis was performed with a BioRad Dcode^{TM} Universal Mutation Detection System, as described in Fischer et al. (*29*). Seventeen µL of PCR product was added to 17 µL of loading buffer, and subsequently loaded directly onto 8% polyacrylamide gels (37.5:1 ratio of acrylamide-bisacrylamide) poured with a 25%-55% gradient of denaturant (100% denaturant is 7M urea, 40% deionized formamide). Electrophoresis was performed at a constant voltage of 130 V and temperature of 60°C for 4 hours in 0.5X TAE buffer (40 mM Tris base, 20 mM sodium acetate, 1 mM EDTA). Gels were stained with EtBr (1 mg/ml) for 15 minutes and destained in de-ionized water for 15 minutes. Bands were then resolved on a Fisher Biotech Electrophoresis Systems Variable Intensity Transilluminator and digital images were captured with an Alpha Imager 2000 Documentation and Analysis System (AlphaInnotech Corp., San Leandro, CA).

Results and discussion

Sulfur speciation

Table I shows many of the redox compounds and ions including some of the sulfur species measurable by the Au/Hg electrode. Sulfide oxidation results in formation of polysulfides, elemental sulfur, thiosulfate, sulfite and polythionates (e.g. tetrathionate) in lab studies (*12-16*) but the presence of such intermediates has been confirmed in few field studies (*10, 30-34*). Sulfite, can be measured (*30*) at pH values <6, which are not common in microbial mats or sediments, where pH values are normally > 7. All of the sulfur species we measured gave a single peak except for polysulfides. At slow scan rates, H_2S, S_8, and polysulfides overlap to give one peak at about −0.60V, and the sum of their contributions is termed S_{red} (*10, 30-35*). However, S_x^{2-} are unique because they exist in two oxidation states and it is possible to discriminate each oxidation state with fast potential scans (*10*). At a potential more positive than −0.6 V, S_x^{2-} reacts to form an HgS_x species at the Au/Hg electrode, which is an electrochemical oxidation of the Hg. Scanning negatively then results in HgS_x reduction to Hg and S_x^{2-} overlapping with, but slightly more positive than H_2S and S_8. The $(x-1)S^0$ atoms

Table I. Measurable redox reactions occurring at the 0.1mm Au/Hg electrode surface vs. the 0.5mm Ag/AgCl reference electrode. Oxygen data were collected by linear sweep voltammetry at a scan rate of 500 mVs^{-1}. Cyclic voltammetry was employed in the absence of oxygen to better measure sulfur species, using 200, 500, or 1000 mVs^{-1}. Square wave voltammetry was used at 200 mVs^{-1} for detection of metal redox species. Potentials can vary with scan rate and concentration.   When applying potential from a positive to negative scan direction, sulfide and S(0) react in a two step process: adsorption onto the Hg surface, and reduction of the HgS film. Polysulfides react in a three step process: adsorption onto the Hg surface, reduction of the HgS$_x$ film, and reduction of the S(0) in the polysulfide. Increasing the scan rate separates electrode reactions 4b and 4c into two peaks because reaction 4c is an irreversible process. Increasing scan rate shifts this signal (36). MDL-minimum detection limit

		E_p (V)	MDL (μM)
1a	$O_2 + 2H^+ + 2e^- \longrightarrow H_2O_2$	-0.30	2
1b	$H_2O_2 + 2H^+ + 2e^- \longrightarrow H_2O$	-1.3	2
2a	$HS^- + Hg \longrightarrow HgS + H^+ + 2e^-$	<-0.60	
2b	$HgS + H^+ + 2e^- \longleftrightarrow HS^- + Hg$	~-0.60	<0.2
3a	$S(0) + Hg \longrightarrow HgS$	<-0.60	
3b	$HgS + H^+ + 2e^- \longleftrightarrow HS^- + Hg$	~-0.60	<0.2
4a	$Hg + S_x^{2-} \longrightarrow HgS_x + 2e^-$	<-0.60	
4b	$HgS_x + 2e^- \longleftrightarrow Hg + S_x^{2-}$	~-0.60	<0.2
4c	$S_x^{2-} + xH^+ + (2x-2)e^- \longrightarrow xHS^-$	~-0.60	<0.2
5	$2 RSH \longleftrightarrow Hg(SR)_2 + 2H^+ + 2e^-$	>H_2S/ HS$^-$	
6	$2 S_2O_3^{2-} + Hg \longleftrightarrow Hg(S_2O_3)_2^{2-} + 2e^-$	-0.15	16
7	$S_4O_6^{2-} + 2e^- \longrightarrow 2 S_2O_3^{2-}$	-0.45	15
8	$FeS + 2e^- + H^+ \longrightarrow Fe(Hg) + HS^-$	-1.1	molecular species
9	$Fe^{2+} + Hg + 2e^- \longleftrightarrow Fe(Hg)$	-1.43	15
10	$Fe^{3+} + e^- \longleftrightarrow Fe^{2+}$	-0.2 to -0.9	molecular species

are reduced to sulfide at a slightly more negative potential. Since the reduction of S^0 atoms in S_x^{2-} is an irreversible process, an increase in scan rate results in a shift of the peak to a more negative potential, permitting a visual separation of the HgS_x reduction form the S_8 reduction (Table I, eqs 4a-c, and Figure 1D) (*10, 36*). The peaks can be resolved at 1000 mVs^{-1}.

Microelectrode gradients

Figure 1 depicts representative voltammograms taken from the profiling of the microbial mat. Scan A, 4 mm above the mat in the overlying water, shows the presence of oxygen, whereas Scan B shows a significant signal due to H_2O_2 at 3 mm above the mat surface, masking any signal due to oxygen (detection limit of 2μM). The magnitude of the H_2O_2 signal is typically equal to that of the oxygen signal because at the electrode surface ambient O_2 reduces to H_2O_2 (Table I, eqs 1a-b), which in turn reduces to H_2O. Here however, there is an excess of H_2O_2 which has also been shown to exist in significant quantities in biofilms (*37*). Scan C, at 2 mm below the surface of the mat, shows 2 peaks. The first, at $E_{1/2}$=-0.15 V represents thiosulfate, and the second, at $E_{1/2}$=-0.60 V represents S_{red}. The single peak at -0.6 V (visible at the fast scan rate of 1000 mVs^{-1}) indicates that polysulfides are not present as part of S_{red}. Thus, S_{red} is, in this case, composed of soluble S_8 because this upper part of the mat is more oxidized due to the oxygen gradient. At 5 mm below the mat surface, Scan D shows a clear double peak at $E_{1/2}$=-0.58 V and -0.68 V, indicative of polysulfide formation. The more positive signal is due to S^{2-} sulfur from H_2S and S_x^{2-} and the more negative signal is due to S^0 sulfur from S_x^{2-}. As the bottom of the mat is passed and the underlying sediment penetrated, Scan E shows only 1 peak at $E_{1/2}$=-0.58 V, indicating only H_2S/HS- is present. Trace amounts of FeS are noted in Scan E and other scans deeper into the mat (data not presented). No scans gave evidence of dissolved Fe or Mn in the mat (detection limit of 10 μM).

Figure 2 shows a concentration versus depth profile of oxygen and sulfur species into the mat. The O_2 profile is typical for microbial mats (*3, 5, 38, 39*). Daytime oxygen levels reach a maximum (~200% saturation) in the water column above the mat/water interface, characteristic of an oxygenic photosynthetic microenvironment. Concentrations of H_2O_2, an intermediate in O_2 formation via water splitting, were observed just above the mat surface. Previous microelectrode studies using membrane electrodes could not determine if H_2O_2 was present, but solid-state microelectrodes used in biofilm work have reported the presence of H_2O_2 (37).

While colorimetric analyses (5, 38, 39) have been employed to measure sulfur species such as thiosulfate, tetrathionate and polysulfides in lab cultures and processed microbial mat cores from thick films that could be cut and analyzed, this is the first report of *in situ* sulfur species determination other than H_2S in a microbial mat. Their occurrences were as predicted, and the overall oxidation appears to be primarily due to biological processes, as no metals (Fe, Mn) which could oxidize sulfide were detected, and O_2 did not co-exist with H_2S.

Fluorescent microscopy

Figure 3 depicts four images typical of 20-micron DAPI-stained sections from within the mat. Picture A is of a 20-micron section taken from 1.20 mm below the mat surface that clearly shows the dominance of sheathed bundles of cells, characteristic of oxygenic photosynthetic cyanobacteria. Picture B shows similar community composition 4.80 mm below the mat surface. Depicted in Picture C, 6.34 mm below the mat surface, is a markedly different community where all sheathed cells are absent, and smaller rods & cocci are dominant. This suggests a striking shift in the community composition occurred between 4.80 and 6.34 mm below the mat surface. It is likely that the shift in dominant organisms is in response to differing metabolic requirements, allowing for reduced substrates diffusing upward from the permanently anoxic sediments to act as electron donors Picture D, taken from 7.23 mm below the surface represents the bottom-most microbial constituents of the mat and further suggests a morphological gradient exists from sheathed cells at the top, oxygenated portion of the mat to smaller rods and cocci in the deeper, more sulfidic zone. Other studies have shown that cyanobacteria, purple sulfur bacteria, and sulfate-reducing bacteria are ultimately responsible for the steep and fluctuating chemical gradients (1).

Molecular analyses

The two methods we compared for DNA extraction yielded differing results. Figure 4 illustrates typical results of extractions performed using CTAB versus extractions performed with an Isoquick kit. The reproducible streaking and multiple band formation in the CTAB extracts were probably not attributable to initial template concentrations, which ranged from 200-800 ng/μL and were

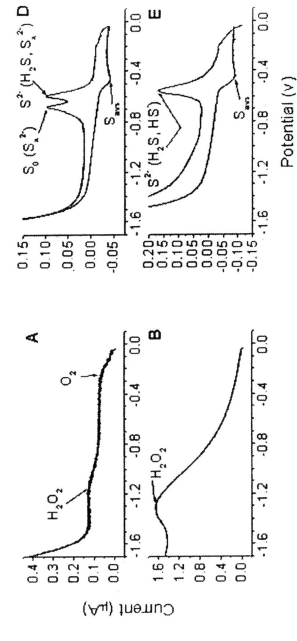

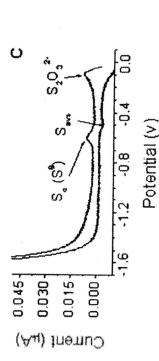

Figure 1. Representative voltammetric scans from varying depths in the mat. A is from 5 mm above the mat surface in the overlying water, B is 3 mm above the mat surface, C is 2 mm below the mat surface, D is 4 mm below the mat surface, and E is 7.55 mm below the mat surface, in the underlying sediment.

A. Linear Sweep Voltammetry @ 500mV/s
B. Linear Sweep Voltammetry @ 500mV/s
C. Cyclic Voltammetry @ 1000mV/s
D. Cyclic Voltammetry @ 1000mV/s
E. Cyclic Voltammetry @ 1000mV/s

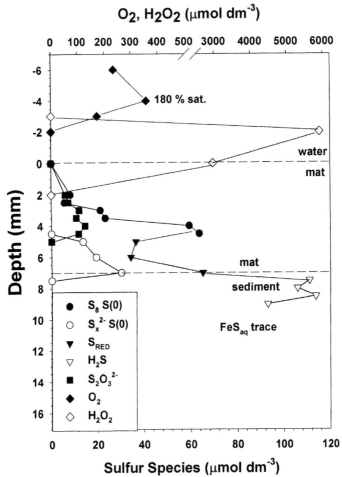

Figure 2. Profile of all chemical components measured in the mat. Once a chemical species is no longer detected it is not plotted. S_{red} represents the sum of all sulfur in the 2- oxidation state from polysulfides and H_2S/HS^- and in the 0 oxidation state from S_8. Note that data points are connected in the figure to show the transistion from S_8 to S_x^{2-} to H_2S.

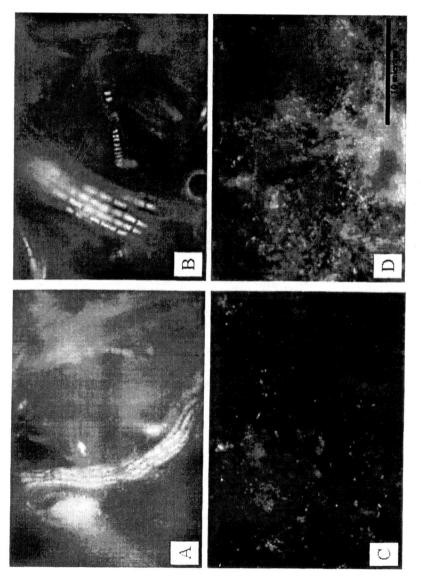

Figure 3. Fluroscent microscopy photographs of mat sections. A. 1.20 mm, B. 4.80 mm, C. 6.34 mm, D. 7.23 mm.

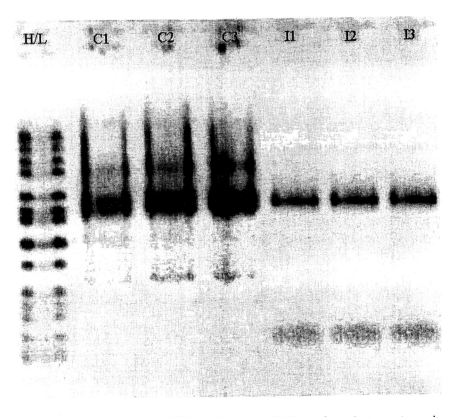

Figure 4. Reverse image of 1% Seakem gel of PCR products (using universal primers Eub A and Eub B) of CTAB (C1, C2, C3) and Isoquick (I1, I2, I3) extracts. Note streaking and multiple bands on C1, C2, and C3.

diluted accordingly prior to PCR, but rather to some unknown contaminant that caused nonspecific binding during the PCR. Although nucleic acids have been extracted from microbial mats with the use of hot phenol in previous experiments (*41, 42*), our comparison found the Isoquick extraction method to yield cleaner DNA, and was subsequently used to obtain the following results.

DNA fingerprints are shown in Figure 5, with each lane of the DGGE gel representing a different depth of the mat, as listed in Table II. A total of 15 different bands could be visualized throughout the profile, however as predicted by the microscopy, each independent section seems to be dominated by two bands (A & B or C & D in Figure 5). Between samples 11 and 12 (depths 3.16 and 3.41 mm), the dominant bands switch from A & B to C & D, suggesting that community composition shifts toward two different organisms, as corroborated by the pattern seen in the microscopy. The microscopy indicated a morphological shift between 4.8 and 6.34 mm, but is within 1.5 mm of the shift shown by the DGGE. The 1.5 mm disparity is most likely due to uneven mat topology and/or sample thickness. The DGGE results came from 60 μm sections while the microscopy used 20 μm sections. Bands representing organisms A & B from Figure 5 are most likely oxygenic photosynthetic cyanobacteria, similar in morphology to the organisms present in panels A & B in Figure 3. Such constituents produce O_2 and organic matter, thus propagating the oxygen gradient, and the surrounding heterotrophic bacteria. Underneath the cyanobacteria, but also exhibiting patchiness throughout the mat (*40*), are purple sulfur bacteria (PSB) and sulfate-reducing bacteria (SRB) (*50*). Bands C & D in Figure 5 and pictured in Figure 3 C & D most likely represent PSB. The PSB use H_2S to produce S_8 (S^0) and organic matter. The sink of organic matter from the cyanobacteria and PSB can then act as fuel for dissimilatory SRB, which reside in the sediments underlying the mat. The SRB use SO_4^{2-} as an electron acceptor, thus releasing H_2S. Sulfate-reducing bacteria are no longer considered to be obligate anaerobes, but have rather been found throughout the mat, including the upper layers where they are in close proximity to cyanobacteria (*54, 43, 44, 51-53*), and may be represented by some of the intermediate bands throughout the mat in Figure 5. Traces of A, B, C, and D can be seen at several non-adjacent depths, as is expected in a non-permanent, stratified, successional mat community. Note also several faint intermediate bands indicating the presence of other species at nearly all depths of the mat.

Conclusions

The data presented clearly show that there is a redox transition between the overlying water, the mat interior, and the underlying sediment, as expected. The transition from reduced sulfide below the mat to partially oxidized sulfide

Table II. Mat sectioning depths and subsequent sample notation. Samples not discussed in the paper yielded insufficient amounts of DNA extract and could not be processed.

Depth (mm)	Sample #	Depth (mm)	Sample #	Depth (mm)	Sample #	Depth (mm)	Sample #
0-0.060	1	1.86-1.92	7	3.72-3.78	13	5.58-5.64	19
0.31-0.37	2	2.17-2.23	8	4.03-4.09	14	5.89-5.95	20
0.62-0.68	3	2.48-2.54	9	4.34-4.40	15	6.2-6.26	21
0.93-0.99	4	2.79-2.85	10	4.65-4.71	16	6.51-6.57	22
1.24-1.30	5	3.1-3.16	11	4.96-5.02	17	6.82-6.88	23
1.55-1.61	6	3.41-3.47	12	5.27-5.33	18	7.13-7.19	24

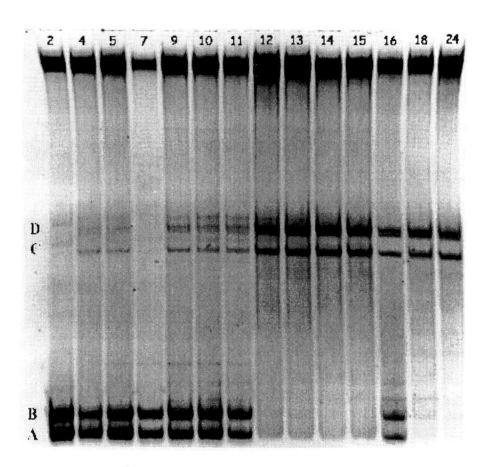

Figure 5. Reverse image of 25/55 DGGE of mat samples. Lanes are numbered by sample, as listed in Table II. Note the distinct shift from A & B to C & D between samples 11 and 12.

species (S_x^{2-}) within the mat to more oxidized sulfur species toward the top of the mat can be explained by the synergy of the chemical and biological processes occurring in the mat as described in Table III. The microscopic data corroborates the patterns present in the DGGE gel (within 1.5 mm), and both combine to form a framework for the observed chemical gradients (also within 1.5 mm). A distinct shift in microbial community is visible with microscopy between 4.8 and 6.3 mm, with DGGE between 3.16 and 3.41 mm, which corresponds to a zone of transition between $S_8/S_2O_3^{2-}$ and S_x^{2-} at 4.2 mm. The offset between each of the methods can be explained by scale. While our electrode profile and DGGE characterization were at roughly the same vertical resolution, lateral resolution could have introduced some disparity between measured chemical and biological gradients. The surface area sampled by the electrode was less than 1 mm^2, while the 20-μm sections used for nucleotide extraction had areas of over 2 cm^2, and were pooled to encompass a thickness of 60 μm. The topography of the mat, variation in organismal stratification, and microbial patchiness all could have contributed to a biological sample less precise than the voltammetric measurements. It is also possible that the freezing of the mat introduced a change in orietation, which could account for differences between the *in situ* voltammetry and biological lab analyses.

Drawbacks to PCR-based analyses of microbial communities have been reviewed at length (55). Two major concerns lie PCR bias and different resolution capabilities of the methods. We attempted to minimize bias by using the same relative amounts of mat material per depth, using the same amount of DNA in the PCR, and using the same amount of PCR product in the DGGE. Additionally, we made comparisons between depths based on presence or absence of certain bands at certain depths, providing an overall picture of community structure shift.

Although the sediments under the mat contained significant quantities of solid phase Fe$_{(II, III)}$, as marsh sediments typically do (*11*), our observations suggest that iron played a very small role in the mat biogeochemistry. Solid phase Fe$_{(II, III)}$ can react with sulfide to form FeS and FeS$_2$, but we did not observe a significant signal for FeS$_{aq}$ in this mat. While the mat surface produces oxidants (O$_2$ and H$_2$O$_2$), which are able to oxidize sulfur compounds, PSB are able to mediate sulfide oxidation directly. The formation of S$_8$ and S$_2$O$_3^{2-}$ within the upper portions of the mat is consistent with the stronger oxidizing characteristics of that section of the mat where oxygenic photosynthesis occurs. Polysulfide formation is occurring in the interior of the mat, most likely as a result of the reaction between HS$^-$ and the S$_8$ produced by PSB and GSB during anoxygenic photosynthesis. Formation of these partially oxidized sulfur species is the result of the oxidation of H$_2$S/HS$^-$ which diffuses from the reducing sediments toward the overlying water. The data set is consistent with laboratory studies of sulfide oxidation (*16*) that show formation of polysulfides early in the

TableIII. Summary of redox reactions that form chemical gradients within the microbial mat environment.

Pathway	Reaction	Contributor
Oxygenic photosynthesis	$CO_2 + H_2O \longrightarrow CH_2O + O_2$	Algae, diatoms, cyanobacteria
Anoxygenic photosynthesis	$CO_2 + 2H_2S \longrightarrow CH_2O + H_2O + 2S^\circ$	Cyanobacteria, purple & green sulfur bacteria
Sulfate reduction	$2(CH_2O) + SO_4^{2-} \longrightarrow 2HCO_3^- + H_2S$	Sulfur reducing bacteria
Fe chemistry	$2Fe^{3+} + H_2S \longrightarrow 2Fe^{2+} + 2H^+ + S^\circ\ (S_x^{2-})$ $Fe^{2+} + H_2S \longrightarrow FeS + 2H^+$ $FeS + H_2S \longrightarrow FeS_2 + H_2$ $Fe^{2+} + S_x^{2-} \longrightarrow FeS + (x-1)\ S_8$	Chemical reaction
Sulfur oxidation	1) $2H_2S + \mu O_2 + Fe^{2+} \longrightarrow$ 2) $H_2S + Fe^{3+} \longrightarrow$ $\left\{\begin{array}{l} S_2O_3^{2-};\ S_8;\ S_x^{2-}; \\ SO_3^{2-},\ SO_4^{2-}; \\ S_4O_6^{2-} \end{array}\right.$	Chemical reaction, sulfur oxidizing bacteria

oxidation and production of thiosulfate later. In Figure 2 polysulfides but not thiosulfate are observed deep in the mat, and thiosulfate but not polysulfides is observed nearer the mat surface.

This was our first attempt to investigate the mat community at high resolution with both voltammetry and molecular biology techniques. Further work is ongoing, including research to produce mat-wide phylogenetic analyses to identify the organisms present, represented by individual DGGE bands. Work is also in progress to create more precise biological sampling techniques, and to provide for a more 3 dimensional profile of redox gradients. Further research is also needed to better understand the effects of the extreme diel fluctuations encountered in the mat. However, it is clear that the microelectrode technique we applied was successful in measuring discrete redox gradients, measuring the absence of dissolved metals, and therefore providing valuable information for predicting variation of the microbial constituents.

Acknowledgements

This work was supported by a grant from the National Oceanic and Atmospheric Administration (NA16RG0162-03). We would like to thank Kathy Coyne, Carol DiMeo, and Matt Cottrell for contributions to our microbial work.

References

1. Stal, L.J. *New Phytol.* **1995**, *131*, 1-32.
2. Jorgensen, B.B.; Revsbech, N.P.; Cohen, Y. *Limnol. Oceanogr.* **1983**, *28(6)* 1075-1093.
3. Van Gemerden, H. *Mar. Geol.* **1993**, *113*, 3-25.
4. Revsbech, N.P.; Jorgensen, B.B.; Brix, O. *Limnol. Oceanogr.* **1981**, *26*, 717-730.
5. Revsbech, N.P.; Jorgensen, B.B. *Limnol. Oceanogr.* **1983**, *28*, 749-756.
6. Visscher, P.T.; Beukema, J.; Van Gemerden, H. *Limnol. Oceanogr.* **1991**, *36*, 1476-1480.
7. Epping, E.H.G.; Jorgensen, B.B. *Mar. Ecol. Prog. Ser.* **1996**, *139*, 193-203.
8. Luther, G.W., III; Brendel, P.J.; Lewis, B.L.; Sundby, B.; Lefrancois, L.; Silverberg, N.; Nuzzio, D.B. *Limnol. Oceanogr.* **1998**, *43*, 325-333.
9. Brendel, P.J.; Luther, G.W., III. *Environ. Sci. Technol.* **1995**, *29*, 751-

10. Rozan, T.F.; Theberge, S.M.; Luther, G.W., III. *Anal. Chim. Acta.* **2000**, *415*, 175-184.

11. Luther, G.W., III; Glazer, B.T.; Hohmann, L.; Popp, J.I.; Taillefert, M.; Rozan, T.F.; Brendel, P.; Theberge, S.M.; Nuzzio, D.B. *J. Environ. Monit.* **2001**, *3*, 61-66.

12. Yao, W.; Millero, F.J. *Geochim. Cosmochim. Acta.* **1993**, *57*, 3359-3365.

13. Yao, W.; Millero, F.J. *Mar. Chem.* **1996**, *52*, 1-16.

14. dos Santos Afonso, M.; Stumm, W. *Langmuir.* **1992**, *8*, 1671-1675.

15. Hoffman, M.R.; *Environ. Sci. Technol.* **1977**, *11*, 61-66.

16. Chen, K.Y.; Morris, J.C. *Environ. Sci. Technol.* **1972**, *6*, 529-537.

17. Jorgensen, B.B. *Science.* **1990**, *249*, 152-154.

18. Chadwell, S.J.; Rickard, D.; Luther, G.W., III. *Aq. Geochem.* **1999**, *5*, 29-57.

19. Vairavamurthy, A.; Mopper, K. In *Biogenic Sulfur in the Environment*, Saltzmann, E.;Copper, W. J., Eds.; American Chemical Society, Washington, D.C., **1989**, *vol. 393*, ch. 15, pp. 231-242.

20. Lalonde, R.T.; Ferrara, L.M.; Hayes, M.P. *Org. Geochem.* **1987**, *11*, 563-571.

21. Garman, E.; Schneider, T.R. *J. Appl. Cryst.* **1997**, *30*, 211-237.

22. Porter, K.G.; Feig, Y.S. *Limnol. Oceanogr.* **1980**, *25*, 943-948.

23. Fry, J.C. In *Methods in Microbiology*, Gigorova, R. Norris, J. R., Eds.; *vol. 22*, Academic Press, London, pp. 41-85.

24. Bloem, J. *Mol. Microbiol. Ecol. Man.* **1995**, *4.1.8* , 1.

25. Sambrook, J.; Fritsch, E.F.; Maniatis, T. *Molecular Cloning: A Lab Manual.* **1989**.

26. Saiki, R.K.; Gelfand, D.H.; Stoffel, S.; Scharf, S.J.; Higuchi, R.; Horn, G.T.; Mullis, K.B.; Erlich, H.A. *Science.* **1988**, *239*, 487-491.

27. Medlin, L.; Elwood, H.J.; Stickel, S.; Sogin, M.L. *Gene.* **1988**, *71*, 491-499.

28. Muyzer, G.; de Waal, E.; Uitterlinden, A.G. *Appl. Environ. Microbiol.* **1993**, *59(3)*, 695-700.

29. Fischer, S.G.; Lerman, L.S. *Proc. Natl. Acad. Sci.* **1983**, *80*, 1579.

30. Luther, G.W., III; Giblin, A.E.; Varsolona, R. *Limnol. Oceanogr.* **1985**, *30*, 727-736.

31. Luther, G.W., III; Church, T.E.; Giblin, A.E.; Howarth, R.W. In *Organic Marine Geochemistry*, Sohn, M., Ed.; American Chemical Society, Washington, D.C., **1986**, *vol. 305*, ch. 15, pp. 231-242.

32. Batina, N.; Cigenecki, I.; Cosovic, B. *Anal. Chim. Acta.* **1992**, *267*, 157-164.

33. Ciglenecki, I.; Cosovic, B. *Electroanalysis.* **1997**, *9*, 775-778.

34. Wang, F.A.; Buffle, J. *Limnol. Oceanogr.* **1998**, *43*, 1353-1361.

35. Jordan, J.; Talbott, J.; Yakupkovic, J. *Anal. Lett.* **1989**, *22*, 1537-1546.
36. Bond, A.M. *Modern Polarographic Methods in Analytical Chemistry.* **1980**, Marcel Dekker, New York.
37. Xu, K.; Dexter, S.C.; Luther, G.W., III. *Corrosion.* **1998**, *54*, 814-823.
38. van den Ende, F.; Van Gemerden, H. *FEMS Microbiol. Ecol.* **1993**, *13*, 69-77.
39. Visscher, P.T.; Nijurg, J.W.; Van Gemerden, H. *Arch. Microbiol.* **1990**, *155*, 75-81.
40. Paerl, H.W.; Pinckney, J.L. *Microb. Ecol.* **1996**, *31*, 225-247.
41. Teske, A.; Ramsing, N.B.; Habicht, K.; Fukui, M.; Kuver, J.; Jorgensen, B.B.; Cohen, Y. *Appl. Environ. Microbiol.* **1998**, *64(8)*, 2943-2951.
42. Bateson, M.M.; Ward, D.M. *Mol. Microbiol. Ecol. Man.* **1995**, *1.1.4*, 1.
43. Jorgensen, B.B. *FEMS Microbiol. Ecol.* **1994**, *13*, 303-312.
44. Minz, D.; Fishbain, S.; Green, S.J.; Muyzer, G.; Cohen, Y.; Rittman, B.E.; Stahl, D.A. *Appl. Environ. Microbiol.* **1999**, *65(10)*, 4659-4665.
45. D'Aquila, R.T.; Bechtel, L.J.; Videler, J.A.; Eron, J.J.; Gorczyca, P.; Kaplan, J.C. *Nucleic Acids Research.* **1991**, *19*, 3749.
46. Murray, A.E.; Hollibaugh, J.T.; Orrego, C. *Appl. Environ. Microbiol.* **1996**, *62(7)*, 2676-2680.
47. Muyzer, G.; Teske, A.; Wirsen, C.O.; Jannasch, H.W. *Arch. Microbiol.* **1995**, *164(3)*, 165-172.
48. Revsbech, N.; Jorgensen, B.B.; Blackburn, T.H. *Limnol. Oceanogr.* **1983**, *28(6)*, 1062-1074.
49. Kuhl, M.; Jorgensen, B.B. *Appl. Environ. Microbiol.* **1992**, *58*, 1164-1174.
50. Ramsing, N. B.; Kuhl, M.; Jorgensen, B.B. *Appl. Environ. Microbiol.* **1993**, *59(11)*, 3840-3849.
51. Canfield, D.E. and D.J. DesMarais. *Science.* **1991**, *251*, 1471-1473.
52. Frund, C. and Y. Cohen. *Appl. Environ. Microbiol.* **1992**, *58*, 70-77.
53. Visscher, P.T., Prins, R.A., Van Gemerden, H. *Appl. Environ. Microbiol.* **1992**, *86(4)*, 283-293.
54. Cypionka, H. *Ann. Rev. Microbiol.* **2000**, *54*, 827-848.
55. Wintzingerode, F.V., Gbel, U.B., Stackebrandt, E. *FEMS Microbiol. Rev.* **1997**, *21*, 213-229.

New Technologies
in Electrochemistry

Chapter 16

Determination of Geochemistry on Mars Using an Array of Electrochemical Sensors

Samuel P. Kounaves[1], Martin G. Buehler[2], Michael H. Hecht[2], and Steve West[3]

[1]Department of Chemistry, Tufts University, 62 Talbot Avenue, Medford, MA 02155
[2]Jet Propulsion Laboratory, Mail Stop 302–231, 4800 Oak Grove Drive, Pasadena, CA 91109
[3]Orion Research, Inc., Beverly, MA 01915

Determining the geochemistry in the remote hostile Martian environment requires sensors specifically designed to meet such a unique challenge. We report here on the initial development, considerations, and a prototype array of electrochemical sensors for measuring in-situ a variety of ionic species in the Martian soil (regolith). The sensor array consists mainly of potentiometric ion selective electrodes but also includes conductivity, and voltammetric microelectrodes for determination of heavy metals. The array functions as an integral unit and is designed to take advantage of data processing systems such as neural networks.

After decades of intensive laboratory and on-site field investigations, we have only just begun to understand the complex chemistry and interactions of the active biogeochemical systems on Earth. Attempting to understand the past or present geochemistry, or any biogeochemistry if it exists, in a remote hostile extraterrestrial environment presents a truly daunting and unique challenge. To even consider such an undertaking, with the slightest hope of obtaining reasonably meaningful analytical data, requires sensors and instrumentation which must meet constraints and withstand rigors far beyond those encountered on Earth. In addition to limits of mass, volume, and power, the sensors and instrumentation have to withstand temperature fluctuations that may range from -120 to 60°C, and attempt to anticipate any unexpected chemistry such an alien environment might present.

There are currently three bodies within our solar system which appear to have had, or still have, the potential to support aqueous biogeochemistry. These include the planet Mars and two of the moons of Jupiter, Europa and Ganymede. Recent data from the Galileo spacecraft confirmed earlier speculations that both Europa and Ganymede are not only covered by frozen water, but may also have thick layers of liquid water beneath their surface. However, it is unlikely that within the next two decades we will be able to land any type of robotic laboratory on their surface capable of taking subsurface samples or performing chemical analyses.

Mars, on the other hand, has been the target of ten successful missions since the early 1970s. Three of these missions included Landers which gave us the first close up look at the surface and chemistry of another world. During the coming decade numerous missions to Mars have been planned, at approximately 18-month intervals, by he National Aeronautics and Space Administration (NASA), the European Space Agency (ESA), and others. Several of these missions will include opportunities for landed exploration of the Martian surface and perhaps subsurface. Our current research efforts are aimed at developing analytical devices, for one or more of these missions, which will provide the maximum data return within the constraints of the transport craft and the planetary environment to be sampled.

The Martian Environment

After decades of astronomical observation and recent lander/orbiter missions, we know that Mars has a cold, desiccated, radiation bathed surface. A barren windswept landscape composed of rocks, soil, sand, and extremely fine adhesive dust. During the past billions of years the wind has swept the finer material into global dust storms, scattering a layer over the entire Martian surface. A detailed description of the geology, composition, mineralogy and structure of Mars can be found in the seminal 1992 compilation by Kieffer *(1)* and more recent findings of the Pathfinder mission in a special section of the Journal of Geophysical Research *(2)*.

A planet's atmosphere also plays a crucial role in its geochemistry. Mars is currently blanketed by an atmosphere which is <1% of standard Earth pressure, and which varies between 6 to 10 torr depending on the season and altitude. However, in stark contrast to Earth, it consists of 95% CO_2. This provides for a concentration of CO_2 that is about thirty times that found on Earth. Other gases include 2.7% N_2, 1.6% Ar, 0.15% O_2, and 0.03% H_2O. With only 40% of the gravitational attraction of Earth, Mars' atmosphere extends about three times as high as Earth's. Consequently, a 1 cm^2 column of air on Mars has a mass of about 15 g compared to 1000 g on Earth. This thin atmosphere allows the soil on Mars to be bombarded with both ultraviolet light and high-energy cosmic rays, probably resulting in both a highly ionized atmosphere and surface.

Based on what is currently known about the evolution of the solar system and Earth, planetary scientists have hypothesized that about 3.5 billion years ago environmental conditions on both Earth and Mars were possibly very similar. Like Earth, Mars would have possessed a warm moist climate and perhaps conditions favorable to life and biogeochemical activity (3, 4). The widespread presence of water on early Mars is clearly evident in photos from the Viking missions in the late 1970s and the Mars Orbital Camera (MOC) aboard the Mars Global Surveyor (MGS) spacecraft which has been orbiting Mars since late 1997 (5-8).

Even though other processes such as volcanic, glacial, or eolian, may have contributed to some of the individual features, taken as a whole there is little doubt that a liquid (most likely water) flowed on the surface of Mars in its past. Photos from the Viking orbiters and the MOC show a variety of outflow channels, sapping, fluvial valley networks, drainage basins, and liquid erosion. Figure 1 shows two such areas on the edge of the northern lowland region of Chryse Planitia, a smooth 2000km wide depression that has the expected characteristics of a large ocean basin and that connects to the even larger basin of Acidalia Planatia. The Tiu Vallis (Fig.1A) and Ares Valles (Fig.1B) both clearly show the erosion of a massive flow into these lower basins (9, 10). Figure 2 (A) shows what appears to be a valley network of tributaries draining into a smooth basin.

There is clear evidence that water still exists on the Martian surface in the form of the north polar ice cap (11, 12), ice cover on many surface features, fogs, clouds in the atmosphere, and possibly in the colder south polar cap, below the crust of CO_2 ice (12, 13). In the low atmospheric pressure of Mars, liquid water eventually phase-separates into vapor and ice, nonetheless a reservoir of water may exist under the surface and could contribute to an underground transport system. Images from the Mars Orbital Camera (MOC), as the one shown in Figure 2(B), suggest the possibility that sources of liquid water at shallow depth are responsible for the runoff features that apparently occurred in the geologically recent past (14). Whether these hydrological features were the result of a few sudden floods or extended periods of moist climate, the presence of water will have had a significant impact on the geochemistry.

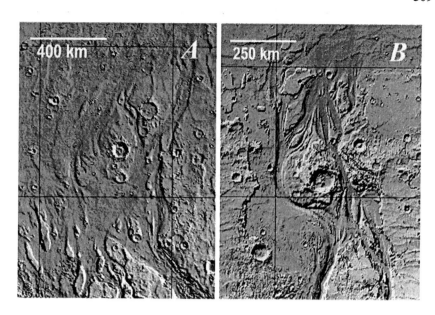

Figure 1.Water flows in the (A)Tiu Vallis and (B) Ares Valles. NASA/MOLA

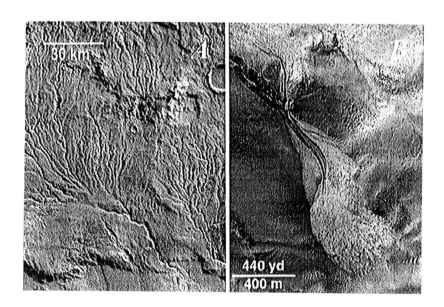

Figure 2.(A) Valley network in the Thaumasia Region 42°S, 93°W. NASA
PIA00185. (B) Recent channels and aprons in East Gorgonum Crater.
NASA/JPL/MSSS MOC2-241

Understanding The Geochemistry of Mars

The geology of Mars shows a clear demarcation of surface topography between the ancient cratered highlands in the south and younger, lower elevation basins in the north. This dichotomy has led to speculation that a significant portion of the planet was once covered by an ocean that was eventually desiccated by some catastrophic event *(9, 15)*. Geochemical indications of this aqueous period in Mars' past should be preserved in the form of salt-rich evaporites resulting from these standing bodies of water and in the geochemical alteration and transport of soluble ionic species. Any primitive biological activity would also have left an imprint on the geochemical differentiation and speciation of many trace metals and inorganic species.

The chemical composition reported by the two Viking and the Pathfinder Landers was very similar. Even though the elemental constituents have been assigned to an assumed set of oxides, it is important to recognize that the raw data from the Sojourner rover's x-ray spectrometer provided elemental composition only. Similar data from Viking has been interpreted as indicating a mix of iron-rich smectite clays, iron oxides, and magnesium sulfate. The iron content is around 15%, with ubiquitous sulfur that could either originate from precipitated salts or volcanic emissions. Although carbonates are expected to be abundant in the Martian soil *(16)*, they appear to be absent on the on the exposed surface *(17)*. However, carbonates can be easily destroyed by UV radiation *(18)* and volcanic sulfuric acid aerosols and sulfates *(19)*, both of which are prevalent on Mars. The Viking experiments also found the surface devoid of any organic molecules *(1)*. This has been attributed to the presence of one or more reactive oxidants in the surface material. Various inorganic oxidants have been proposed *(20, 21)*, most recently superoxide radical ion formation by UV on the soil has been suggested as a most likely mechanism *(22)*.

Data from the Viking and Pathfinder missions suggest that the Martian surface soil consists of approximately 10% salts (dominated by sulfur- and chlorine-containing salts, presumed to be sulfates and chlorides). Widespread salt evaporites can form and accumulate in small enclosed basins such as craters, or on a large scale associated with lakes *(5)* or a large northern ocean *(15)*. Salts can also accumulate wherever volcanic gases act upon the soil, and also in areas where microbial activity might have existed. Since characteristic salts are formed by each of the above processes, a chemical analysis of the salts present at a given location can provide information on the geochemical history of Mars, and in particular the history of liquid water on its surface.

Several models have been developed to theoretically determine likely salt and mineralogical compositions *(23, 24)*. Most recently Catling *(25)* has developed a sedimentation model and demonstrated its use in calculating the evaporite mineral sequence that would be expected in such closed basins.

Figure 3 shows the expected layering using water soluble components derived from weathered igneous rock similar to Martian meteorite basalts, assuming an atmospheric P_{CO2} level of >0.75 torr, and that conditions are neither strongly reducing or oxidizing. The first major carbonate to precipitate would be siderite ($FeCO_3$), followed by magnesian calcites ($Ca_XMg_YCO_3$), hydromagnesite ($Mg_5(CO_3)_4(OH)_2$-$4(H_2O)$), gypsum ($CaSO_4$·$2(H_2O)$), and finally by highly soluble salts such as NaCl.

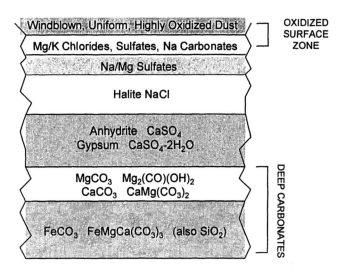

Figure 3. Theoretically derived layering in an evaporite basin.

The Legacy of the Mars Environmental Compatibility Assessment (MECA) Instrument Package

The MECA instrument package, shown in Figure 4, was originally designed and assembled for inclusion on NASA's Mars 2001 Lander Mission. The mission was cancelled in 2000 due to the loss of the Mars Polar Lander in December of 1999. MECA, which was completed and flight-qualified prior to being cancelled, was designed to evaluate potential geochemical and environmental hazards that might confront future Mars explorers, and to guide NASA scientists in the development of realistic Mars soil simulants. In addition to the NASA objectives, MECA had the potential to return data that would be directly relevant to basic geology, geochemistry, paleoclimate, and exobiology.

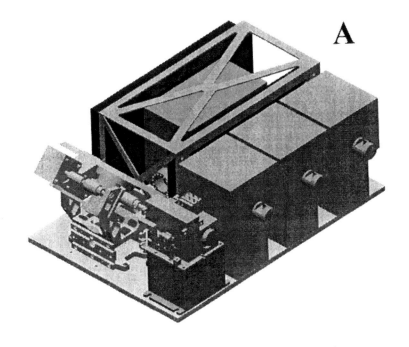

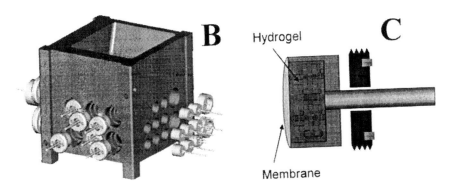

Figure 4. (A) The MECA instrument box showing the four wet chemistry cells and the microscopy station. (B) Exploded view of a wet chemistry cell showing lateral placement of ISE sensors. (C) Diagram of typical ion selective electrode.

The MECA instrument package contained a wet chemistry laboratory, an optical and atomic force microscope, an electrometer to characterize the electrostatics of the soil and its environment, and an array of material patches to study the abrasive and adhesive properties of soil grains. Due to launch vehicle limits, the MECA package was limited to; a mass of 10 kg, peak power of 15 W, and a 35×25×15cm volume. In addition to these flight constraints, MECA had to withstand temperatures ranging from -120 to 60°C, a near vacuum atmosphere, and any unexpected chemistries such an alien environment might present.

The development of MECA, and its use to analyze the surface material in a remote hostile environment, posed a unique set of challenges for remote chemical analysis and more specifically for electrochemical analysis. MECA served as a rigorous test bed for next-generation electrochemical sensors for in-situ planetary chemical analysis.

The MECA Wet Chemistry Laboratory (WCL)

The WCL consisted of four thermally insulated, single-use, independent analysis cells. Each cell was capped with a water reservoir/actuator assembly. The actuator assembly consisted of a sealed water tank with a puncture valve, a sample loading drawer, a stirrer motor with impeller, and a solid pellet dispenser. The pressurized tank contained 30 mL of a "leaching" solution. The solution, which contained several ions at 0.01 millimolar concentrations corresponding to the ISE, served both to extract the soluble components from the soil and as a calibration standard for the reference and ion selective electrodes. The sample loading "drawer" would receive the soil from the lander robotic arm, remove excess soil, deposit it in the chamber, and effect a crude chamber seal. The sample drawer sealed with enough force to maintain a chamber overpressure sufficient to prevent boiling at 27°C (less than 25 torr). The drawer loading compartment held approximately 1.0 cm^3 of soil, and the base was a spring-loaded flap, which would retract to allow the soil to fall into the cell as the drawer was closed. A sieve or screen prevented particles >0.5 mm from falling into the receptacle, while a gap between the receptacle and the seal allowed excess soil to fall off. A scraper or leveling tool removed excess soil as the drawer closed.

Each rectangular cell, fabricated from an epoxy resin and designed to be inert in a range of environments, was 4×4 cm wide and 5 cm deep, with an internal volume of about 35 mL. The cells were designed to lose < 0.5 thermal watts of power against a 40°C temperature gradient. A Viton sealing surface insured a leak rate of < 0.1 cm^3/minute of water vapor at 30 torr (corresponding to 4 micrograms per minute) against an outside pressure of 5 torr, over the operating temperature range, despite contamination with dirt and dust. This leak rate corresponds to a partial pressure of water ten times lower than that of the Mars ambient contribution at 1 cm from the leak.

The WCL Electroanalytical Sensor Array

Arrayed around the perimeter at two levels were 26 sensors. A description of the sensors included in the MECA WCL can be found in Table 1 and shown in Figure 4B-C. The array included both voltammetric and potentiometric based sensors. The *ion selective electrodes* (ISEs) were based on commercially available polymer membrane and solid pellet technology (ThermoOrion Research, Inc.). The *anodic stripping voltammetry* (ASV) for measuring concentrations of heavy metals such as lead, copper, mercury, and cadmium at part-per-billion levels was performed at an array of 10μm microfabricated ultramicroelectrodes *(26-28)*. Specially designed and configured metal electrodes were used to measure conductivity and redox potential.

Table 1. Species and Parameters Measured by the MECA Wet Chemistry Cell Sensor Array.

Parameter/Species	Sensor Configuration &Method
Total Ionic Content.	Conductivity cell, 4-electrode planar chip
H^+	pH, polymer membrane, potentiometric
H^+	pH, iridium dioxide, potentiometric
Dissolved O_2	Au electrode 0.25-mm, membrane-covered, 3-electrode, Au cathode, using CV.
Dissolved CO_2	ISE gas permeable membrane, potentiometric
Redox Potential.	Pt electrode, 1.0-mm disc, potentiometric
Oxidants and Reductants	Au electrode, 0.25-mm disc, using CV
Cu^{2+}, Cd^{2+}, Hg^{2+}, & Pb^{2+}	Au MEA, chip, 512 10-μm elements, ASV
Ag^+, Cd^{2+}, Cl^-, Br^-, I^-,	ISE, solid-state pellet, potentiometric
Li^+, Na^+, K^+, Mg^{2+}, Ca^{2+}, NH_4^+, NO_3^-, ClO_4^-, HCO_3^-	ISE, polymer membrane, potentiometric

CV: Cyclic Voltammetry, ISE: Ion Selective Electrode, ASV: Anodic Stripping Voltammetry, MEA: Microelectrode Array

In some cases the sensors and/or the sensing function of the WCL were duplicated. This was necessitated by; (a) the relative importance of the measured parameter; and (b) the expected reliability of the individual sensor. After being subjected to a series of freeze/thaw cycles during the equipment qualification stage, the solid-state sensors proved to be very reliable. Thus, only one sensor each for ORP, conductivity, CV, ASV, silver, and cadmium were included. There are also several redundant measurements. For example, oxygen and other oxidants or reductants are determined not only by the bare and membrane-covered CV electrodes but also by the ORP sensor; heavy metals are determined by ASV as well as by silver and cadmium ISEs. Because the pH and reference electrodes are inherently less reliable than solid-state sensors and these parameters are so important, three pH sensors and three reference electrodes were included. The reference electrodes were critical for all measurements except CV, CO_2 and conductivity. Since the chloride ISE could have also served as a reference if all the others had failed, a second chloride ISE was also included.

The most critical electrochemical sensors were designed to suffer minimal effects from exposure to a dry evacuated environment. Several of these, the solid-state ISEs, ORP, and conductivity sensors, contain no fluid and could tolerate a vacuum environment well. The gel-backed polymer ISEs were tested in simulated flight environments and were shown to be sufficiently robust and even tolerant of complete dehydration. Ion-selective electrodes (ISEs) were chosen as the predominate sensor because they possess several desirable characteristics. These include, a wide dynamic detection range, availability for a substantial number of ions, and intrinsic simplicity compared to most other analytical tools. Many ISEs can be made compact, rugged, capable of surviving harsh chemical and physical environments, and resistant to radiation damage. In the WCL they were used for determining soluble ions such as sodium, potassium, magnesium, calcium, the halides, pH, dissolved CO_2, and O_2 levels. Figure 5 shows the typical response of an array of ISEs to a calibration solution containing 10^{-5}M of $KHCO_3$, NH_4NO_3, $NaCl$, $CaCl_2$, and $MgCl_2$, and 2.5×10^{-3}M of Li_2SO_4. Each ion was calibrated minus its own ion. For example, the Li^+ ISE was calibrated in the presence of all the above ions except Li^+. Only the perchlorate ISE was calibrated by itself in de-ionized water. All ISEs were calibrated with a primary ion concentration ranging from 1M to 10^{-6}M. Serial dilutions were performed using a 1M stock of each solution to the desired concentration. The calibration was performed using several beakers containing the range of primary ion to be monitored. Stock solutions of the other leaching solution components were added to each solution so the proper background of 10^{-5}M or 2.5×10^{-3}M would be maintained in relation to the primary ion.

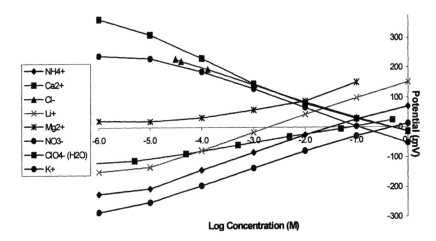

Figure 5. Typical ISE response to nine ions dissolved in multi-analyte solution

The Electronic Tongue – An Integrated Electrochemical Sensor Array

The MECA WCL demonstrated both the untapped potential of electrochemical sensing, and the challenges that still need to be met. Rarely have electrochemical sensors been required to operate and survive under the conditions imposed by an in-situ planetary analysis, including the heat and humidity of the Cape Canaveral launch pad, the desiccating vacuum of space, and the extreme temperatures found on the surface of Mars. During the MECA evaluations the sensors were subjected to a full regimen of shock, vibration, and environmental cycling tests. Even though some sensors such as the glass pH electrode have been used in laboratories for decades, they were eliminated early in the program because of cracking when subjected to temperatures ranging from −100 to 60 °C. But even more troublesome was the degradation of the insulation caused by the repeated expansion and contraction, and the resulting electrical leakage paths between the test solution and the wire lead.

One important finding of the MECA project was that polymer-based ISEs proved to be more resilient under such harsh conditions than originally expected. The MECA ISEs survived despite fears that temperature cycling might cause the membranes to undergo glass transition at low temperature and become brittle. Dehydration of the hydrogel also proved to be of minor concern.

As a next generation of MECA-type WCL sensors, we have been developing an integrated, rugged, low mass/power, electroanalytical sensor composed of an array of *ion selective electrodes (ISE)* and *microelectrodes* (MEA). This *ion sensor array* (ISA) (also referred to as electronic-tongue or e-tongue), builds on the MECA legacy but also extends it based on several new transduction and fabrication concepts. Figures 6 through 8 show the general configuration of the sensor assembly on the ceramic substrate and the analysis chamber with the controller electronics board. The microfabrication, integration, and multiplexing of a large number of individual ISEs and MEAs on a single substrate had not been previously attempted. The key to fabrication is the customization of each sensor by incorporating within it a species selective carrier (ionophore) using several special electrochemical processes to immobilize the ionophore within the sensing membrane.

Recent developments have made it practical to decrease the size of the ISE sensor to a point were a large number of them can be fabricated within a small area, thus, allowing a broad spectrum of analytes to be detected and measured. To fabricate such a device requires the efficient and reproducible deposition of the desired ions or ionophores onto small 50-200μm diameter areas. The novel concept in this work is the electrochemical modification of an individual element in an array of ISE sensor substrates, by electronically addressing it. By applying a voltage to the desired element, the appropriate ionophore or counter-ion is immobilized via an electropolymerization process. The cell allows a rapid flow-through exchange of reagents such so that every electrode pad can be customized in rapid succession. In additon to polymers, other materials may also be deposited on any electrode pad.

Unlike pattern-dependent sensor array devices, this electrochemical sensor will provide both identification and reliable quantitative chemical data. The use of an integrated set of species-semispecific, ionophore-based ISEs, has required that we simultaneously address several fundamental scientific questions, including the scaling of the transduction mechanisms, the doping process, the selection and deposition mechanisms of the appropriate ionophore matrices and polymer substrates, the processing of the array of electroanalytical signals, and the chemometric analysis for interpreting and generating the chemical speciation of the sampled environment.

Such integrated sensor array would have a dramatic impact by enabling a small, low-power, cubic-centimeter-sized device that could be used in-situ on remote planets. This sensor could determine the inorganic constituents of the soil, elucidate the speciation of trace metals, and identify possible chemical biosignatures. The successful implementation of this new ISE fabrication technique will also allow the development of other microsensor arrays that can be used for the in situ determination of the chemical composition of water and water-solvated soil samples in terrestrial oceans, waters, and other aqueous systems.

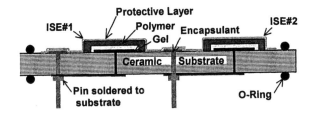

Figure 6. Schematic diagram of two ISE sensor array elements fabricated on a ceramic substrate.

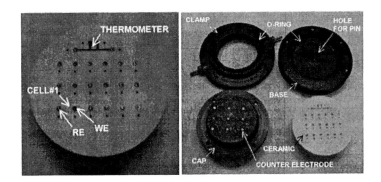

Figure 7. Left: The 4.5-cm diameter ceramic substrate showing the working (WE) and reference (RE) electrodes and a four-terminal thermometer. Right: Chamber parts including ceramic substrate (lower right) and three chamber units: Base (upper right), Clamp (upper left), and Cap (lower left).

Figure 8. Deposition/Sensor chamber mounted on electronics board.

Acknowledgments

MECA was developed at the Jet Propulsion Laboratory, supported by the National Aeronautics and Space Administration (NASA), Human Exploration and Development of Space (HEDS) Enterprise. The contributors to the original MECA instrument can be found at http://mars.jpl.nasa.gov/2001/lander/meca/. The work on the ion sensor array was supported by NASA's Advanced Environmental Control and Monitoring Program.

References

1. Kieffer, H.H.; Jakosky, B.M.; Snyder, C.W.; Matthews, M.S. (Eds) *Mars*; University of Arizona Press: Tucson, AZ, 1992.
2. Results from Mars Pathfinder *J. Geophys. Res.*, **1999**, *104*, 8521-9081.
3. Baker, V. R.; et al. *Nature* **1991**, *352*, 585-94.
4. Jakoskey, B. M.; Jones, J. H. *Nature* **1994**, *370*, 328-29.
5. Carr, M. H. *Water on Mars*; Oxford University Press: New York, **1996**.
6. Smith, P. H.; et al. *Science* **1997**, *278*, 1758-64.
7. Ward, A. W.; et al. *J. Geophys. Res.* **1999**, *104*, 8555-71.
8. Golombek, M. P. *Science* **1999**, *283*, 1470-71.
9. Head, J. W. et al. *Science* **1999**, *286*, 2134
10. Nelson, D.M.; Greeley, R. . *J. Geophys. Res.* **1999**, *104*, 8653
11. Zuber, M.T. et al. *Science* **1998**, *282*, 2053
12. Smith, D.E., et al. *Science* **1999**, *284*, 1495
13. Reference 1, pp 767-95.
14. Malin, M. C.; Edgett, K. S. *Science* **2000**, *288*, 2330-35.
15. Parker, T.J. et al. *J. Geophys. Res.* **1993**, *98*, 11,061-67
16. Reference 1, pp 626-51.
17. Reference 1, pp 594-625.
18. Mukhin, L.M. et al. *Nature* **1996**, *379*, 141-3
19. Bell, J.F. et al. *J. Geophys. Res.* **2000**, *105*, 1721-55
20. Klein, H.P. *Icarus* **1977**,*34*, 666-72
21. Zent, A.P.; McKay, C.P. *Icarus*, **1994**, *108*, 146-57
22. Yen, A. S.; Kim, S. S.; Hecht, M. H.; Frant, M. S.; Murray, B. *Science* **2000**, *289*, 1909-12
23. Clark, B. C.; VanHart, D. C. *Icarus* **1981**, *45*, 370.
24. DeBraal, J. D.; Reed, M. H.; Plumlee, G. S. *LPI Tech. Rep. 92-04,* **1992**, 10
25. Catling, D. C. *J. Geophys. Res.*, **1999**, *104*, 16,453-69.
26. Feeney, R.; Kounaves, S. P. *Electroanalysis*, **2000**, *12*, 677-84
27. Nolan, M.; Kounaves, S. P. *Anal.Chem.*, **1999**, *71*, 3567-3573
28. Kounaves, S. P. et al. *Anal.Chem.* , **1994**, *66*, 418-423

Chapter 17

Integrating an Ultramicroelectrode in an AFM Cantilever: Toward the Development of Combined Microsensing Imaging Tools

Christine Kranz[1], Boris Mizaikoff[1], Alois Lugstein[2], and Emmerich Bertagnolli[2]

[1]School of Chemistry and Biochemistry, Georgia Institute of Technology, Atlanta, GA 30332–0400 (email: christine.kranz@chemistry.gatech.edu)
[2]Institute für Festkörperelektronik und Mikrostrukturzentrum der TU-Wien, Floragasse 7, A–1040 Wien, Austria

A novel approach to integrate a frame microelectrode or ring microelectrode in a standard AFM-tip is presented. Based on microfabrication and Focused Ion Beam (FIB) techniques, an electroactive area can be reproducibly integrated in an exactly defined distance above of the apex of a scanning probe tip. The apex of this modified AFM-tip is then sharpened with FIB, in order to ensure an imaging quality comparable to a non-modified AFM-cantilever. Thus, topographical and electrochemical information can be obtained simultaneously during AFM-imaging. With the demonstrated approach, a precisely defined and constant distance between the micro-electrode and the sample surface can be maintained. The feasibility of this approach is demonstrated by simultaneous topographical and electrochemical imaging of a gold grating on gallium arsenide.

Introduction

Recent developments in scanning probe microscopy are aiming at information on the surface topography, and simultaneously at the physical and chemical properties at the sub-micron scale. A promising technology for the investigation of biological/biochemical systems is scanning electrochemical microscopy (SECM), which was first introduced by the groups of Bard and Engstrom (1-3). Investigations with SECM can be performed in buffered solutions and, therefore, allow the examination of biological species.

In the feedback mode, changes of the diffusion limited Faraday current generated at the microelectrode occur due to hemispherical diffusion of a redox mediator. The microelectrode is scanned in constant height in the x,y-plane within a distance of a few electrode radii above the sample surface. The current response as a function of the microelectrode position is mainly influenced by (i) the morphology of the investigated surface, (ii) the reactivity of the sample surface and (iii) the distance between microelectrode and sample (4).

In generation/collection (G/C) experiments, either substrate-generated species are collected at the tip or, vice versa, tip generated species are collected at the substrate electrode (5,6). Potentiometric tips can be used to detect and image electro-inactive species, which are generated e.g. by biological species at the sample surface. However, positioning of potentiometric tips is more sophisticated due to the absence of redox active species (7-9).

The feedback mode as well as the generation/collection mode were applied for SECM experiments, in order to investigate redox processes at biological surfaces, such as immobilized enzymes (10) and enzyme-labelled immunosensing systems (11,12), transport pathways in membranes (13) and permeability measurements of biological tissues (14). Recently, the redox activity and the detection of metabolism processes of individual cells as well as single exocytosis events were investigated with SECM (15,16).

Still, one of the major drawbacks of SECM is the lateral resolution, which is limited compared with the spatial resolution achievable for Scanning Tunneling Microscopy (STM) or Atomic Force Microscopy (AFM). Hence, further progress in information quantification and qualification has to address the issue of sub-micrometer and nanometer-sized electrodes (*nanoelectrodes*), in order to obtain improved lateral resolution. The integration of current independent height information and the precise knowledge of the distance between the electrode and the sample surface is crucial for accessing the nanometer region. The working distance of the microelectrode is dependent on the electrode size. The current measured at the electrode is usually evaluated for for positioning the electrode in an optimum constant height above the sample surface. Resulting, a significant decrease of the electrode area with an improved lateral resolution would lead to tip crashes in a conventional SECM experiment, due to the diminished working distance.

Several approaches in literature describe the separation of the electrochemical and the topographical information (17-19). One possible solution is the integration of the shear force mode adopted from scanning near field optical microscopy (SNOM) into SECM. Either using an optical detection system (20-22) or a tuning fork (23,24), enables to independently record topographical and electrochemical images. In another approach, electrodes shaped like AFM cantilevers, have been realized by applying a metal coating to a needle shaped AFM-tip, with subsequent insulation (25,26).The combination of SECM with other scanning probe techniques, such as STM, AFM, SNOM, etc., is of particular interest in order to overcome the current limitations in spatial resolution and to retrieve complementary topographical and chemical information. In particular, the trend towards nanoelectrodes for enhanced spatial resolution and the use of micro/nanosensors as SECM tips demands novel approaches for manufacturing and positioning the probe in SECM.

In this contribution, we describe the first integration of *frame-* and *ring-*shaped sub-micro- and nanoelectrodes in AFM-tips, by applying a Focused Ion Beam (FIB) technique. FIB is a microfabrication tool with increasing relevance for maskless processes. The presented fabrication procedure allows the integration of an electroactive area in a precisely defined and exactly known distance beneath the apex of an AFM-tip. With this approach, SECM functionality can be integrated into any standard AFM without further major modification of the core instrument. This is demonstrated in this work with simultaneous topographical and electrochemical contact mode imaging of a gold grating on a gallium arsenide wafer, representing conductive and non-conductive sample areas.

Experimental

The electrochemical experiments were performed in potassiumchloride/potassium-ferrocyanide solution (Fluka, Neu-Ulm/Germany). Aqueous ferrocyanide solutions contained 0.010 mol l^{-1} potassium ferrocyanide trihydrate $[Fe(CN)_6]^{4-}$ with 0.5 mol l^{-1} potassium chloride as a supporting electrolyte. Standard silicon nitride (Si_3N_4) cantilevers were obtained from Digital Instruments (Santa Barbara, CA). The cantilevers have been electrically connected by the gold spring of the cantilever mount, which was subsequently coated with insulation varnish (RS Components, UK).

Cyclic voltammograms were recorded in a Faraday cage with a two-electrode set-up using a silver/silverchloride reference electrode and a potentiostat (27).

Apparatus

Simultaneous AFM/electrochemical imaging was performed with a DI Nanoscope III atomic force microscope (Digital Instruments, Santa Barbara, CA) equipped with a sample cell for measurements in the liquid phase. A scanning head with a maximum imaging range of 120 x 120 μm was used. Silicon nitride cantilevers (length: 200 μm, nominal spring constant: 0.06 Nm^{-1}) with pyramidal tips (base: 4x4 μm, height: 2.86 μm) were used for the reference measurements.

For combined electrochemical and topographical measurements, the AFM was shielded with a Faraday cage using the contact mode base. A three electrode set-up was used for these experiments, with a silver wire operating as an AgQRE, a platinum wire as auxiliary electrode and the integrated microelectrode serving as the working electrode. The potential control and the current measurements were done with a bipotentiostat PG10 (IPS-Jaissle, Münster, Germany). The output signal was directly fed to the data acquisition board of the AFM, which enables the simultaneous correlation of the electrochemical and the topographical image.

Integrated tip preparation

The gold layer was sputtered onto the silicon nitride cantilevers (Parameters: working pressure: 5 x 10^{-3} Torr, power: 172 W, Ar-pressure), which were previously coated with a 5 nm chromium adhesion layer (Sputtering apparatus: Von Ardenne, LS 320 S, Dresden/Germany). Gold layers were deposited homogeneously with a thickness up to 100 nm.

The mixed silicon oxide and silicon nitride insulation layers were subsequently deposited onto the gold-coated AFM cantilever by plasma enhanced chemical vapor deposition (PE-CVD, PlasmaLab80 Plus, Oxford Instruments, Oxford/UK) (Parameters: temperature: 300 °C, pressure:1 Torr, power: 10 W, reaction gases: tetrahydrosilane + ammonia for silicon nitride and tetrahydrosilane + N_2O for silicon oxide). The cantilevers have been tempered at 300 °C. Processing times up to 10 min led to a layer thickness up to 500 nm.

Alternatively, the insulation of the tips was done by anodic deposition of polyphenyleneoxide (2-allylphenole in methanole/water) onto the gold-coated cantilevers (28). A potential of 4 V was applied for 15 min in a two-electrode set-up. The resulting insulation layer thickness was < 200 nm.

3-dimensional pre-shaping of the pyramidal AFM-tip and structuring of the coated modified AFM-tip was performed with a Focused Ion Beam (FIB) system (Micrion 2500, Peabody, MA) equipped with a liquid gallium ion source.

Operation parameters of the FIB system: acceleration voltage: 15 - 50 kV, current: 1 pA - 40 nA.

Results and Discussion

The main focus of this work is dedicated towards novel, reproducible fabrication techniques for sub-micro- and nanoelectrodes, utilizing micromachining technologies. This approach ensures that the electrode shape, geometry and position can be varied maintaining low fabrication tolerances and high reproducibility.

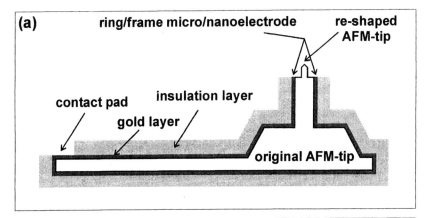

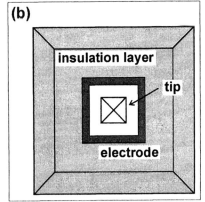
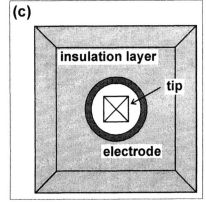

Figure 1: (a) Schematic cross-section of the modified and shaped AFM-tip with an integrated electrode, (b) top view of an integrated frame micro/nanoelectrode and (c) integrated ring micro/nanoelectrode.

Standard microfabrication processes, such as sputtering and PE-CVD, have been used to coat commercially available Si_3N_4 AFM cantilevers, as discussed in the experimental section. For facile tip-shaping and in order to achieve thin, insulating films, polyphenyleneoxide layers have been deposited as an alternative. Figure 1 (a) shows a schematic cross-section of a modified cantilever after the coating and cutting procedures.

FIB as milling and cutting tool with a resolution less than 25 nm, allows to modify small structures, such as the present AFM-tips, in three dimensions. Using several cutting and milling steps with the focused ion beam, rectangular (*frame*) and circular (*ring*) integrated elecrodes have been fabricated, as schematically demonstrated in Figure 1 (b) and (c).

Figure 2 shows the realized electrode geometries, ranging from an electrode diameter of 2 μm to 500 nm. On the left hand side the cantilevers are depicted after the sputtering and coating step. The right hand side shows different electrode geometries, achieved after several FIB cutting/milling steps. Furthermore, these images depict different insulation coatings, with polyphenyleneoxide (Figure 2 (c)) as the thinnest achieved insulation layer (< 200 nm).

The FIB images on the left hand side in Figure 2 (a) and 2 (c) show pyramidal shaped AFM-tips. Both tips are sputtered with a gold layer (100 nm) and are subsequently insulated with alternating Si_3N_4/SiO_2 layers (2 (a)) and polyphenyleneoxide (2 (c)), respectively. The images on the right hand side depict the integrated tips after FIB modification and reveal square-shaped electrodes (*frame electrode*), due to the initial pyramidal tip geometry.

In contrast, Figure 2 (b) shows an AFM-tip, which was cylindrically preshaped by FIB, prior to the metal/insulation coating procedures. This modification step allows the integration of a *ring electrode*, as indicated schematically in Figure 1.

The characterization of the fabricated integrated electrodes was performed by cyclic voltammetry. Several semi-analytical approximations can be found in literature for the description of the steady state current of a ring microelectrode (*29-32*). A comparison of the experimental results and the theoretical values is described elsewhere (*27*).

AFM imaging with FIB-modified cantilevers

An important prerequisite for simultaneous electrochemical and high resolution topographical imaging is the quality and stability of the re-modeled AFM-tip. Hence, at different fabrication steps contact mode AFM images in air and in aqueous solution were recorded. As an example, surfaces of self-assembled monolayers on a (100) p-doped silicon wafer with an island size of 1-2 μm and a height of 0.26 nm (*33*) were used. These studies aim at the

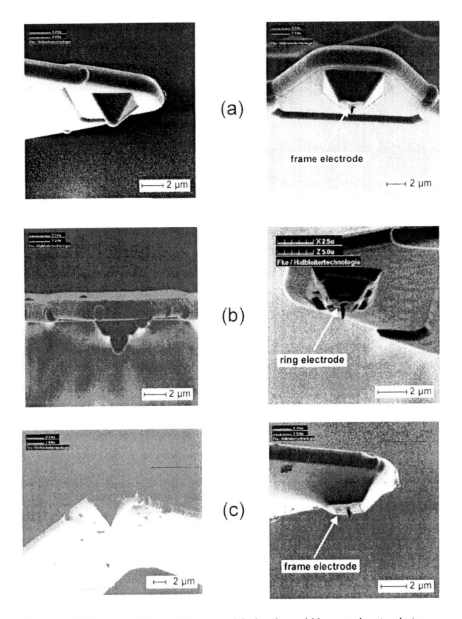

Figure 2: FIB Images of (left) AFM-tips modified with a gold layer and an insulation layer and (right) FIB-cutted AFM-tips. (a) integrated frame nanoelectrode (gold) with siliconnitride/siliconoxide insulation (edge length: 500 nm); (b) integrated ring microelectrode (gold) with siliconnitride insulation (inner diameter: 900 nm, outer diameter: 1000 nm); (c) integrated frame microelectrode (gold) with polyphenyleneoxide insulation (edge length: 2 μm).

feasibility of using modified cantilevers as force sensors. Although up to 1 μm of material (gold, silicon nitride) was deposited on the flexible cantilevers, only 1 of the modified cantilevers showed undesired bending for a total amount of more than 50 modified cantilevers. The demands on the re-modeling procedure, especially the re-sharpening of the original tip, is demonstrated in Figure 3. Due to charging effects during the PECVD process the very end of the tip shows a comparatively large radius of the tip curvature, in contrast to the sharp tip on an unmodified AFM-pyramid. This results in a broadening effect of the imaged structures, due to a curvature up to 600 nm for an 800 nm thick, insulating layer. Figure 3 demonstrates this effect by imaging a self-assembled monolayer with a pillar shaped tip (curvature: 300 nm, pillar height: 2 μm), compared to a non-modified AFM-tip.

In order to ensure high resolution imaging, the original tip is re-shaped by FIB, with a curvature similar to the initial Si_3N_4-tip. Even smaller curvatures than usual for conventional Si_3N_4-tips can be realized with this technique.

Simultaneous AFM imaging and electrochemical mapping with an intergrated ring sub-microelectrode

Figure 4 shows the simultaneously recorded AFM/SECM image of a gold grating, with a periodicity of 4.3 μm and a height of 0.2 μm. The images were recorded with an integrated ring sub-microelectrode (outer diameter: 900 nm, height of the re-shaped pillar: 1.5 μm), as shown in Figure 2. The imaging quality achieved with the re-modeled AFM-tip is comparable to an original Si_3N_4 cantilever with a nominal appendix of 20 nm. The SECM image was recorded in the feedback mode, involving the oxidation of ferrocyanide to ferricyanide in a potassium chloride solution at the integrated ring sub-microelectrode. The electrode was held at a potential of 600 mV vs. AgQRE. The periodicity of the conducting/non-conducting sections in the electrochemical image corresponds well to the topographical image of the gold grating on the GaAs substrate. The simultaneously recorded topographical and electrochemical image in Figure 4 demonstrates clearly the feasibility of integrating SECM functionality into a commercial AFM-tip.

Environmental applications of SECM

SECM is a non-invasive in-situ method and, hence, particularly suitable for the investigation of biological specimen or biomolecules. Recent literature reveals that ongoing work in the field of SECM is markedly focused towards cell biology and cell signaling processes (*34*). Potentiometric and amperometric

328

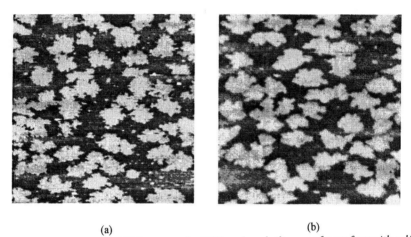

(a)   (b)

Figure 3: Contact mode AFM images of a (100) p-doped silicon wafer surface with self-assembled monolayer islands (diameter: 1-2 µm, height: 0.26 nm) in air. (a) AFM-image with a non-modified, standard AFM-tip (scan area:
10 x 10 µm, scan rate: 5 Hz). (b) AFM-image with the pillar-shaped tip of a modified cantilever before final re-sharpening (scan area: 10 x 10 µm, scan rate: 5 Hz).

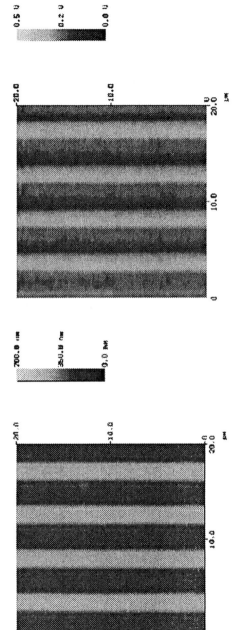

Figure 4: Height and current images of a model gold grating, simultaneously recorded in 0.01 mol l⁻¹ [Fe(CN)6]⁴⁺ contained in 0.5 mol l⁻¹ KCl. The integrated ring sub-microelectrode used for this experiment had an inner diameter of 900 nm and a re-shaped tip height of 1.5 μm. Scan parameters: scan area: 20 μm x 20 μm, scan rate: 2 Hz. The tip was scanned from left to right and top to bottom.

(left)   Top view of the AFM image (scale: 0 – 700 nm).
(right)   Top view of the simultaneously recorded SECM image obtained with the integrated electrode (scale: 0 – 0.5 V) The electrode was held at a potential of + 0.6 V versus AgQRE.

330

microelectrodes are applied, in order to detect neurotransmitters, metal cations and free radicals. Morever, due to the variety of electrochemical processes involved in biological processes, SECM is nowadays deployed for the investigation of enzyme activities, imaging of antibodies, electrochemical detection of metabolites, investigation of cell communication and the characterization of cultivated cell reactions to external stimuli. Biomedical applications are focused on e. g. the investigation of ionophoretic transport phenomena and measurements of redox activities of normal and cancerous cells.

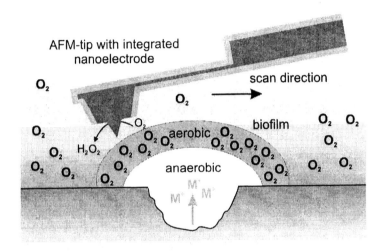

Figure 5: Scheme of simultaneously recorded microbial topography and electrochemical activity in seawater. The depletion of O_2 due to the presence of microbes can be imaged in a generation/collection mode of the integrated electrode, by reducing O_2 to H_2O_2. The recorded current would decrease above the microbial cells.

As advanced environmental analysis is closely related to biological and biochemical processes, imaging methods providing laterally resolved chemical information are gaining substantial importance. One of the most interesting fields of application for SECM can be envisaged in the field of marine environmental sciences and marine biology. In particular, surface phenomena involving electrochemical interactions, such as biocorrosion and biofouling, can be investigated with high spatial resolution. The growth and behavior of microorganisms and their metabolism in aqueous solution can be monitored, in order to gain deeper insight on surface corrosion processes.

As an example, the physical presence of microbes on a surface exposed in a marine environment can cause microbially induced corrosion (MIC) or biocorrosion (35). This effect is a significant problem for a wide variety of industrial applications relying on the use of metal surfaces in a marine environment. Corrosion prevention measures still have to be optimized, as the precise mechanisms of MIC are not fully understood. The initial states of the microbial growth may be mapped with SECM by measuring the oxygen depletion due to the oxygen consumption of microorganisms with high spatial resolution, as schematically shown in Figure 5 using the scanning probe technology presented in this work. In principle, even single cell activity can be addressed based on the oxygen reduction of the surrounding environment (36). Furthermore, oxygen depletion promotes the growth of anaerobic organisms, such as sulfate reducing bacteria, which are among the most frequent reasons for biocorrosion. Additionally, pH-changes resulting from microbial metabolism can be imaged with integrated pH-microelectrodes.

The lateral resolution of electrochemical imaging experiments can be improved significantly by integrating the electrode into a scanning probe tip. This novel technology provides a tool for the improved in-situ investigation of biocorrosion processes and similar phenomena.

Conclusion

We present a novel approach, which enables the integration of frame micro/nanoelectrodes and ring micro/nanoelectrodes into AFM-tips, using microfabrication and micromachining techniques. This technology ensures the reproducible and accurate fabrication of such combined tips.

As demonstrated, simultaneously recorded topographical and electrochemical images can be obtained by using these integrated tips in a commercial AFM instrument.

The distance of the electroactive area to the sample surface can be exactly correlate by adapting the length of the topographical probe with a micromachining technique like FIB. Thus, an optimized and defined working distance of the electrode is ensured, without theoretical fitting of current/distance approach curves.

Ongoing work is focused on integrated nanoelectrodes and different electrode geometries, in order to further improve the lateral resolution of the electrochemical image. The desribed fabrication of the integrated tips is the first step towards the integration of electrochemical micro- and nanosensors into an imaging scanning probe technique. Since the positioning of the tip is based on the AFM-cantilever and not on the current, integrated microsensors can be used as imaging tools also for electroinactive species, which is important in many biological processes.

Acknowledgement

C. Kranz acknowledges financial support from the *Deutsche Forschungsgemeinschaft (DFG)* under research grant KR1797/1-1. B. Mizaikoff thanks the *Austrian Science Foundation (FWF)* for support within project P14122-CHE. Finally, the authors thank Prof. E. Gornik (Institut für Festkörperelektronik, TU-Wien) and his staff for the opportunity to use the clean room facilities.

1. Engstrom, R. C.; Weber, M.; Wunder, D. J.; Burgess, R.; Winquist, S. *Anal. Chem.* **1986**, *58*, 844-848.
2. Engstrom, R. C.; Meany, T.; Tople, R.; Wightman, R.M. *Anal. Chem.* **1987**, *59*, 2005-2010.
3. Liu, H. Y.; Fan, F.-R. F.; Lin, C. W.; Bard, A. J. *J. Am. Chem. Soc.* **1986**, *108*, 3838-3839.
4. Kwak, J.; Bard, A. J. *Anal. Chem.* **1989**; *61,* 1221-1227.
5. Zhou, F.; Bard, A. J. *J. Amer. Chem. Soc..* **1994**, *116*, 393-395.
6. Bard, A. J.; Fan, F.-R. F.; Mirkin, M. V. in *Electroanalytical Chemistry*, Bard, A. J.; Ed.; Marcel Dekker: NY, 1994 Vol. 18, 287-293.
7. Horrocks, B. R.; Schmidtke, D.; Heller, A.; Bard, A. J. *Anal. Chem.* **1993**; *65*, 3605-3614.
8. Horrocks, B. R.; Mirkin, M. V.; Pierce, D. T.; Bard, A. J.; Nagy, G.; Toth, K. *Anal. Chem.* **1993**, *65*, 1213-1224.
9. Wei, C.; Bard, A. J.; Kapui, I.; Nagy, G.; Toth, K. *Anal. Chem.* **1995**, *67*, 1346-1356.
10. Wittstock, G.; Schuhmann, W. *Anal. Chem.* **1997**, *69*, 5059-5066.
11. Shiku, H.; Hara, Y.; Matsue, T.; Ochida, I.; Yamauchi, T. *J. Electroanal. Chem.* **1997**, *438*, 187-190.
12. Wittstock, G.; Yu, K. J.; Halsall, H. B.; Ridgway, T. H.; Heinemann, W. R. *Anal. Chem.* **1995**, *67*, 3578-3582.
13. Scott, E. R.; Phipps, J. B.; White, H. S.; *J. Invest. Dermatol.* **1995**, *104*, 142-145.
14. Macpherson, J. V.; Beeston, M. A.; Unwin, P. R.; Hughes, N. P.; Littlewood, D. *J. Chem. Soc. Farady Trans.* **1995**, *91*, 1407-1410.
15. Yasukawa, T.; Kaya, T.; Matsue, T. *Anal. Chem.* **1999**, *71*, 4637-4641.
16. Hengstenberg, A.; Blöchl, A.; Dietzel, I. D.; Schuhmann, W. *Angew. Chem.* **2001**, *in press*.
17. Wipf, D. O.; Bard, A. J. *Anal. Chem.* **1992**, *64*, 1362-1367.
18. Borgwarth, K.; Ebling, D. G.; Heinze, J. *Ber. Bunsen-Ges. Phys. Chem.* **1994**, *98*, 1317-1321.
19. Jones, C. E.; Macpherson, J. V.; Barber, Z. H.; Somekh, R. E.; Unwin, P. R. *Electrochem. Commun.* **1999**, *1*, 55.

20. Ludwig, M.; Kranz, C.; Schuhmann, W.; Gaub, H. E. *Rev. Sci. Instrum.* **1995**, *66*, 2857-2860.

21. Hengstenberg, A.; Kranz, C.; Schuhmann, W.; *Chem. Europ. J.* **2000**, *6*, 1547-1554.

22. Kranz, C.; Ludwig, M.; Gaub, H. E.; Schuhmann, W. *Adv. Mater.* **1996**, *8*, 634-637.

23. James, P. J.; Garfias – Mesias, L. F.; Moyer, P. J.; Smyrl, W. H. *J. Electrochem. Soc.* **1998**, *145*, L64-L66.

24. Büchler, M.; Kelley, S. C.; Smyrl, W. H. *J. Electrochem. Solid-State Lett.* **2000**, *3*, 35-38.

25. Macpherson, J. V.; Unwin, P. R. *Anal. Chem.* 2000, **72**, 276-285.

26. Macpherson, J. V.; Unwin, P. R. *Anal. Chem.* 2001, **73**, 550-557.

27. Kranz, C.; Mizaikoff, B.; Friedbacher, G.; Lugstein, A.; Smoliner, J.; Bertagnolli, E.; *Anal. Chem.* **2001**, in press.

28. Potje-Kamloth, K.; Janata, J.; Josowicz, M. *Ber. Bunsenges. Phys. Chem.* **1989**, *93*, 1480-1485.

29. Szabo, A. *J. Phys. Chem.* **1987**, *91*, 3108-3111.

30. Fleischman, M.; Bandyopadhyay, S.; Pons, S. *J. Phys. Chem.* **1985**, *89*, 5537.

31. Fleischman, M.; Pons, S. *J. Electroanal. Chem.* **1987**, *222*, 107.

32. Smythe, W. R. *J. Appl. Phys.* **1951**, *22, 1499.*

33. Vallant, T.; Brunner, H.; Mayer, U.; Hoffmann, H.; Leitner, T.; Resch, R.; Friedbacher, G. *J. Phys. Chem. B* **1998**, *102*, 7190-7195.

34. Wittstock, G; Horrocks, B. R. in *Scanning Electrochemical Microscopy*, Bard, A. J.; Mirkin, M. V., Ed.; Marcel Dekker: NY, **2001**.

35. Borenstein, S. W. Microbiologically Influenced Corrosion Handbook, Industrial Press Inc.: NY, **1994**.

36. Yasukawa, T.; Kaya, T.; Matsue, T. *Chem. Lett.* **1999**, 975-976.

Trace Metal Complexation

Chapter 18

A Review of Competitive Ligand-Exchange–Voltammetric Methods for Speciation of Trace Metals in Freshwater

HanBin Xue[1] and Laura Sigg[2]

[1]Swiss Federal Institute for Environmental Science and Technology (EAWAG),
CH–6047 Kastanienbaum, Switzerland
[2]Swiss Federal Institute for Environmental Science and Technology (EAWAG),
CH–8600 Dübendorf, Switzerland

The competitive ligand exchange /voltammetry techniques are very useful to indirectly determine free aqua metal ion concentrations and to detect specifically strong ligands at the extremely low metal levels in natural waters. A number of criteria must however be fulfilled to obtain accurate results. We review in this chapter the techniques under freshwater conditions, in terms of principles of the ligand-exchange reactions, of theoretical criteria and of practical assumptions. Ligand-exchange kinetics have been studied as a prerequisite for equilibrium speciation. We also discuss and give recommendations for determining optimal working windows and for selecting the various types and amounts of ligands used under freshwater conditions. Accuracy and validation of these methods are discussed. Finally, some results on the speciation of. Cu, Zn, Cd, Co and Ni in Swiss lake, river and ground waters are presented.

Introduction

Anthropogenic inputs of trace metals have increased concentrations of metal ions over background levels in many rivers, lakes and groundwaters. Metal ions occur in natural waters in a variety of chemical forms (i.e., speciation), namely in solution as free aqua ions, as complexes with inorganic or organic ligands, as well as in particulate or colloidal phases. These various chemical species possess unique reactivities, which in turn affect their environmental fate. For example, the transport of metal ions in natural waters and between different compartments in the environment is controlled by metal speciation. The distribution of metal ions between the particulate and the solution phase is an important factor that affects the fate of metals in the environment. This distribution process in turn is influenced by solution phase complexation reactions. All these reactions can be linked to the free ion concentrations of the metals.

The bioavailability of metal ions to organisms (both as nutrients and as toxic compounds) is strongly dependent upon chemical speciation. In many instances, these effects are related to the concentrations of the free aqua ions (1, 2). The concentration of free aqua ions in natural waters is regulated by the complex interactions between the various ligands present in solution, particle surfaces and biota. Many studies have provided evidence for the significance of the complexation of various metals by natural organic ligands in solution. This phenomenon has been observed in the oceans (3-10), and is also likely to occur in freshwater (11-15) where dissolved organic carbon (DOC) levels are generally higher (16).

Complexation of metal ions by organic ligands cannot be modelled in a simplistic manner (e.g., models used in calculation of inorganic speciation), because natural organic matter (NOM) is heterogeneous with respect to its composition and to the metal-binding functional groups. Humic and fulvic acids (macro-organic ligands that occur ubiquitously in natural waters) are important components of NOM. Complexation of metal ions by humic or fulvic acids has been widely investigated and characterized by complex relationships involving the ratio of metal ions to ligand concentrations, ionic strength and pH. Various models have been developed to interpret the experimental humic and fulvic acid-metal complexation data (16-21). Moreover, a number of studies have indicated in seawater and in freshwater the presence of strong ligands for various metal ions, which may be freshly produced by organisms (6, 11, 12, 22-26). In addition to these natural organic ligands, synthetic organic ligands are also introduced by anthropogenic inputs into natural waters. For instance, EDTA and NTA have been detected in Swiss rivers in a similar concentration range as that of trace metals (27). Sulfide has recently been shown to be an important ligand of trace metals, even in oxic waters (28, 29).

The general concepts of metal speciation and bioavailability are currently being evaluated to define water quality criteria for metals in freshwater in the USA and in Europe (*30, 31*). Significant progress has been made in modelling the interactions of metals with dissolved organic matter and with suspended particles (*19*). In order to develop and apply such models to water quality criteria, reliable data sets of free ion concentrations and of complexation parameters for trace metals are requested.

A number of analytical approaches have been developed to measure metal speciation in natural waters, according to various criteria. Labile metal concentrations are defined with respect to a certain technique, e. g. in the case of direct voltammetric measurements, such as of a voltammetric in situ probe with microelectrodes (*32*). The recently developed technique of diffusive gradients in thin film (DGT) measures operationally defined DGT-labile metal species (*33, 34*). Metal ion-selective electrodes have been used to directly measure ambient free metal concentrations in river or lake water, but they are in general limited by their sensitivity and selectivity (*35-37*). To determine representative complexation parameters (ligand concentrations, conditional stability constants), it is essential that an analytical method can be used at the natural low range of trace metal concentrations (typically in the nanomolar range).

The indirect methods of competitive ligand-exchange (CLE) followed by voltammetric measurements (cathodic stripping voltammetry CSV or anodic stripping voltammetry ASV, CLE-CSV or CLE-ASV) have been applied to metal speciation at ambient levels in seawater and in freshwater (*4, 5, 7, 9, 11-15, 24, 36, 38-42*). In these methods, a competing ligand with known complexation properties (complex stoichiometry and stability constants) and known concentration is added to a water sample (Figure 1). A new equilibrium between the original metal species and the complexes with this competing ligand (Figure 1, MR_z) is established. The concentration of the complexes with the competing ligand is measured specifically by CSV or ASV. Free aqua metal ion concentrations are obtained by equilibrium calculations and calculated for the original sample. By titrating a water sample with a metal M, complexation parameters (ligand concentrations and conditional stability constants) are determined.

In this paper, we review the techniques of competing ligand-exchange coupled with voltammetry for their use under freshwater conditions, in terms of principles of the ligand-exchange reactions, of theoretical criteria and of practical assumptions. Special attention is given to studies related to ligand-exchange kinetics, as a prerequisite for equilibrium speciation. We also discuss and give recommendations for determining optimal working windows and for selecting the various types and amounts of ligands used under freshwater conditions. Accuracy

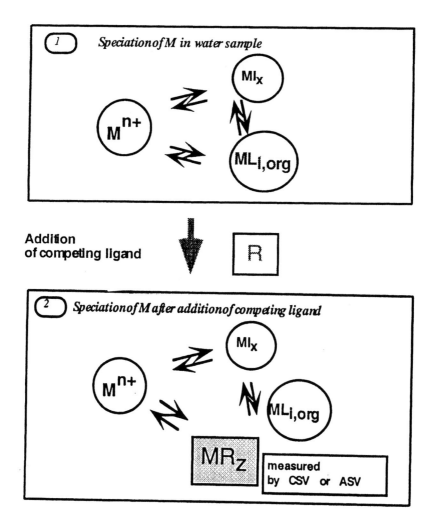

Figure 1. Scheme of a competing ligand exchange method coupled to voltammetric measurement. In a water sample, the dissolved metal M is originally distributed among free aquo metal ions (M^{n+}), inorganic complexes (MI_x, e. g., MOH^+, MCO_3^0) and organic complexes ($ML_{i,org}$). After addition of a competing ligand R and equilibration, the complexes MR_z are measured specifically by CSV or ASV. The mass balance of M in the case 1 is given by eq 1, in the case 2 by eq 4 in the text.

and validation of these methods are discussed. Finally, some results on speciation of Cu, Zn, Cd, Co and Ni in Swiss lake, river and ground waters are presented.

Principles of competitive ligand-exchange /voltammetric methods

Calculation of ligand-exchange equilibrium

In the following section we present a series of examples to illustrate the theory and data treatment of ligand-exchange equilibria. An essential assumption in these calculations is that equilibrium has been reached after addition of the competing ligand. Ligand-exchange kinetics are discussed in a subsequent paragraph.

Example 1. Ligand-exchange and DPCSV for Cu, Co or Ni speciation

The ligands used include catechol, oxine, tropolone, salicyladoxime and benzoylacetone for Cu, and dimethylglyoxime for Co and Ni (*7, 9, 12, 15, 38, 40, 42, 43*). The complexes formed with these ligands are suitable for measurement by DPCSV (differential pulse cathodic stripping voltammetry). They are accumulated by adsorption on the mercury electrode at constant potential and then measured by reduction in DPCSV mode. In the following case, the complexes of a metal M with the added ligand R are measured by DPCSV after adsorption of MR_z on the electrode. A titration of a sample with increasing concentrations of a metal ion (M^{2+}) is conducted in the presence of the added ligand (R).

The following mass balance is valid in the original water sample:

$$[M]_T = [M^{2+}] + \sum [MI_{i,x}] + \sum [ML_i] \qquad (1)$$

where I_i and L_i respectively stand for the ith inorganic ligand and organic ligand; $[M]_T$, $[M^{2+}]$, $[MI_{i,x}]$ and $[ML_i]$ represent the concentrations of total dissolved metal, free aqua metal ion, inorganic and organic complexes.

The stability constant for a complex with a number x of inorganic ligand I_i is defined as

$$\beta_{Iix} = \frac{[MI_{i,x}]}{[M^{2+}][I_i]^x} \qquad (2)$$

For each natural organic ligand a conditional stability constant may be defined, assuming a 1:1 stoichiometry and omitting the unknown charge of the ligand L_i :

$$K_{L_i} = \frac{[ML_i]}{[M^{2+}][L_i]} \qquad (3)$$

After addition of a competing ligand R, the mass balance of the metal (M) at any point of the titration is

$$[M]_T = [M^{2+}] + \sum [MI_{i,x}] + \sum [ML_i] + \sum [MR_z] \qquad (4)$$

where $[MR_z]$ is the concentration of the competing ligand complexes (Figure 1). The stability constant for a complex with a number z of added ligand R is:

$$\beta_{R_z} = \frac{[MR_z]}{[M^{2+}][R]^z} \qquad (5)$$

For most competing ligands, only 1:1 (MR_1) and 1:2 (MR_2) complexes need to be taken into account (z = 1 or 2). The sum of natural metal species ($[M]_{nat}$) is the sum of the free metal and of the inorganic and natural organic complexes at the new equilibrium:

$$[M]_{nat} = [M^{2+}] + \sum [MI_{i,x}] + \sum [ML_i] = [M]_T - \sum [MR_z] \qquad (6)$$

Based upon equilibria, the concentrations of inorganic and competing ligand complexes are calculated, using their respective stability constants (β_{Iix} or β_{Rz}) and the free ligand concentrations ($[I_i]$ or $[R]$).

$$\sum [MI_{I,x}] = [M^{2+}] \sum (\beta_{Iix} [I_i]^x) = [M^{2+}] \alpha_{in} \qquad (7)$$

$$\sum [MR_z] = [M^{2+}] (\sum \beta_R [R]^z) = [M^{2+}] \alpha_R \qquad (8)$$

where the complexation coefficient of inorganic ligands (α_{in}) is the sum of the products of complex formation constant (β_{1i}) and free ligand concentrations [I_i], and the complexation coefficient of the competing ligand (α_R) is the sum of the products of complex formation constant β_{Rz} and free ligand concentrations [R].

The inorganic coefficient α_{in} is easily calculated if alkalinity, pH and major ion composition are known for a water sample. The free ligand concentration [R] is calculated from the proton and major ion equilibria. If the stability constants of the competing ligand complexes for the metal, proton and major cations are available, the coefficient α_R can easily be calculated. If these data are not available, the conditional stability constants and α_R must be determined in solutions identical to the composition of the natural water samples (*7, 15, 38, 42-44*).

Using DPCSV, the sum of the complexes $\sum[MR_z]$ is measured and natural organic complexes (ML_i) supposedly are non-electrochemically labile. An internal calibration in each water sample is used, in which a linear relationship between the total metal concentration [M]$_T$ and the voltammetric peak current i_p is established at high metal concentrations, at which the natural organic ligands are saturated:

$$i_p = S \left([M]_T - \sum[ML_i]\right) = S [M]_T - S\sum[ML_i] \qquad (9)$$

where S is the slope or the CSV sensitivity (e.g., nA M^{-1}), and $\sum[ML_i]$ is the sum of the metal-organic ligand complex concentrations which keeps constant with addition of the metal and can be obtained operationally from the intercept of the linear eq (9). In this case:

$$[M]_T - \sum[ML_i] = [M^{2+}] + \sum[MI_{ix}] + \sum[MR_z] = [M^{2+}](1 + \alpha_{in} + \alpha_R) \quad (10)$$

The free ion concentration of the metal is obtained from the peak current i_p and the CSV sensitivity S after correction with the factor $(1 + \alpha_{in} + \alpha_R)$:

$$[M^{2+}] = \frac{i_p}{S(1 + \alpha_{in} + \alpha_R)} \qquad (11)$$

In most cases the factor $(1 + \alpha_{in})$ is negligible in comparison to α_R.

Finally, the sum of natural species is determined by mass balance from eqs (6) and (8),

$$[M]_{nat} = [M]_T - [M^{2+}] \alpha_R \qquad (12)$$

Using eqs 11 and 12, a data set of $[M_{nat}]$ and $[M^{2+}]$ is thus acquired over the titration range, for which the following relationship is valid in the absence of the added ligand:

$$[M]_{nat} = [M^{2+}] (1 + \alpha_{in} + \alpha_L) \tag{13}$$

where α_L is the complexation coefficient of natural organic complexes, equal to the sum of the products of conditional stability constants and free ligand concentrations:

$$\alpha_L = \sum K_i[L_i] \tag{14}$$

Because of the heterogeneous nature of natural ligands, the coefficient α_L continuously varies with the metal to ligand ratio, and thus with the added metal concentration. When the concentration of $[M]_{nat}$ approaches the ambient total metal concentration $[M]_{T,amb}$, the corresponding coefficient $(\alpha_L)_{amb}$ or $(\sum K_i[L_i])_{amb}$ reflects the complexation by natural organic ligands at the ambient level. Data fitting and interpolation are needed in the titration range close to the ambient total metal concentration. Because of their simplicity, discrete ligand models with 1:1 complexes are used for this purpose. The actual number of ligands needed to fit the data depends upon the chemical composition of the water samples and the titration window (see below for detail). Typically, one or two ligands provide satisfactory fits to many data sets with a titration range over $1 - 2$ orders of magnitude of metal concentrations. The relevant complexation parameters (i.e., conditional stability constant and ligand concentration) of the natural organic ligands are either obtained graphically (45-47) or are optimized numerically using fitting programs like FITEQL (48-50). The stability constants obtained are conditional with respect to pH and concentration of major ions. The ambient free metal ion concentration $[M^{2+}]_{amb}$ are interpolated from the fitted conditional stability constants K_i and the total available ligand concentrations $[L_i]_T$.

A typical titration curve in terms of free ion concentration as a function of natural metal concentration is illustrated for Co speciation in lake water (Figure 2). The regression line at higher concentrations (inset Figure 2) was used for calibration of electrochemically labile Co using Eq 9, while the data at lower concentrations (large plot Figure 2) were used to determine the amount of metal binding by natural organic ligands in the water sample with Eq 13. The curve in Figure 2 was fitted by a one-ligand model over the narrow titration range, using the equilibrium program FITEQL (48).

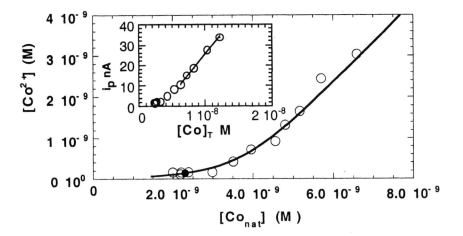

Figure 2. Titration curves of a lake water at pH 7.95 in terms of [Co²⁺] as a function of [Co_{nat}] and DPCSV peak current ip vs. [Co]_T (insert). The open dots were measured and the solid dots ([Co²⁺] = 0.14 nM) are for ambient [Co]_T (2.3 nM). The curve was calculated with the fitting parameters: log K = 10.0 and [L]_T = 3.8 nM. The regression line in the insert is for the calibration of labile Co. The sample was taken in Lake Biel (Switzerland) at 25 m depth, on Jan. 8, 1997.

Example 2. Ethylenediamine-exchange and DPASV for Cd speciation

Cadmium (Cd) complexation in natural water samples is evaluated by ligand-exchange with ethylenediamine (EN). The formed CdEN complex is measured by DPASV (differential pulse anodic stripping voltammetry), since it is electrochemically labile on the time scale of DPASV. The addition of EN increases the concentration of labile Cd, and thus the signal of the DPASV measurement. Some natural weak organic Cd complexes may however also contribute to the labile Cd, whereas the stronger organic Cd complexes and colloidal Cd species are not labile under these conditions. The following calculations are done under the assumption that the CdEN complexes represent approximately the total labile organic complexes in the presence of a large EN concentration.

The Cd mass balance in a water sample is defined as in eq (1). The mass balance for Cd after addition of EN is:

$$[Cd]_T = [Cd^{2+}] + \sum[CdI_{i,x}] + \sum[CdL_i] + \sum[CdEN_i]$$
$$= [Cd]_{lab} + \sum[CdL_i] \qquad (15)$$

where all the symbols are identical to the example in eqs (1) and (4). In this case the metal, M, is Cd and EN is the competing ligand, R. The Cd organic complexes CdL_i are in this case the stronger, DPASV non-labile complexes. The labile complexes. The labile Cd is measured by DPASV using an internal calibration at high metal concentrations, as described by eq (9) only ASV instead of CSV. Then free cadmium at any titration point is:

$$[Cd^{2+}] = [Cd]_{lab} / (1 + \alpha_{in} + \alpha_{EN}) \qquad (16)$$

$$[Cd]_{nat} = [Cd]_T - \sum[CdEN] = [Cd]_T - [Cd^{2+}] \alpha_{EN} \qquad (17)$$

$$\sum[CdL_i] = [Cd^{2+}] (\sum K_i[L_i]) = [Cd]_T - [Cd]_{lab} \qquad (18)$$

By fitting the data set of $[Cd_{nat}]$ and $[Cd^{2+}]$ (or $\sum[CdL_i]$ and $[Cd^{2+}]$) with the program FITEQL, we obtain the conditional complexation parameters of Cd complexes with the stronger natural ligands. The ambient free cadmium ion concentration $[Cd^{2+}]_{amb}$ is computed at the ambient total Cd concentration, using these complexation parameters. Using this technique, however $[Cd^{2+}]_{amb}$ may be overestimated because of the presence of some labile (weak) organic complexes.

Example 3. EDTA-exchange and DPASV for Zn Speciation

In the case of Zn, EDTA is used as a competing ligand which forms DPASV non-labile complexes (ZnEDTA). Total dissolved Zn in freshwater systems comprises electrochemically labile and non-labile species on the time scale of DPASV. In a similar way as for Cd, non-labile Zn ($\sum[ZnL_i]_{nl}$) comprises Zn in strong organic complexes and Zn in colloidal species. The labile species include free metal ions ($[Zn^{2+}]$), inorganic ($\sum[ZnI_{ix}]$) and weak organic complexes ($\sum[ZnL_i]_{lab}$). The ZnEDTA complexes increase the concentration of the non-labile complexes. Exchange between the strong organic complexes and EDTA cannot be assessed. This method does therefore not directly give access to the stability of the strong organic Zn complexes.

The total dissolved Zn is distributed in an original sample as follows:

$$[Zn]_T = [Zn^{2+}] + \sum[ZnI_{ix}] + \sum[ZnL_i]_{lab} + \sum[ZnL_i]_{nl} \tag{19}$$

$$[Zn]_T = [Zn]^0_{lab} + \sum[ZnL_i]_{nl} \tag{20}$$

where $[Zn]_T$ represents the concentration of total dissolved Zn, $[Zn]^0_{lab}$ the sum of labile species in the absence of EDTA.

The addition of EDTA modifies the mass balance by forming additional non-labile species:

$$[Zn]_T = [Zn]_{lab} + \sum[ZnL_i]_{nl} + [ZnEDTA] \tag{21}$$

If the added EDTA concentration just competes with the labile inorganic and organic complexes, the strong organic complexes $\sum[ZnL_i]_{nl}$ remain constant, and [ZnEDTA] is measured at any given total Zn concentration as:

$$[ZnEDTA] = [Zn]^0_{lab} - [Zn]_{lab} \tag{22}$$

The relationship (22) is illustrated in Figure 3. The free Zn ion concentration can be calculated from the [ZnEDTA] at any given total Zn concentration:

$$[Zn^{2+}] = [ZnEDTA] / \alpha_{EDTA} \tag{23}$$

where the complexation coefficient α_{EDTA} is the product of the calculated free EDTA concentration and the complex stability constant of ZnEDTA. This coefficient is usually in the range $\alpha_{EDTA} = 6 - 30$ with the addition of $1 \times 10^{-8} - 1 \times 10^{-7}$ M EDTA and in the presence of 1- 2 mM Ca^{2+}. Subsequently, the corresponding concentration of natural Zn species $[Zn]_{nat}$ is calculated as:

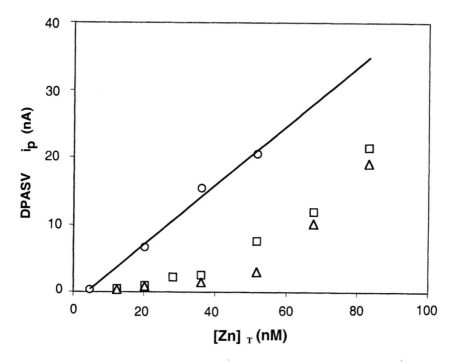

Figure 3. Zn titration curves of a lake sample in terms of DPASV peak current i_p vs. total Zn concentration. The circles represent the results in the absence of EDTA, the squares in the presence of 20 nM EDTA, the triangles in the presence of 30 nM EDTA. The regression line is for calibration of labile Zn. The differences between curves in the absence and in the presence of EDTA give [ZnEDTA].The sample from Lake Greifen (Switzerland) at 2.5 m depth, 12 July 1999, with original total dissolved Zn = 4.6 nM, determined ambient free Zn concentration $[Zn^{2+}]_{amb}$ = 0.58 nM and Zn in strong organic complexes $\sum[ZnL_i]$ = 3.7 nM.

$$[Zn]_{nat} = [Zn]_T - [ZnEDTA] \qquad\qquad (24)$$

Ambient free Zn ionic concentration $[Zn^{2+}]_{amb}$ is estimated by interpolating $[Zn]_{nat}$ to the ambient total Zn concentration with the linear relationship between $[Zn^{2+}]$ (Eq 23) and $[Zn]_{nat}$ (Eq 24). The inorganic complex concentration is estimated from the product of $[Zn^{2+}]_{amb}$ and the inorganic complexation coefficient, and the weak (labile) organic complex concentration from the mass balance of labile Zn, $[Zn]^0_{lab}$ (Eq 19). The strong organic complexes $\sum [ZnL_i]_{nl}$ are simply equal to the non-labile Zn concentration in the original water sample without EDTA.

With this technique, $[Zn^{2+}]_{amb}$ may be underestimated due to potential exchange between the strong organic complexes and EDTA. However, these reactions are not very significant because α_{EDTA} is quite low. This method does not directly give accesss to the stability of the strong organic Zn complexex. The stability constant of the strong organic Zn complexes has therefore been determined by competition of Cu and Zn for strong ligands (51).

Selection of competing ligands and of their optimal concentration

A suitable competing ligand for a given metal is well characterized with respect to the formed complexes (stoichiometry, complex stability and acidity constants), and the metal ligand complexes of interest can be accurately and selectively measured by voltammetry (DPASV or DPCSV). To efficiently compete with the natural ligands, the complexing coefficient (product of stability constant and ligand concentration) of the complexes with the added ligand must match that of the natural organic ligand complexes. Under these conditions, the equilibrium relationships can be accurately computed. The added ligand concentration should generally be in excess over that of the specific metal, in order to maintain a constant level of the ligand during the titration with the metal. Thus, the stability constant of the competing ligand must be on average lower than that of the natural ligands to match the complexing coefficients. In general, the complexing coefficient (α_R) of the added ligand should be just between 0.1 and 10 times that (α_L) of the natural ligands at the ambient specific metal concentration.

To determine practically the optimal ligand concentration, it is recommended to first titrate a water sample with the competing ligand (Figure 4A and 4B). At the optimal concentration, the ligand competes with the natural ligands and forms accurately measurable complex concentrations, by binding about 20-80% of the total metal. This concentration is situated in the steeply rising section of the titration curve (Figure 4A and B). The curves in Figure 4

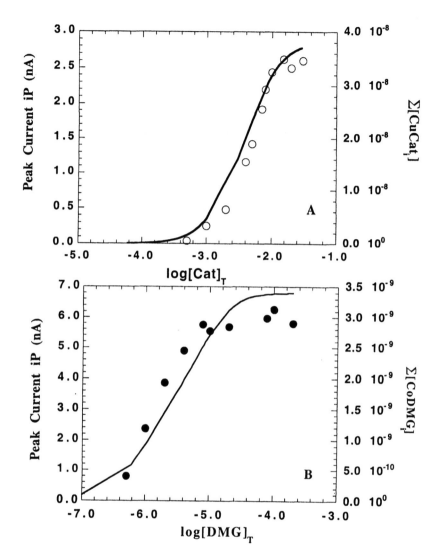

Figure 4. Dependence of DPCSV peak current and of calculated concentration of complexes ($\sum[MR_z]$) on added ligand concentration in natural samples. (A): for Cu-catechol complexes at pH 8.0 in Lake Greifen water (collected at 5 m depth on 21 March, 1990) with addition of 1.6×10^{-8} M Cu (O) ; the curve is calculated for the competition of catechol with the natural ligands. (B): for Co-DMG complexes at pH 7.98 in samples from the Aare River near Niederried (July 17, 1996), with 1.4 nM of original dissolved and 1.0 nM of additional Co(II)(●). The curve is calculated for the competition of DMG with the natural ligands.

have been calculated for the competition between the added ligand (catechol for Cu in A, and DMG for Co in B) and the natural ligands, by taking into account all relevant equilibria.

If a too low ligand concentration is added, the formed complexes may not be measurable. If a too high ligand concentration or a too strong ligand is used, the added ligand may outcompete all natural ligands for the specific metal ion and bind close to 100% of the total metal. Because the natural ligand-metal complexes are determined indirectly from mass balance relationships, the data interpretation may then be difficult or erroneous. If close to 100% of the total metal concentration $[M]_T$ are bound to the competing ligand complex concentration ($[MR]$), the calculated value of the natural organic complexes $[ML]$ cannot be determined with sufficient precision. The titration points would produce a nearly linear curve for free metal ion versus total metal, and only very strong natural ligands may be detected by extrapolation, which play an important role only at lower metal concentrations than at the ambient level. Within the suitable range, different ligand concentrations should give the same results on free metal ion concentration and on complexing coefficients. In contrast, an unsuitable range of added ligand concentrations may lead to erroneous results.

If several competing ligands are available for a specific metal, various factors must be considered for their selection. For example, when ligand exchange and DPCSV are used for Cu speciation, catechol, tropolone, oxine or salicylaldoxime (SA) are possible choices for competing ligands. Salicylaldoxime is an ideal ligand with good sensitivity under seawater conditions (38), but it may not be a suitable ligand for freshwater samples. To match the working window for freshwater, its affinity is in many cases too strong and restricts the Cu titration range. The sensitivity and stability constant of catechol are suitable for Cu titrations in most freshwater samples, but catechol is easily oxidized in presence of O_2. Therefore, caution must be exercised in some water samples with slower rates of ligand exchange, because only short equilibration times can be used with catechol.

Various voltammetric techniques have been compared for the speciation of Cu in seawater (52). The results obtained with various competing ligands were in good agreement with respect to the determined free $[Cu^{2+}]$.

Accuracy and validity of ligand-exchange techniques

The accuracy of metal complexation data using ligand exchange methods depends on a number of criteria:

i) Ligand exchange equilibrium must be established between the added ligand and the natural ligands in the sample. Knowledge of ligand-exchange kinetics is therefore needed to select suitable equilibration times.

ii) The concentration of the specific metal complexes with the competing ligand must be accurately measured.

iii) Constant pH must be maintained using adequate non-complexing pH buffers.

iv) The complexation parameters of the competing ligand R with M must be accurately known.

Each of these criteria is reviewed in the next sections and a discussion on the validity of the ligand-exchange techniques is provided.

Ligand-exchange kinetics

The competitive ligand-exchange methods rely on equilibrium relationships. It is thus essential that equilibrium is achieved under the experimental conditions. Ligand-exchange reactions involving the exchange of a metal ion from a strong natural complex to a complex with an added ligand may occur over various pathways (53, 54). The kinetics of such reactions cannot be readily predicted and should therefore be determined experimentally to define the conditions under which equilibrium is reached. Some case studies are presented below.

Example 1. Catechol (or tropolone)-exchange kinetics for Cu speciation

The competing ligand catechol offers a good sensitivity and a favorable working window for the analysis of Cu speciation by natural ligands in lake water. Using catechol, only short equilibration times can however be used because of its rapid oxidation in the presence of O_2. Short equilibration times (5 minutes) have therefore been used in our experiments. Control studies in the absence of O_2 (deoxygenated water) showed no variations of Cu-catechol reduction peak heights for samples from Lake Greifen over equilibration times of 3-120 minutes at constant pH (12).

The ligand-exchange kinetics of Cu in lake water was also examined using tropolone as a competing ligand, which is more stable than catechol. We found rapid equilibration with tropolone in Cu titrations of Lake Sempach water over time periods of a few minutes to 24 hours. Indeed we observed no variation in measured pCu using different equilibration times and competing ligands . The agreement between $[Cu^{2+}]$ results with catechol and those with tropolone or oxine (for which equilibration times were much longer (14-18 h)), suggests that the short equilibration time with catechol was sufficient to achieve ligand-exchange equilibrium in those samples (36). This assumption, however, may not be valid for every sample. The ligand-exchange kinetics of Cu with catechol and with other competing ligands thus needs to be further examined.

Example 2. EDTA exchange for Zn speciation

Metal exchange with a strong ligand such as EDTA, which reacts with major ions as well as with trace metals, is often very slow due to complicated pathways of cation exchange (55-57). Furthermore, low EDTA concentrations are used in these ligand-exchange experiments. For the speciation of Zn in Lake Greifen water using competitive ligand-exchange with EDTA, we found that the minimum time needed to reach equilibrium was about 15 hours (13).

Example 3. Competitive ligand-exchange with DMG for Ni speciation

The equilibration time in ligand-exchange techniques is especially critical for Ni because of its generally slow coordination kinetics (53).

In order to validate a method of competitive ligand- exchange with DMG for Ni speciation, the kinetics of $Ni(DMG)_2$ complex formation were examined (42). Using the optimal DMG concentrations of $5 - 6$ μM, the observed pseudo-first order rate is log $k \approx -5$ (s^{-1}), i. e. a half-life of $4 - 7$ hours for this reaction. The complex formation rate constant k_f ($4x10^3$ to $1.5x 10^4 s^{-1} M^{-1}$) appears to be pH-dependent. The rate constants estimated in our study for DMG-exchange agree well with the calculated rate constant of water loss by Ni^{2+} (56, 58), and with the kinetics reported for similar ligands (53).

The observed first order constants indicate very slow rates of exchange of Ni between natural ligands and DMG (logk_{obsd}, -5 to -6 (s^{-1})) and long half-lifes up to 100 hours (Figure 5). The overall second order constants of the tested samples are between $5x10^2$ to $7x10^3 s^{-1} M^{-1}$. These data suggest that ligand-exchange kinetics for Ni must be individually examined prior to conducting equilibrium experiments. The speciation results after long equilibrations only reflect the Ni speciation close to equilibrium, in particular with regard to the occurrence of strong ligands. Ni speciation in natural waters may never really reach equilibrium. The steady state of a water system may be perturbed by new inputs, which may persist for a long time (days) in their original speciation. Slow exchange of strong Ni-binding ligands has also been observed in wastewater effluents (59).

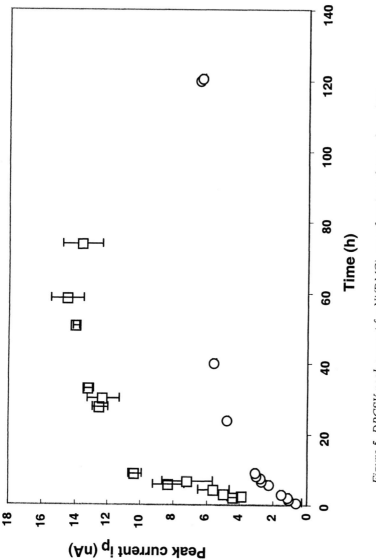

Figure 5. DPCSV peak current for Ni(DMG)₂ as a function of time after addition of DMG, indicating the ligand-exchange kinetics of Ni between natural ligands and DMG in lake waters. (O) for sample from Lake Greifen at 2.5 m depth collected on 12 July 1999, [Ni]ₜ = 11.9 nM, [DMG] = 6 μM and pH = 7.95; (□) for sample from Lake Lucerne at 2 m depth collected on 7 Jan. 1998, [Ni]ₜ = 9.1 nM, [DMG] = 4.6 μM and pH = 7.70.

Accuracy of voltammetric measurement

The accuracy of the voltammetric measurement critically depends on the calibration of the voltammetric signal, which is outlined in eqs (9) to (11). As mentioned above, an internal calibration has to be used, in order to take into account the effects of natural organic matter and of other components of natural water samples. The internal calibration is however also problematic because of the assumption that the relevant organic ligands have been completely titrated. This problem and the resulting errors in the measured concentrations have been discussed thoroughly in (60).

pH buffers

Accurate measurements of free ion concentration and of complexation by natural organic ligands in freshwaters must be conducted at exactly constant pH close to the in situ value. All complexation and ligand-exchange reactions are very sensitive to pH. Generally, a pH buffer is needed because the pH of a sample may change during the measurement. Morever, the buffer should not complex the specific metal, and at least exhibit no obvious effects on the resulting metal speciation under the working conditions with the used concentration. HEPES (N-2-hydroxyethylpiperazine-N'-2-ethanesulfonic acid) has been widely used as a pH buffer for seawater (7). For some lake and river samples, lower concentrations (5 –6 mM) of HEPES have also been used to keep the pH constant in the range pH = 7.5 –8.3, without any observable effects on the resulting metal speciation (12, 15, 24). Some water samples are however difficult to buffer with these HEPES concentrations (samples with high levels of DOC (> 20 mg/L) or with high CO_2 content). Moreover, experimental evidence for weak complexing effects of HEPES on Cu has been found (61-63). MOPS (3-morpholinopropanesulfonic acid) and MES (2-[N-morpholino] ethanesulfonic acid) were found to be non-reactive to copper and are therefore more adequate buffers (62-64). MOPS can be used for a pH range of 6.5 - 8.3, while MES buffers at a much lower pH (5.5 - 6.7). The measured Cu speciation is strongly pH-dependent (Figure 6). A relatively small change in pH results in large differences in $[Cu^{2+}]_{amb}$ (Figure 6B) and in the calculated speciation It is therefore very important to maintain the pH constant and close to the original value.

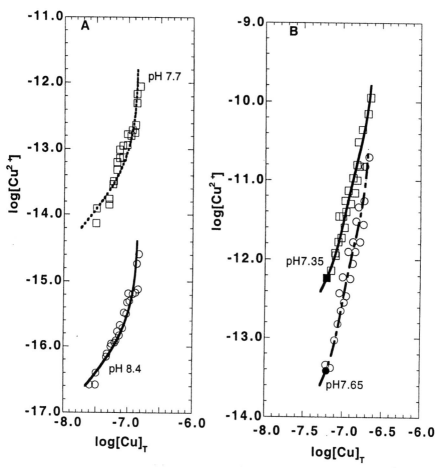

Figure 6. Comparison of titration curves of natural waters at differently buffered pH values in terms of log[Cu²⁺] vs. log[Cu]ₜ. The water samples were collected at 2.5 m depth of Lake Greifen on 12 July 1999 (A) and from infiltrated groundwater (Glattfelden) on 25 May 1999 (B). The dots were determined by catechol exchange and DPCSV, the curves were calculated using the fitting parameters of log K =15.9 and [L]ₜ = 144 nM at pH 8.4, and log K = 13.4 and [L]ₜ = 142 nM at pH 7.7 in (A); of log K₁ = 13.7, logK₂ = 11.6, [L₁]ₜ = 94 nM and [L₂]ₜ = 126 nM at pH 7.65, and log K₁ = 12.4, logK₂ = 10.5, [L₁]ₜ = 101 nM and [L₂]ₜ = 148 nM at pH 7.35 in (B).

Determination of conditional stability constants or complexation coefficients of competing ligands

For seawater, a conditional stability constant for the complexation of a specific metal with a competing ligand is measured in artificial seawater or in UV-oxidized seawater (*7, 38, 44*). This conditional stability constant or the corresponding complexation coefficient can reliably be used in seawater because of its stable composition, ionic strength and pH. Freshwater samples have a much more variable composition. Conditional complexation coefficients have therefore to be determined for each sample. They can be calculated based on equilibrium relationships for a given major ion composition and pH, if the intrinsic stability constants of the competing ligand with the specific metal, with protons and with major cations are known. In many cases, this information is not available. Under these circumstances, the complexing coefficient must be determined individually in synthetic solutions or in organic-free natural water, using the techniques recommended for seawater (*43, 44*). A well-characterized ligand like EDTA is used as a model ligand to determine the conditional stability constant of a competing ligand with a specific metal. Using the conditional stability constants measured under various conditions (e.g. pH, ionic strength), the intrinsic stability constants are obtained.

The determination of the intrinsic stability constant of the $Ni(DMG)_2$ complex provides an example. Literature values of the stability constants of the Ni-DMG-complexes vary widely (*65, 66*). Moreover, the stability constants of DMG with major cations are not available. Therefore, we measured the conditional stability constant $\beta^*_{Ni(DMG)2}$ in several series of synthetic solutions as well as in an UV-irradiated lake water sample by ligand competition with EDTA by DPCSV (*42*). After correcting for the effect of ionic strength, a final average value of $\log\beta_{Ni(DMG)2} = 22.9\pm0.7$ was obtained for the intrinsic stability constant. No significant differences were found in the stability constants obtained in natural water samples and in synthetic solutions, nor in solutions with different Ca concentrations (1- 2.5 mM). We conclude that complexation of Ca and DMG can be neglected under freshwater conditions with $[Ca] \leq 2$ mM.

The determined intrinsic stability constant of $Ni(DMG)_2$ is few orders of magnitude higher than that ($\log \beta = 17.84$, I=0) selected by (*66*), but close to the value which can be calculated from conditional stability constants determined in seawater by (*43*). It is also close to a value ($\log \beta = 21.8$, I = 0.1) reported earlier by (*67*) and compiled by (*65*). This case illustrates that stability constants must be carefully selected and in some cases checked experimentally.

Validation of ligand-exchange methods

It can be problematic to accurately know and compute the equilibrium relationships. A number of assumptions have to be made, which may not always be valid in actual natural water samples. These assumptions include that chemical reactions such as adsorption or oxidation do not significantly decrease the concentration of added ligand in solution, that mixed complexes are not formed between the added ligand and natural ligands, and that competition with other cations does not affect the used equilibrium relationships.

The assumptions involved may be validated in various ways. First of all, the determined speciation can be compared with the calculated speciation in synthetic solutions with well-defined ligands (inorganic and organic), as we did for Cu, or Zn speciation (*12, 13*). Furthermore, speciation results may be compared by using different techniques or different competing ligands in the same natural water samples (*36, 39, 52, 64*). To examine ligand exchange/DPCSV for Cu speciation, we compared the determined $[Cu^{2+}]$ in lake water samples using various competing ligands, namely catechol, oxine or tropolone (*36*). Despite limitations of each ligand, relationships between $[Cu^{2+}]$ and $[Cu]_T$ determined with each of these three ligands were in good agreement, as shown in the titration curve of a lake water sample (Figure7A). The average error in calculated $[Cu^{2+}]$ at lower $[Cu]_T$ (1×10^{-8} - 4×10^{-8}), using these three competing ligands, was less than 0.2 pCu units. This error is also observed if different concentrations of a single competing ligand are used. This agreement indicates that the above criteria are met, in particular that competition between copper and other trace metals for either the natural or added ligands have not led to significant biases. Reliable measurements of pCu in a water sample should be independent of the added ligand or its concentration.

It is also useful to compare different techniques, such as ligand exchange / DPCSV and ISE (ion selective electrode). Because of its limited sensitivity, ISE (CuS solid membrane) can however not be used in freshwater with very low $[Cu]_T$, due to non-ideal electrode behavior resulting from a lack of equilibration between the electrode membrane and the bulk solution. As an example, we have determined ambient pCu by catechol/DPCSV and by Cu-ISE in surface waters from Lake Orta, an Italian lake with high total copper concentrations and lower pH resulting from inputs of industrial pollutants (*36*). Good agreement of the determined pCu was observed in this case, with values of pCu = 9.8 –10.0 by CSV and pCu = 9.4 – 9.8 by ISE, with $[Cu]_T$ 65 –71 nM at pH 7.0 –7.2.

Figure 7B exemplifies a titration curve over a larger concentration range, to compare the relationships between $[Cu^{2+}]$ and $[Cu]_T$ measured by ligand-exchange with catechol/DPCSV and Cu-ISE in a sample from Lake Lucerne. Good agreement was observed between the two methods at total copper

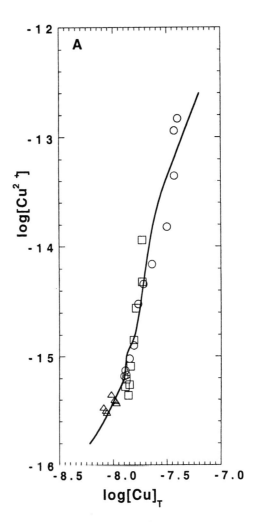

Figure 7. Relationship between log [Cu²⁺] and log [Cu]_T or complexing coefficient log α_L in titrations of lake water samples. In (A): the data were determined by ligand-exchange / DPCSV using different added ligands: catechol (O), oxine (□), and tropolone (Δ). The curves were fitted to the catechol data by FITEQL using a two-ligand model (log K₁ = 15.5, log K₂ = 12.6, and [L₁]_T =19.0 nM, [L₂]_T =100 nM). The water sample was collected from a depth of 2.5 m of Lake Sempach on 11 October 1994. In (B): the data were determined by catechol-exchange / DPCSV as log [Cu²⁺] (O) and log α_L (□), and by Cu-ISE as log [Cu²⁺] (●) and log α_L (■). The curves were calculated by fitting all the data measured by DPCSV plus that measured by ISE at [Cu]_T > 0.5 µM to a three-ligand model (logK₁ = 14.3, logK₂ = 10.8, logK₃ = 8.6 and [L₁]_T = 18.5 nM, [L₂]_T = 100 nM, [L₃]_T = 6.0µM). The water samples were collected on 22 July 1991 at 20 m depth from Lake Lucerne.

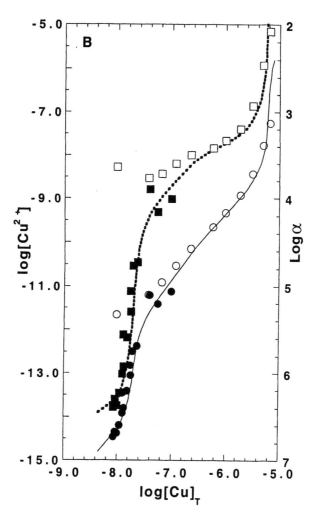

Figure 7. *Continued.*

concentrations above 0.3 μM, but the ISE measurements of [Cu^{2+}] deviated progressively upward from the DPCSV values at lower copper concentrations. The ISE measurements are however not reliable at low total Cu concentrations (about < 0.1 μM). The agreement of the two methods within their overlapping range (0.3-1.6 μM [Cu]$_T$) for the Lake Lucerne sample (Figure 7B) provides further support for the validity of both methods within their respective windows.

Interpretation of metal speciation in natural waters

Conditional stability constants and ligand concentrations

At any point of a metal titration, the complexation coefficient ($\alpha_L = \sum K_i[L_i]$, product of stability constant and available ligand concentration) is directly obtained by a competitive ligand-exchange technique. This coefficient depends on the working window and varies continuously, even within one titration. As an example plotted in Figure 7B, lower values of the complexation coefficient were obtained by ISE, since this technique has a lower detection window than that of catechol-exchange/DPCSV. Natural waters contain a range of complexing ligands with widely varying complex stability. Various complexes may be detected in function of the working window in a metal titration (*60, 68, 69*). If a discrete ligand model is used to fit the data, the required number of ligands depends not only on the heterogeneity of the sample, but also on the titration range and the desired fitting error. More ligands are needed to fit a wider titration window, as in the example shown in Figure 7B. In this case, three ligands are needed to fit the titration in the range of log[Cu]$_T$ = -8.1 to -4.7, including data from ISE and DPCSV. If only the titration data obtained by ligand-exchange/ DPCSV are fitted in the range of log[Cu]$_T$ = -8.1 to -7.0, two ligands are sufficient. In general, one or two ligands are often satisfactory for natural freshwater samples in a titration range over one to two orders of magnitude of total concentrations. The fitted parameters, conditional stability constant and ligand concentration are dependent of each other in a given detection window. The product (α_L) of the two parameters reflects the complexation within a given window (Figure 7B). To obtain a meaningful comparison of two or more water samples, the range of total metal concentrations must be constrained within the same window.

Graphical methods with linear transformation are only suitable for narrow titration ranges with one or two ligands, and are difficult to use for titrations over a wider range. These methods have been discussed on the basis of model data in (*69*).

Comparison of complexation by shift of titration curves

To compare the complexation properties of various samples, it is recommended to consider the shift of titration curves (Figure 8). A shift of a titration curve to the right with respect to the x-axis (total metal concentration) indicates that an increase in total Cu concentration is needed to have the same free Cu^{2+} concentration and therefore stronger complexation.

In figure 8, the Cu titration curves of Thur river water samples are plotted in terms of $log[Cu^{2+}]$ vs. $log[Cu]_T$. Ultrafiltrate samples (<10kD) are compared with conventional filtrates (<0.45 μm) at two different sampling sites (1 and 4). We can see that Cu complexation in the samples from the upstream site (No.1) was weaker than that in the samples from the downstream site (No. 4), considering shifts of the titration curves for both the filtrate and the ultrafiltrate samples. At both sites, the titration curves of the ultrafiltrate are shifted to the left of those of the filtrate. This shift indicates that Cu complexation in the ultrafiltrate was weaker than that in the filtrate. For the upstream sample, the shift of the curves is parallel over the whole titration window, indicating the similar stability of the natural organic ligands in both fractions and less concentration of the ligands in the ultrafiltrate than in the filtrate (Figure 8 caption). The organic ligands at the upstream site originate mostly from soil runoff and are thus likely to be mostly of the humic and fulvic acid type, which occur both in the fractions with similar properties, < 10kD solution and >10kD colloidal. Only in a parallel section of titration curves, one may compare the complexation using one of the fitting parameters, which are conditional average, and dependent of each other's within the limited range. The shift of the curves for the downstream sample is not completely parallel. The curves for the both fractions are almost overlapped in the lower section (close to ambient copper concentrations) and upwards separate. This kind of shift implies that strong Cu binding ligands other than natural souses (as in the upstream sample) appear to be mostly in the < 10 kD range and to bind copper more stronger in the lower concentration than in the higher section. At the downstream site, organic ligands may still originate partly from natural soil runoff, but the increase of DOC indicates other sources, namely sewage inputs, inputs from agricultural soils and biotic production in the river itself. The Cu binding ligands originating from these various sources appear to be mostly in the < 10 kD range and to form stronger complexes than those at the upstream site (*70*).

Stability constants obtained from different working windows may differ, because different components of the heterogeneous organic ligands are

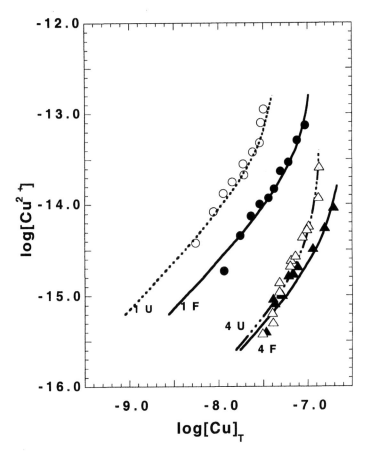

Figure 8. Titration curves of <0.45 μm-filtrate and <10 kD ultrafiltrate samples with Cu (log [Cu²⁺] versus log [Cu_T]). Samples are collected from the Necker site, upstream of Thur river (1F, ●) for filtrate < 0.45 μm and 1U, O for ultrafiltrate < 10 kD), and from the Andelfingen site, downstream of Thur river (4F, ▲ for filtrate < 0.45 μm and 4U, △ for ultrafiltrate < 10 kD) on 12 Nov. 1997. The dots are measured and the curves calculated using the parameters: log K = 13.55, 13.57, 14.68 and 14.45; [L]_T = 40.7, 120, 143, and 270 nM respectively for 1U, 1F, 4U and 4F. A shift of the titration curves to the left indicates less complexation of Cu.

determined (71). The titration curves calculated by the models may show the relative complexation within their respective window. Figure 9 shows Cu titration curves of fulvic acids (FA) and of a lake water sample calculated by different models, using fitting parameters obtained by different techniques at different detection windows. Except for the WHAM model, all curves are plotted in the respective working windows. Our results for FA using ligand-exchange / DPCSV have been obtained at much lower total Cu concentrations than those measured by ISE ($[Cu_T] > 10^{-7}$ M and $[Cu^{2+}] > 10^{-10}$ M), from (17). Therefore, the complexing parameters from these two techniques are hardly comparable. To further compare our results with literature data, the WHAM model with universal data base (72) was used to calculate relations of $[Cu^{2+}]$ and FA bound Cu in the range of natural water. The curve calculated by WHAM for FA is located above that by CSV in the low range of total Cu (around 10^{-8} M). The slopes of the curves however differ over a wider range. This discrepancy in the lower range of total Cu is probably due to the database included in the WHAM model, which relies mostly on studies performed at higher metal concentrations.

Thus, we may also compare the titration curves of FA with those of lake water, using all the data obtained with the same technique and in the same concentration range, as shown in Figure 9. The direct comparison of the titration curves is more reliable than a comparison of the stability constants or ligand concentrations. This comparison shows that complexation of Cu by lakewater ligands was stronger than that by FA at the ambient concentrations. However, the titration curves of ligands in the lake sample and of FA approach each other closer at higher copper concentrations, indicating that FA may play a role at higher Cu concentrations as weaker ligands in lakewaters (25).

Speciation of Cu, Zn, Cd, Co, and Ni in freshwater

An overview of the results of equilibrium speciation of Cu, Zn, Cd, Co, and Ni in Swiss lakes and rivers is presented in Table 1. All these results have been determined by competitive ligand exchange / voltammetry methods. Only the complexation parameters of the strongest ligands are listed in Table 1, because this class of strong ligands plays the most important role in complexing the metals at ambient level. The studied trace metals appear to be mostly bound in strong organic complexes at their natural concentrations in surface freshwater. The complex stability decreases in the order Cu > Ni > Cd > Co > Zn, which is in accordance with the general order of chemical coordination stability.

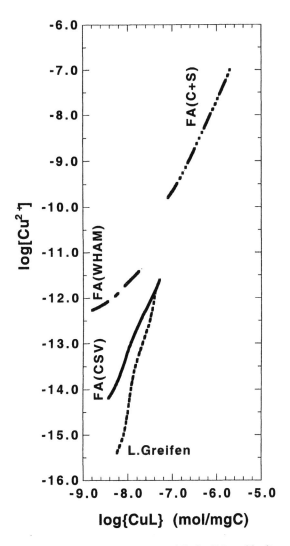

Figure 9. Comparison of the complexation of Cu by FA and by ligands of lake water (pH 8.0±0.1) in terms of log[Cu²⁺] (M) vs. bound Cu {CuL} (mol (mgC)⁻¹). The complexation by FA (CSV) is calculated with the fitting parameters: log K_1 = 14.1, log K_2 = 11.7, $[L_1]_T$ = 6.46 μmol (gC)⁻¹ and $[L_2]_T$ = 77.4 μmol (gC)⁻¹, the curve for Lake Greifen with the fitting parameters of log K_1 = 15.5, logK_2 = 12.6, $[L_1]_T$ 30 nM and $[L_2]_T$ 100 nM.. The lake water, with 2.8 mg L⁻¹ DOC, was collected at 2.5 m depth of Lake Greifen on 26 April 1996. The curve for the FA (C+S) was calculated using the parameters of the 5-site model measured with ISE (17). The curve for FA (WHAM) was computed by the WHAM model with its database (71). Except for the WHAM model, all curves are plotted in the respective working windows.

Table 1 Equilibrium Speciation of Trace Metals in Swiss lakes and Rivers determined by Ligand-exchange and Voltammetry[1].

Metals	$[M]_T$	$pM= -log[M^{2+}]_{amb}$	$LogK_l$	$[L_l]_T$	Complexed Fraction[2]
	nM			nM	%
Cu	6-20	13.1-15.6	13.5-15.5	10-100	>99.9
	(to 70)	(to 16)	(to 16.2)	(to 280)	
Zn	10-40	8.6-10.0	7.8 –9.6	10-100	~50
	(4 - 80)				(15-90) [3]
Cd	0.03-0.1	11.5-12.5	10.1-10.5	1.6-6.0	>98
	(to 0.4)	(to 10.4)	(to 9.5)		
Co	1.0-6.5	9.3-10.0	9.5-11.5	1.0-8.0	80-90
					(59-96)
Ni	4.0-13.0	13.1-14.5	12.1-14.1	45-100	>99.9
	(to 30)				

1) pH of water samples were in the range 7.7 – 8.5; speciation was determined at pH close to the original value. DOC ranged from 0.9 to 4.3 mgL^{-1}, with few exceptions up to 6.2 mgL^{-1}. Data in brackets represent extreme values. All results were obtained by ligand exchange/voltammetry, using catechol/DPCSV for Cu, EDTA/DPASV for Zu, EN/DPASV for Cd and DMG/DPCSV for Co and Ni.

2) The complexed fraction indicates the fraction of dissolved metals bound in organic complexes.

3) The complexed fraction of Zn dynamically varies due to relatively weak and pH-dependent complexation, high background concentrations, and uptake by phytoplankton.

Fulvic and humic acids (HA) are likely to play an important role for metal complexation in systems with high DOC and relatively high metal concentrations, especially if DOC mostly originates from soil or wetland. In systems with high biological productivity and relatively low DOC, such as eutrophic lakes, it appears probable that some specific stronger ligands occur. Furthermore, these strong ligands are present at low concentrations (about 1 - 300 nM) and thus represent only a tiny fraction of DOC in a typical sample. The concentrations of specific Cu or Cd binding ligands in relationship to DOC ($1x10^{-6}$ -$1x10^{-4}$ mol (g DOC)$^{-1}$) are 1 - 3 orders of magnitude lower than the typical total number of binding groups in HA or FA (1 - $10x10^{-3}$ mol (g C)$^{-1}$). This observation indicates that those specific stronger ligands are only present at low concentrations in freshwater, and that humic substances would play a role as weaker ligands with higher concentrations.

With regard to the origin of these ligands, the differences in metal complexation observed amoung different lakes supply more evidence that the occurrence of strong ligands may be linked to biological productivity. Complexation of Cu, Ni, Co or Cd by specific ligands was stronger in eutrophic lakes than that in oligotrophic lakes (*14, 15, 24, 25, 42*). Cu and Ni titration curves in samples from lakes with different trophic states are compared in Figure 10, illustrating the differences in complexation of the metals by shifts of the titration curves. Seasonal variations of Cu complexation have also been observed to follow the variations in primary productivity, chlorophyll-a concentration and C-14 assimilation in a eutrophic lake (*12*). In addition, Cu in the euphotic zone of a deep eutrophic lake showed stronger complexation than that in the hypolimnion, whereas no differences with depth were observed in an oligotrophic lake (*24*). All these observations indicate that these stronger ligands are probably of recent biological origin.

No systematic differences in Zn complexation in various lakes were observed, probably due to relatively weak complexation, higher background concentrations, and large variations due to uptake by phytoplankton. However, Cu competition for strong natural ligands with Zn has been observed in a eutrophic lake (*51*). Competition for ligands among various trace metals has not yet been widely examined and is expected to depend on the relative binding strength, on exchange kinetics, and on the original speciation of metal inputs.

The observed strong complexation of trace metals in freshwater is relevant with regard to bioavailability and toxicity of these elements, as well as for their mobility. Decrease of free metal ion concentrations is expected to decrease metal toxicity. For example, the dissolved concentrations of Cu, Zn, Ni and Co in Thur river increased downstream, but the concentrations of the free $[Cu^{2+}]$ did not obviously increase because of increased inputs of complexing organic compounds (Figure 8) (*70*). Only a few examples of relationships between free metal ions and interactions of metals with organisms have been studied in freshwater. We have related the Cu or Cd (total cellular or intracellular) contents in algae to $[Cu^{2+}]$ or $[Cd^{2+}]$ in several lakes, in which $[Cu^{2+}]$ and $[Cd^{2+}]$ have been determined by competitive ligand exchange / voltammetry (*73*). In another study, the cadmium content in insect larvae has been related to $[Cd^{2+}]$ in various lakes (*74*). In this case, $[Cd^{2+}]$ has however not been directly measured, but estimated by the WHAM model. Further field evidence for the relationship between determined free metal concentrations and organism uptake or toxicity is needed in order to evaluate the general concepts of speciation and bioavailability under realistic conditions, and to develop better models for water quality standards.

Metal distribution among different phases or compartments is also related to complexation in solution. Competition between complexation in solution and binding to solid surfaces may affect the metal mobility in groundwater, as well as

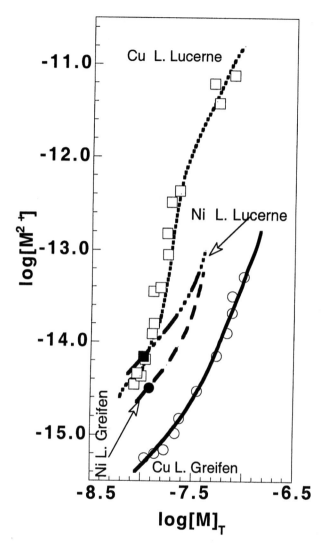

Figure 10. Comparison of Cu or Ni complexation between natural ligands from different lakes by the shifts of the titration curves in terms of $\log[M^{2+}]$ vs. $\log[M]_T$. The curve with open circles for Cu speciation in a Lake Greifen sample (13 Oct. 1993) is fitted by the parameters: $\log K_1 = 14.6$, $\log K_2 = 13.0$, $[L_1]_T = 56.9$ nM, and $[L_2]_T = 150$ nM; the curve with open squares for Cu speciation in a Lake Lucerne sample (22 July 1991) is calculated by the fitting parameters: $\log K_1 = 14.3$, $\log K_2 = 11.0$, $[L_1]_T = 19$ nM and $[L_2]_T = 110$ nM. The curve with a solid circle (ambient position) for Ni speciation in a Lake Greifen sample (12 July 1999) is fitted by the parameters: $\log K = 14.05$ and $[L]_T = 45.6$ nM; the curve with a solid square (ambient position) for Ni speciation in a Lake Lucerne sample (13 July 1999) is calculated by the parameters: $\log K = 13.55$ and $[L]_T = 53.7$ nM.

the removal of metals by settling particles in lakes. Natural groundwater ligands were shown to decrease the adsorption of Cu, Ni, Cd to a model solid phase (75). As another example, cobalt was not efficiently adsorbed and transferred to aquatic sediments after discharge by a nuclear power station, probably due to competition by dissolved organic ligands (15).

References

1. Sunda, W. *Biol. Oceanogr.* **1988/89**, *6*, 411-442.
2. Morel, F. M. M.; Hudson, R. J. M.; Price, N. M. *Limnol. Oceanogr.* **1991**, *36*, 1742-1755.
3. Bruland, K. W. *Limnol. Oceanogr.* **1989**, *34*, 267-283.
4. Bruland, K. W. *Limnol. Oceanogr.* **1992**, *37*, 1008-1017.
5. Donat, J. R.; Bruland, K. W. *Mar. Chem.* **1990**, *28*, 301-323.
6. Coale, K. H.; Bruland, K. W. *Limnol. Oceanogr.* **1988**, *33*, 1084-1101.
7. Van den Berg, C. M. G. *Anal. Chim. Acta* **1984**, *164*, 195-207.
8. Van den Berg, C. M. G. *Mar. Chem.* **1995**, *50*, 139-157.
9. Moffett, J. W.; Brand, L. E.; Croot, P. L.; Barbeau, K. A. *Limnol. Oceanogr.* **1997**, *42*, 789-799.
10. Kozelka, P. B.; Bruland, K. W. *Marine Chemistry* **1998**, *60*, 267-282.
11. Achterberg, E. P.; Van den Berg, C. M. G.; Boussemart, M.; Davison, W. *Geochim. Cosmochim. Acta* **1997**, *61*, 5233-5253.
12. Xue, H. B.; Sigg, L. *Limnol. Oceanogr.* **1993**, *38*, 1200-1213.
13. Xue, H. B.; Sigg, L. *Anal. Chim. Acta* **1994**, *284*, 505 - 515.
14. Xue, H. B.; Sigg, L. *Anal. Chim. Acta* **1998**, *363*, 249-259.
15. Qian, J.; Xue, H. B.; Sigg, L.; Albrecht, A. *Environ. Sci. Technol.* **1998**, *32*, 2043-2050.
16. Buffle, J. *Complexation reactions in aquatic systems: an analytical approach*; Ellis Horwood limited: Chichester, 1988.
17. Cabaniss, S. E.; Shuman, M. S. *Geochim. Cosmochim. Acta* **1988**, *52*, 185-193.
18. Tipping, E.; Hurley, M. A. *Geochim. Cosmochim. Acta* **1992**, *56*, 3627-3641.
19. Tipping, E. *Aquat. Geochem.* **1998**, *4*, 3-48.
20. Benedetti, M. F.; Milne, C. J.; Kinniburgh, D. G.; Van Riemsdijk, W. H.; Koopal, L. K. *Environ. Sci. Technol.* **1995**, *29*, 446-457.
21. Kinniburgh, D. G.; Milne, C. J.; Benedetti, M. F.; Pinheiro, J. P.; Filius, J.; Koopal, L. K.; Van Riemsdijk, W. H. *Environ. Sci. Technol.* **1996**, *30*, 1687-1698.
22. Bruland, K. W.; Donat, J. R.; Hutchins, D. A. *Limnol. Oceanogr.* **1991**, *36*, 1555-1577.

23. Moffett, J. W.; Brand, L. E. *Limnol. Oceanogr.* **1996**, *41*, 388-395.
24. Xue, H.; Oestreich, A.; Kistler, D.; Sigg, L. *Aquatic Sci.* **1996**, *58*, 69 - 87.
25. Xue, H.; Sigg, L. *Aquatic Geochemistry* **1999**, *5*, 313-335.
26. Buckley, P. J. M.; Van den Berg, C. M. G. *Mar. Chem.* **1986**, *19*, 281-296.
27. Giger, W.; Schaffner, C.; Kari, F. G.; Ponusz, H.; Reichert, P.; Wanner, O. *EAWAG news* **1991**, *32*, 27-31.
28. Rozan, T. F.; Benoit, G.; Luther, G. W. *Environ. Sci. Technol.* **1999**, *33*, 3021-3026.
29. Rozan, T. F.; Lassman, M. E.; Ridge, D. P.; Luther, G. W. I. *Nature* **2000**, *406*, 879-882.
30. Allen, H. E.; Hansen, D. J. *Water Env. Res.* **1996**, *68*, 42-54.
31. Peijnenburg, W. J. G. M.; Posthuma, L.; Eijsackers, H. J. P.; Allen, H. E. *Ecotoxicol. Environ. Safety* **1997**, *37*, 163-172.
32. Tercier-Waeber, M.-L.; Buffle, J. *Environ. Sci. Technol.* **2000**, *34*, 4018-4024.
33. Davison, W.; Zhang, H. *Nature* **1994**, *367*, 546-548.
34. Zhang, H.; Davison, W. *Anal. Chem.* **1995**, *67*, 3391-3400.
35. Sunda, W. G.; Hanson, P. J. In *Chemical modeling in aqueous systems*; Jenne, E. A., Ed.; American Chemical Society: Washington, 1979, pp 147-180.
36. Xue, H.; Sunda, W. G. *Environ. Sci. Technol.* **1997**, *31*, 1902-1909.
37. Camusso, M.; Tartari, G.; Zirino, A. *Eviron. Sci. Technol.* **1991**, *25*, 678-683.
38. Campos, M. L. A. M.; Van den Berg, C. M. G. *Anal. Chim. Acta* **1994**, *284*, 481-496.
39. Donat, J. R.; Lao, K. A.; Bruland, K. W. *Anal. Chim. Acta* **1994**, *284*, 547-571.
40. Donat, J. R.; Van den Berg, C. M. G. *Mar. Chem.* **1992**, *38*, 69-90.
41. Nimmo, M.; van den Berg, C. M. G.; Brown, J. *Estuarine, Coastal and Shelf Sci.* **1989**, *29*, 57-74.
42. Xue, H.; Jansen, S.; Prasch, A.; Sigg, L. *Environ. Sci. Technol.* **2001**, *35*, 539-546.
43. Van den Berg, C. M. G.; Nimmo, M. *Sci. Total Environ.* **1987**, *60*, 185-195.
44. Zhang, H.; Van den Berg, C. M. G.; Wollast, R. *Mar. Chem.* **1990**, *28*, 285-300.
45. Scatchard, G. *Ann. N. Y. Acad. Sci.* **1949**, *51*, 660-672.
46. Ruzic, I. *Anal. Chim. Acta* **1982**, *140*, 99-113.
47. Van den Berg, C. M. G. *Anal. Chim. Acta* **1982**, *11*, 307-322.
48. Westall, J. C. ; Department of Chemistry, Oregon State University: Corvallis, 1982.
49. Cernik, M.; Borkovec, M.; Westall, J. C. *Environ. Sci. Technol.* **1995**, *29*, 413-425.

50. Westall, J. C.; Jones, J. D.; Turner, G. D.; Zachara, J. M. *Environ. Sci. Technol.* **1995,** *29,* 951-959.

51. Xue, H. B.; Kistler, D.; Sigg, L. *Limnol. Oceanogr.* **1995,** *40,* 1142-1152.

52. Bruland, K. W.; Rue, E. R.; Donat, J. R.; Skrabal, S. A.; Moffett, J. W. *Anal. Chim. Acta* **2000,** *405,* 99-113.

53. Margerum, D. W.; Cayley, G. R.; Weatherburn, D. C.; Pagenkopf, G. K. In *Coordination chemistry*; Martell, A. E., Ed.; Am. Chem. Soc.: Monogr., 1978; Vol. 2, pp 1-220.

54. Hering, J. G.; Morel, F. M. M. *Environ. Sci. Technol.* **1990,** *24,* 242-252.

55. Hering, J. G.; Morel, F. M. M. *Environ. Sci. Technol.* **1988,** *22,* 1469-1478.

56. Hering, J. G.; Morel, F. M. M. In *Aquatic Chemical Kinetics- Reaction Rate of Processes in Natural Waters*; Stumm, W., Ed.; John Wiley & Sons, Inc.: New York, Chichester, 1990, pp 145 - 171.

57. Xue, H. B.; Sigg, L.; Kari, F. G. *Environ. Sci. Technol.* **1995,** *29,* 59 - 68.

58. Eigen, M.; Wilkins, R. G. In *Mechanisms of Inorganic Reactions (ACS Symposium Series No.49)*; American Chemical Society: Washington, DC,, 1965, pp 55-80.

59. Sedlak, D. L.; Phinney, J. T.; Bedsworth, W. W. *Environ. Sci. Technol.* **1997,** *31,* 3010-3016.

60. Kogut, M. B.; Voelker, B. M. *Environ. Sci. Technol.* **2001,** *35,* 1149-1156.

61. Vasconcelos, M. T. S. D.; Azenha, M. A. G. O.; Lage, M. O. *Analytical Biochemistry* **1996,** *256,* 248-253.

62. Yu, Q.; Kandegedara, A.; Xu, Y.; Rorbacher, D. B. *Analytical Biochemistry* **1997,** *253,* 50-56.

63. Mash, H. Ph. D., The Ohio State University, Columbus, 2000.

64. Rozan, T. F.; Benoit, G.; Mash, H.; Chin, Y. P. *Environ. Sci. Technol.* **1999,** *33,* 1766-1770.

65. Sillen, L. G.; Martell, A. E. *Stability Constants of Metal-Ion Complexes*; The Chemical Society, Burlington House, W.1: London, 1964.

66. Martell, A. E.; Smith, R. M. *Critical Stability Constants*; Plenum Press: New York and London, 1989.

67. Vlacil, F. *Collection Czechoslov. Chem. Commun. (in Germ.)* **1961,** *26,* 658-666.

68. Van den Berg, C. M. G.; Nimmo, M.; Daly, P.; Turner, D. R. *Anal. Chim. Acta* **1990,** *232,* 149-159.

69. Miller, L. A.; Bruland, K. W. *Anal. Chim. Acta* **1997,** *343,* 161-181.

70. Sigg, L.; Xue, H.; Kistler, D.; Schönenberger, R. *Aquatic Geochem.* **2000,** *6,* 413-434.

71. Town, R.W.; Filella, M. *Limnol. Oceanogr.* 2000, 45, 1341-1357.

72. Tipping, E. *Comput. Geosci.* **1994,** *20,* 973-1023.

73. Knauer, K.; Ahner, B.; Xue, H.; Sigg, L. *Env. Toxicol. Chem.* **1998,** *17,* 2444-2452.

74. Hare, L.; Tessier, A. *Nature* **1996,** *380,* 430-432.

75. Weirich, D. Ph. D., Swiss Federal Institute of Technology, Zürich,2000.

Chapter 19

Voltammetric Evidence Suggesting Ag Speciation Is Dominated by Sulfide Complexation in River Water

Tim F. Rozan and George W. Luther, III

College of Marine Studies, University of Delaware, Lewes, DE 19958

Using voltammetric methods, free HS^-/H_2S, acid volatile sulfide (AVS), and only Cr(II) reducible sulfide (OCRS) fractions were measured in seven rivers spanning the mid-Atlantic region of the United States to determine the effect of sulfide complexation on dissolved Ag (Ag_{diss}) speciation in freshwaters. In all waters, the AVS: Ag_{diss} ratio was observed to be greater than 600. This increased to a range of 1000 – 5000 for rivers that received sewage treatment plant (STP) effluent. In laboratory experiments, Ag was observed to quickly (< 1 min.) replace both Cu and Zn in metal sulfide solutions, indicating that the measured AVS concentrations in natural waters provide an excess pool of available sulfide for complete Ag complexation.

Introduction

In natural waters determining the speciation of a trace metal, such as Ag, is an important, if not daunting task. Yet knowledge of the exact speciation is critical for understanding the geochemical behavior and bioavailabilty of trace metals in aquatic environments. Unfortunately, most current analytical techniques are not able to identify specific metal complexes or directly measure the free metal ion (M^+), which is thought to be the most toxic form, at real world concentrations (*1,2*). To circumvent this problem, thermodynamic equilibrium models have been used to calculate the expected speciation of trace metals in natural waters (*3,4*). However, the success of these models lies in the ability to accurately measure potential complexing ligand concentrations and determine their respective stability constants. While most previous trace metal research has focused on natural organic matter (NOM) complexation, recent data have shown that inorganic sulfides can play an important role in the trace metal speciation in oxic waters (*5-7*).

As a class B metal, Ag(I) will preferentially form complexes with soft bases, such as Cl, O, and S. The high affinity for reduced sulfur (sulfides) is reflected in the large stability constants that have been experimentally determined for inorganic silver sulfides (e.g., $K_{AgHS^0} = 10^{11.6}$ (*8*)). These large stability constants suggest that if sulfides are present and available in natural systems significant Ag sulfide complexation will occur. The resulting Ag sulfide complexes would be extremely stable, which could dramatically reduce the bioavailabilty of Ag to aquatic microorganisms, especially for those that uptake trace metals through their cell membranes (*9*).

In fresh waters, dissolved silver (Ag_{diss}) concentrations have been reported spanning a range of concentrations from < 10 pM to 620 pM (*10,11*). The elevated Ag_{diss} levels have been directly linked to inputs of treated sewage plant (STP) effluent, which is believed to be the primary source of Ag_{diss} to most natural waters (*12*). In one river system, a mass balance showed that the Ag_{diss} contained in effluent from 5 STPs accounted for > 90% of the total mass of Ag_{diss} measured in downstream river water samples (*13*). While these recent studies have provided some of the first reliable Ag_{diss} concentrations for surface waters, they lack specific data on potential ligands for Ag_{diss} complexation and only speculate on the Ag_{diss} speciation in the water column (*14*).

Conversely, studies (*12, 15-17*) have reported inorganic sulfide concentrations in oxic waters, but have not measured the corresponding sulfide speciation. While inorganic sulfide has been measured in surface waters and STP effluent at concentrations ranging from < 500 pM to 500 nM (*12,15-17*), these measurements have been conducted using the non-specific methylene blue (or Cline) method (18). This spectrophotometric method uses H_2SO_4 (HCl for Cline), which results in the measurement of both free bisulfide and acid labile metal sulfides, such as Fe and Zn sulfides. In oxic river waters, metal (Cu, Fe, and Zn) sulfide complexation has been shown to account for > 90% of the total dissolved sulfide concentration (as measured by Cr(II) reduction (*5*)). However, no data exist about the effect of sulfide complexation on other less abundant class B metals, including Ag, which form strong sulfide complexes.

Due to the extremely low Ag_{diss} concentrations observed in most natural systems, direct observations of specific Ag complexes may not be possible with current analytical techniques. However, evidence may be inferred from the presence of only Cr(II) reducible sulfides (OCRS = CRS – AVS). Cr(II) is one of the few reducers capable of reducing Ag^+ in the Ag_2S to Ag^0, releasing all of the complexed sulfide as free HS^-/H_2S for subsequent measurement (eq 1-2).

$$E_{red} \quad Ag_2S + 2H^+ + 2e^- \rightarrow H_2S + 2Ag \qquad -0.060 \text{ V} \qquad (1)$$

$$E_{ox} \quad Cr^{2+} \rightarrow Cr^{3+} + 1e^- \qquad +0.408 \text{ V} \qquad (2)$$

$$Ag_2S + 2H^+ + 2Cr^{2+} \rightarrow 2Cr^{3+} + H_2S + 2Ag \qquad +0.756 \text{ V} \qquad (3)$$

This contrasts with direct acid leaches, which measure the acid volatile sulfides (AVS) fraction, and may not be sufficiently strong enough to cause dissociation of the Ag sulfide (especially on the time scale, $t_{diss} < 20$ min.) used for recovery and measurement (17). However, the presence of a resistant OCRS fraction in natural waters would be indicative of the presence of Ag sulfides (and other resistant class B metals, such as Hg sulfides).

The goal of this research was to provide evidence showing that Ag speciation in fresh waters is controlled by sulfide complexation. To accomplish this objective, the different sulfide fractions (free HS^-/H_2S, AVS, OCRS) were measured and compared with the dissolved Ag concentrations (Ag_{diss}). Seven rivers covering a range of watershed development in the mid-Atlantic region (Figure 1) were sampled. In addition, effluents from two different size STPs were also measured to provide "natural" water samples with elevated Ag_{diss} concentrations and help define the role of STPs cycling Ag in the environment.

To accurately measure the low sulfide fractions found in natural systems, several different voltammetric techniques were employed. Voltammetry not only provided a simple analytical method with excellent sensitivity (10 nM without a deposition step), but also allowed for discrimination among specific sulfide species (e.g. HS^-, S_x^{2-}, and S^0 (19)). Additionally voltammetry was used in a series of laboratory experiments, to determine if Ag_{diss} would replace Zn in Zn sulfides (representing an AVS) and Cu in Cu sulfides (representing CRS). The extent of Ag replacement was determined by direct electrochemical measurement of released Cu^{2+} and Zn^{2+} ions back into the laboratory solutions.

Methods

All laboratory solutions and standards were prepared using N_2 purged Nanopure waters (O_2 content $< 0.2 \mu M$). Sulfide standards and solutions were prepared from reagent grade $Na_2S \bullet 9 \ H_2O$ (Aldrich). Metal standards and

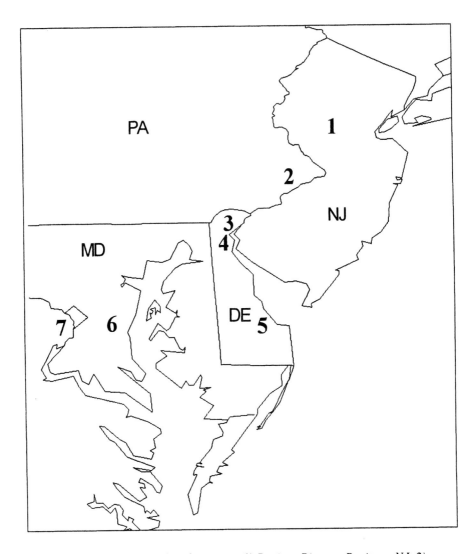

Figure 1. Field sampling locations. 1) Raritan River at Raritan,, NJ, 2) Schuylkill River at Philadelphia, PA, 3) Brandywine River at Wilmington, DE, 4) Christina River at Elsmere, DE , 5) Broadkill River at Milton, DE, 6) Patuxent River, Leeland, MD, 7) Potomac River at Washington, DC.

solutions were prepared using reagent grade $AgNO_3$, $Cu(NO_3)_2$, and $Zn(NO_3)_2$ (Aldrich). All metal standards were stored at pH = 2.0 in Teflon bottles (opaque for Ag^+). Free HS^-/H_2S were determined by the current produced from the reduction of HgS (eq 4) using cathodic stripping square wave voltammetry (CSSWV) [initial potential – 0.05 V, final potential – 1.4 V, scan rate 200mV/s]. A 60 s deposition step @ - 0.5 V was added to increase sensitivity, which resulted in a picomolar detection limits (D.L. = 450 $\pm$ 18 pM). For replacement experiments, free Cu^{2+} and Zn^{2+} concentrations were determined by metal amalgam oxidation (eq 5 and 6) using anodic stripping square wave voltammetry (ASSWV) [initial potential – 1.25 V, final potential – 0.1 V, scan rate 200mV/s]. A 30 s deposition step @ - 1.25 V was used to allow for sub-micromolar detection

$$\text{Ep}$$

$$HgS + 2e^- + H^+ \rightarrow HS^- + Hg \qquad -0.66 \text{ V} \qquad (4)$$

$$CuHg \rightarrow Cu^{2+} + Hg^0 + 2e^- \qquad -0.15 \text{ V} \qquad (5)$$

$$ZnHg \rightarrow Zn^{2+} + Hg^0 + 2e^- \qquad -1.05 \text{ V} \qquad (6)$$

limits. All electrochemical measurements were conducted with a PAR 303B hanging drop mercury electrode (HDME) using a DLK-100 voltammetric analyzer (Analytical Instrument Systems). Replacement reactions were conducted in oxygenated, pH = 7.5 water (buffered with $NaHCO_3$) to simulate natural fresh waters. The formation and stability of the Zn sulfide clusters used for these experiments were confirmed by their UV spectra signal (18), which was measured using a multi-diode array UV-vis spectrophotometer (HP8486).

Natural water samples were collected following clean protocol and filtered in a Class 100 clean room using 0.2 μm Nuclepore filters. Dissolved sulfide fractions were determined by CSSWV on a HDME following stepwise sulfide leaching (20). First, the presence of any free HS^-/H_2S in the sample was measured directly. Next, acid leached sulfides (AVS) were liberated with 3 M HCl. The HCl was added to samples releasing sulfide as H_2S, which was then purged from the samples and trapped in 1 M NaOH. The isolated sulfides were then measured by CSSWV. Finally, acidified Cr(II) was added to the samples to determine the OCRS fraction. A Jones reduction column (21) was used to reduce Cr (III) [1 M $CrCl_3$(in 1N HCl)] to Cr(II). Ag_{diss} in the water samples was measured using an inductively coupled plasma mass spectrometer (ICP-MS).

Quantification of Cu sulfides and S_x^{2-} was performed on separate sub-samples using previously reported methods (7,22). In summary, an HCl acid titration was performed on the water sample to induce dissociation of the metal sulfide complexes (7). Free sulfide released was electrochemically measured at discrete pH values, which correspond to Cu sulfide (pH < 5), Zn sulfide (pH <

6.7), and $FeSH^+$ (pH > 6.7) dissociation. After each measurement at a specific pH (pH = 6.8, 5.2, and 2.8), the free sulfide was purged prior to the next acid addition. S_x^{2-} and S_8 were measured directly by their discrete peak potentials at ambient pH to determine their contribution to final sulfide concentration (7,20). Cu sulfides could be determined semi-quantitatively from the acidified (pH = 2.5) sulfide concentration after subtracting off the contribution polysulfides and S_8 from the final electrochemical sulfide signal.

Ag speciation calculations were performed using the computer program $MINEQL^+$ (23), which was modified to account for potential multi-nuclear metal sulfide clusters (5).

Results

Laboratory solutions

Initial acid titrations of micromolar Ag sulfide solutions showed a strong resistance to acid dissolution and release of sulfides. Thus we hypothesized that the AVS fraction would not include Ag sulfides to any great extent. To test this hypothesis, both a 3M HCl leach and Cr(II) reduction were performed on sub-samples of Ag sulfides from laboratory prepared aqueous solutions, fresh precipitates, and ground acanthite (Ag_2S). Aqueous micromolar level solutions were shown to be completely dissociated < 1 min, by measuring the free sulfides released on Cr(II) reduction, whereas complete dissociation from acid leaching required ≈ 1 hr (Figure 2). If the laboratory Ag sulfide solution was allowed to equilibrate for 1 day, the time needed for complete dissociation to occur increased by 20 min., suggesting that "aged" natural samples may be more resistant to acid leaching than laboratory prepared solutions (Figure 2). For the fresh Ag sulfide precipitate and acanthite, Cr(II) reduction required > 15 min (precipitate) and > 25 min (mineral) before dissociation was completed. However, acid leaching required > 5 hr (precipitate) and > 10 hr (mineral) before any free sulfides were even detected.

The inertness of all Ag sulfides to acid leaching demonstrated an extreme stability for the Ag sulfide complex and suggested that Ag should be able to displace other more numerous trace metals (e.g. non-class B metals: Cu, Fe, Pb, and Zn) that were already complexed with sulfide. To test this assertion, replacement experiments were conducted to determine whether Ag could replace the Zn in Zn sulfide complexes. Since Zn sulfides have been shown to form highly stable multi-nuclear clusters in laboratory solutions (19), metal sulfide clusters were formed prior to any laboratory replacement experiment.

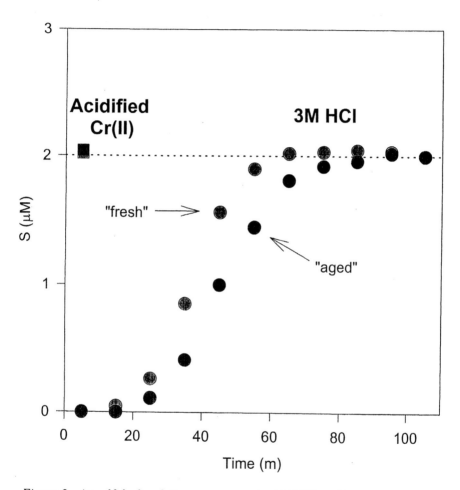

Figure 2. Ag sulfide dissolution as measure by 1M HCl acid leaching (circles) and acidic Cr(II) reduction (squares). Ag sulfides were "aged" either 1 hr (filled symbols) or 1 day (open symbols) prior to measurement.

UV-vis spectra were used to observe the formation and stability of Zn sulfide clusters over time (*19*). Zn sulfide complexes produced two peaks at 210 nm and 290 nm, which were separated from the Zn^{2+} peak at 214 nm and HS^- peak at 230 nm. As the Zn sulfide solution was allowed to age, the magnitude of the 290 nm peak remained constant, while the magnitude of the 210 nm peak increased, indicating growth of lower molecular weight Zn sulfide clusters. In Figure 3, the growth of the 210 nm peak over time is shown for three different concentrations of Zn (0.1 μM, 0.6 μM, and 2.0 μM) and sulfide (0.2 μM, 1.2 μM, and 4.0 μM). The range of concentrations used is typical of those found in natural waters and STP effluent. Additionally, it was observed that at higher concentrations (Zn = 10 μM and HS^- = 20 μM), the 210 nm peak rapidly decreased ($t_{1/2}$ = 2 hr, Figure 4a), probably due to increased clustering and precipitation. For the lower concentrations, the 200 nm peak was found to stabilize in ≈ 2 days and have a half-life > 30 days (Figure 4b).

One of the Zn sulfide solutions (Zn = 0.6 μM and HS^- = 1.2 μM) was sampled daily from day 2 to day 10 to determine the stability of the Zn sulfides using mass recoveries. At each time step, Zn^{2+} and HS^- were analyzed using square wave voltammetry before and after acidification to pH = 6.4, which caused the Zn sulfide solution to dissociate into Zn^{2+} and free HS^-/H_2S. In none of the samples was free Zn^{2+} observed before acidification, indicating no breakdown in the Zn sulfide. After acidification, recoveries were Zn = 0.58 ± 0.03 μM and HS^-/H_2S decreased to a stable concentration of 0.82 ± 0.07 μM. The loss of free HS^-/H_2S was attributed to oxidation the non-Zn complexed reduced sulfur.

Using stable Zn sulfide clusters, as in fig. 4b, the Zn sulfide solutions were titrated with free Ag^+ aliquots (0.1 μM). Following a 15 s mixing and 15 s equilibrium period, replaced Zn was measured by ASSWV. Free Zn^{2+} was measured due to the stability and inertness of Ag sulfides. For each Ag addition, voltammetric scans were conducted until the free Zn^{2+} peak was observed to stabilize, indicating a completion of the replacement reaction. Ag was observed to replace Zn in a 1:1 ratio (eq 7,8) and on extremely short time scales, with complete replacement for each

$$ZnS + Ag^+ -> AgS^- + Zn^{2+} \qquad (7)$$

$$Zn_3S_3 + 3Ag^+ -> Ag_3S_3^{3-} + 3Zn^{2+} \qquad (8)$$

addition occurring within one minute (Figure 5). The replacement of Zn in Zn sulfide clusters (eq 8) is similar to the results generated from Ag ion doping of laboratory prepared Zn sulfide colloids, where the Ag was thought to be incorporated into the Zn sulfide lattice (*24*). These results suggest that the Ag

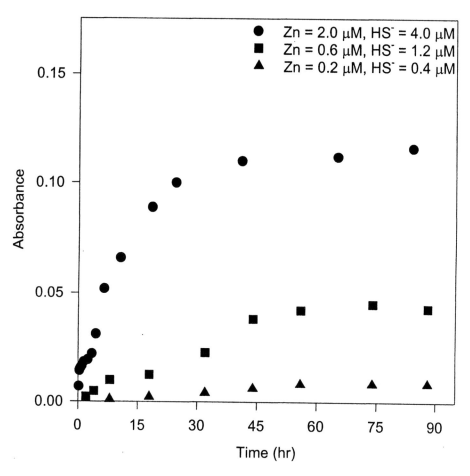

Figure 3. UV-vis analyses of the Zn sulfide cluster formation at pH = 8.0 as determined from the 210 nm peak. Stabilization occurs after 30 min for the 2.0 μM Zn and 4.0 μM HS⁻ solution (circles), after 50 min. for the 0.6 μM Zn and 1.2 μM HS⁻ solution (squares), and after 60 min. for the 0.2 μM Zn and 0.4 μM HS⁻ solution (triangles).

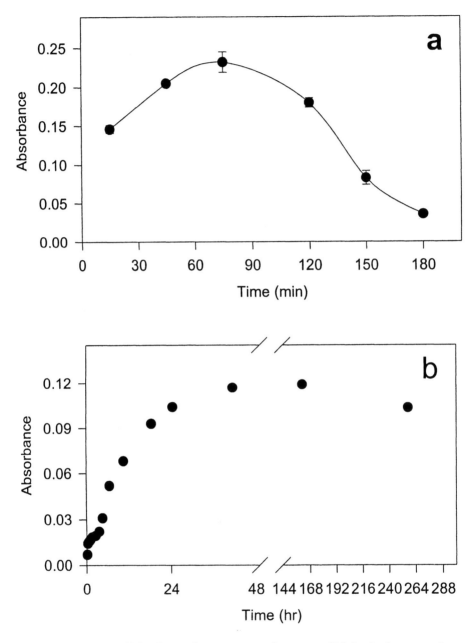

Figure 4. Zn sulfide cluster dissociation with time at pH 8.0. As determined from the 210 nm peak for a) Zn = 10 μM, HS⁻ = 20 μM and b) Zn = 0.6 μM, HS⁻ = 1.2 μM.

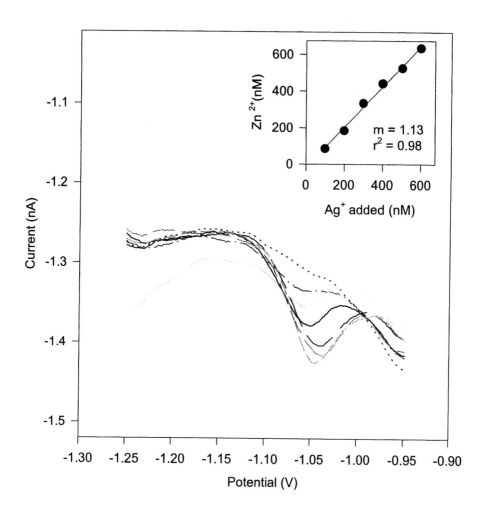

Figure 5. ASSWV voltammetric scans of replaced Zn^{2+} during Ag titrations of a 48 hr "aged" Zn sulfide solution. Zn^{2+} concentrations reflect a 1:1 replacement ratio with the titrated Ag. Insert shows amount of Zn^{2+} released during each Ag titration step.

382

can be forming either multi-nuclear Ag sulfide clusters or Ag-metal-sulfides (eg Ag/ZnS).

Ag titration of Cu sulfide solutions was also conducted to determine if Ag would replace both the Cu(II) and Cu(I). Peak potentials produced by reduction of Cu(II) at - 0.15 V and Cu(I) at 0.05 V were measured (7) to determine how Ag attacked a Cu sulfide complex. For the first two Ag additions, only Cu(II) was observed (Figure 6). However, when additional Ag was added, both Cu(II) and Cu(I) were measured (Figure 6). Both Cu(I) and Cu(II) were replaced on short time scales (< 1 min.) after Ag$^+$ additions, indicating that the Cu complexed sulfide in OCRS fraction is also available for Ag complexation.

Field samples

With the information gathered from laboratory experiments, river water and STP effluent were collected and analyzed for the presence of Ag sulfides. All the rivers, which received STP effluent, were observed to have elevated Ag$_{diss}$ concentrations (Table I). In the two rivers that were not impacted by STP effluent, the Christina and the Broadkill, the Ag concentrations were the lowest (< 30 pM). At the most urbanized sampling locations (the Brandywine River at Wilmington, DE, the Potomac River at Washington, D.C. and the Schuylkill River at Philadelphia, Pa.), Ag$_{diss}$ concentrations were found to be higher, but only moderately so. The relatively small increases in Ag$_{diss}$ in the Schuylkill and the Potomac, relative to the other rivers, are due to the high dilution from the larger river discharge volume. This dilution effect is not as pronounced in the smaller Brandywine River, which had an Ag$_{diss}$ concentration of 320 pM. AVS and OCRS concentrations in the river waters sampled generally followed a similar pattern to the Ag$_{diss}$, with OCRS concentrations only being observed in significant quantities (> 15 nM) in rivers which received STP effluent. One notable exception to this sulfide pattern was the Christiana River whose extensive fresh water marsh system produced larger concentrations of AVS.

Analysis of STP effluents revealed the nanomolar concentrations of Ag$_{diss}$, and micromolar concentrations of AVS and OCRS (Table I). The effluent from the smaller STP (0.5 MGD) was observed to have a Ag$_{diss}$ concentration of 250 pM. In comparison, the effluent from the larger STP (90 MGD) had a Ag$_{diss}$ concentrations of 470 pM. While these concentration differences appear relatively minor, the total mass of Ag$_{diss}$ being discharged by each STP is quite substantial. The mass differences in Ag$_{diss}$ between the STPs reflect the larger commercial inputs from the City of Wilmington, DE versus the more residential inputs from the town of Lewes, DE. However, in both STP effluents, significant differences were observed between the AVS and OCRS concentrations. This is a direct result of the large increase in the concentrations of S$_x^{2-}$ (and Cu sulfides),

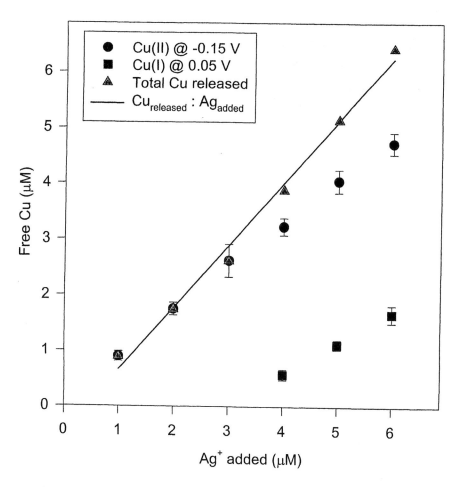

Figure 6. Replaced Cu(II) and Cu(I) from Cu sulfides during Ag titrations in 0.01 M NaClO₄. Cu(II) was measured by ASSWV at a peak potential of – 0.15 V, while Cu(I) was measured at a peak potential of 0.05 V. Regression for Cu$_{released}$ to Ag$_{added}$ gives a slope = 1.09 (r^2 = 0.97).

which appears to be a key indicator of the presence of STP effluent in river water.

Table I. Metal sulfide data measured in river water from the Mid-Atlantic region of the United States. All samples are from the (< 0.2 μm) filtrate fraction. (AVS) represents the concentrations liberated using a 3M HCl acid leach. (OCRS) represents only the Cr(II) reducible sulfides including the contribution from Cu sulfides (CuS) and polysulfides (S_4^{2-}). CuS concentrations were corrected to account for 30% polysulfide, while S_4^{2-} was corrected to represent only the S^0 contribution. Errors are based on triplicate analysis of a single sample. Ag_{diss} instrumentation (ICP-MS) analysis error < 2.5%.

Location	pH	AVS	OCRS	Ag_{diss}	CuS	S_4^{2-}	OCRS/ Ag_{diss}	AVS/ Ag_{diss}
		(nM)	(nM)	(nM)	(nM)	(nM)		
Delaware								
Broadkill	7.8	20.8±1.1	1.7±0.08	0.031	1.3±0.21	-	13	680
Brandywine	7.36	382±15	8.4±0.09	0.323	6.7±0.52	-	6	1200
Christiana	7.18	473±22	7.0±0.04	0.016	5.3±0.58	-	125	29600
Wil. STP	7.76	2850±98	325±15	0.464	26±2.2	270±22	640	6150
Lewes STP	7.82	1220±62	450±38	0.242	52±5.2	365±52	1250	5050
Maryland								
Potomac	7.78	522±27	29±1.1	0.13	6.2±0.44	18±2	175	4000
Patuxent	7.46	34.8±11	0.9±0.01	0.027	0.5±0.06	-	15	1300
New Jersey								
Raritan	7.38	223±10	7.7±0.6	0.047	4.1±0.2	-	65	4750
Pennsylvania								
Schuylkill	6.99	153±8	37 + 1.3	0.056	7.8±0.75	22±4	500	2700

Discussion

In all the waters sampled, the AVS was found to exceed the Ag_{diss} concentrations by greater than 600 times (Table I). In most of the rivers sampled, the AVS/Ag_{diss} ratio was found to range from 1000 – 5000. In contrast, the AVS/Ag_{diss} in STP effluent was found to be > 5000, which suggests that both AVS and Ag_{diss} in river water originate from STPs. Irrespective of the source, the available pools of sulfides from AVS provide sufficient excess to allow for complete Ag sulfide complexation.

The presence of OCRS in all samples provides corroborating evidence that Ag sulfide complexes can dominate Ag speciation. Unfortunately both polysulfides (S_x^{2-}) and Cu sulfides are partially resistant to acid leaching, making precise sulfur speciation in the OCRS fraction somewhat complicated to deconvolute. At pH < 6.0, S_x^{2-} breaks down into one H_2S and $(x-1)S^0$ molecules. The S^0 is further reduced by Cr(II) to H_2S, which is included in the OCRS fraction. However, S_4^{2-} can be directly measured (and standardized) by voltammetry and then subtracted from the OCRS fraction, assuming only S_4^{2-} is present (19). Similarly, a portion of the total Cu sulfides contributes to the OCRS fraction. During Cu sulfide formation, some Cu(II) is reduced to Cu(I), with an equal amount of sulfides being oxidized to polysulfides. Since extent of this redox reaction is not known for natural systems (estimated to be ~ 30% from laboratory prepared low micromolar level solutions (7)), the contribution of Cu(I) polysulfides to the CRS fraction in a natural water sample can only be estimated.

In the waters sampled, the OCRS concentrations were still greater than the Ag_{diss} concentrations even after the contribution from both electrochemically measured S_4^{2-} (corrected for the S^0 contribution) and Cu sulfides (assuming 30% polysulfide formation) was subtracted (Table I). Highly urbanized and STP impacted rivers were found to contain high Cu sulfides (and S_x^{2-} observed in the Schuylkill and Potomac rivers) and correspondingly high OCRS concentrations. However, after removing the contribution from Cu sulfides, the OCRS concentrations still were in excess to the entire measured Ag_{diss} concentration. On average the OCRS: Ag_{diss} > 100 (Table I). In contrast, less urbanized (and non-STP impacted) waters, with low Cu sulfide concentrations and no measurable S_x^{2-}, the OCRS: Ag_{diss} ~ 10 (Table I). While this still left sufficient quantities of OCRS to account for all of the Ag_{diss}, it must be noted that this excess is only slightly greater than total experimental error, which included purge and trap recovery variation (< 10%) and Cu sulfide measurements (10 – 15%) (see Table I). Even with the potential analysis problems due to low OCRS and Cu sulfide concentrations in less urbanized waters, the implication of complete sulfide complexation on free Ag^+ concentrations in river water is quite plausible, especially since STP effluent inputs appeared to be the primary source of Ag_{diss} to river water.

To amplify this assertion, a simple MINEQL$^+$ speciation calculation performed to determine the effect of sulfide complexation on free Ag^+ concentrations that would be found in typical river water. The speciation calculation was conducted using the Ag_{diss} concentrations, AVS concentrations (representing available sulfides), and ambient water conditions (e.g. pH and ionic strength) found in the Raritan River sample. First, only the formation of

Ag bisulfide species ($K_{AgHS^0} = 10^{11.6}$ (8)) was assumed to be occurring. Using this assumption, the free Ag^+ concentration in the river water was calculated to be 1.4×10^{-16} M, with the majority (> 99%) of the Ag being complexed to sulfides. When the speciation was re-calculated assuming the formation of multi-nuclear metal sulfide clusters with stoichiometries of either 2:1 or 3:3, ($K_{Ag_2S} = 10^{29.6}$ and $K_{Ag_3S_3} = 10^{23.2}$ (24)), the free Ag^+ concentration dramatically decreased to 2.0×10^{-27} M.

In either case, the existence and availability of AVS in oxic river dramatically decreased the free Ag^+ concentration, implying an equally dramatic effect on Ag toxicity. However, the effect of sulfide complexation on reducing Ag toxicity in fresh waters is only suggested by the large stability constants and needs to be directly measured in future research, especially when one considers the presence of multi-nuclear metal sulfide clusters in fresh waters (5).

Acknowledgements

This research was made possible in part by a NSF post-doctoral grant to T.F. Rozan and funding provided by the photographic imaging and manufacturing association to G.W. Luther. The authors wish to thank Rob Sherrell for the Ag_{diss} measurements on the ICP-MS.

References

1. Bruland, K.W.; Rue, E.L.; Donat, J.R.; Skrabal, S.A.; Moffet, J.W. *Anal. Chimi. Acta* **2000**, *405*, 99-113.
2. Xue, H. ;Sunda W. *Environ. Sci. Technol.* **1997**, *31*, 1902-1909.
3. Buykx, S.E.J.; Cleven, R.F.M.J.; Hoegee-Wehmann, A.A.; van den Hoop, M.A.G.T. *Fresen, J.Anal. Chem.* **1999**, *363*, 566-602.
4. Breault, R.; Colman, J.; Akien, G.; McKnight, D. *Environ. Sci. Technol.* **1996**, *30*, 3477-3486.
5. Rozan, T.F.; Lassman, M., Ridge, D., Luther III, G.W. *Nature*, **2000**, *406*, 879-882.
6. Rozan, T.F. ;Benoit, G. *Geochim. Cosmochim. Acta* **1999**, *19/20*, 3311-3319.
7. Rozan, T.F; Benoit, G.; Luther III, G.W. *Environ. Sci. Technol.* **1999**, *33*, 3021-3026.
8. Al-Farawati, R. van den Berg, C.M.G.*Mar. Chem.***1999**, *63*, 331-352,
9. Peters, J.W.; Lanzilotta, W.N.; Lemon, B.J.; Seefeldt, L.C. *Science* **1998**, *282*, 1853-1858.

10. Wen, L.; Santschi, P.H; Gill, G.A.; Paternostro, C.L.; Lehman, R.D. *Environ. Sci. Technol.* **1997**, *31*, 723-731.
11. Hurley, J.P.; Shafer, M.M.; Cowell,S.E.; Overdier, J.T.; Hughes, P.E.; Armstrong, D.E. *Environ. Sci. Technol.* **1996**, *30*, 2093 – 2098.
12. Shafer, M.M.; Overdier, J.T.; Armstrong, D.E. *Environ. Toxicol. Chem.* **1998**, *17*, 630 – 641.
13. Rozan, T.F. ;Hunter, K.S. *Sci. Total Environ.*, **2001**, in press.
14. Adams, N.W.H. ; Kramer, J.R. *Environ. Toxicol. Chem* **1999**, *18*, 2674 - 2680.
15. Adams, N.W.H.; Kramer, J.R. *Environ. Toxicol. Chem.* **1999**, *18*, 2667 – 2673.
16. Bell, R.A.; Kramer, J.R. *Environ. Toxicol. Chem.* **1999**, *18*, 9-22.
17. Radford-Knoery, J. ;Cutter, G.A. *Anal. Chem.* **1993**, *65*, 976-982.
18. *Standard Methods for the examination of Water and Wastewater.*19th edition, Amer. Public health Assoc., Washington, D.C., 1992.
19. Luther III, G.W.; Theberge, S.M.; Rickard, D.T. *Geochim. Cosmochim. Acta* **1999**, *63*, 3159-3166
20. Rozan, T.F.; Theberge, S.M.; Luther III. G.W. *Anal. Chimi. Acta* **2000**, *415*, 175-184.
21. Henneke, E.; Luther III, G.W.; deLange, G.J.; Hoefs, J. *Geochim. Cosmochim. Acta* **1996**, *61*, 307-321.
22. Luther III, G.W.; Rickard, D.T; Theberge, S.M.; Olroyd, A. *Environ. Sci. Technol.* **1996**, *30*, 671-679.
23. Schecher, W.D. ;McAvoy, D.C. MINEQL$^+$: A chemical equilibrium program for peronal computers. Environmental Research Software, Hallowell, ME, 1989.
24. Hao, E.; Sun, Y; Yang, B.; Zhang, X.; Liu, J.; Shen, J. *J. Colloid. Inter. Sci.* **1998**, *204*, 369-373.
25. Rozan, T.F. ; Luther III. G.W. Ag sulfide complexation: stoichiometries and stability constants. *Environ. Sci. Technol*, submitted.

INDEXES

389

Author Index

Subject Index